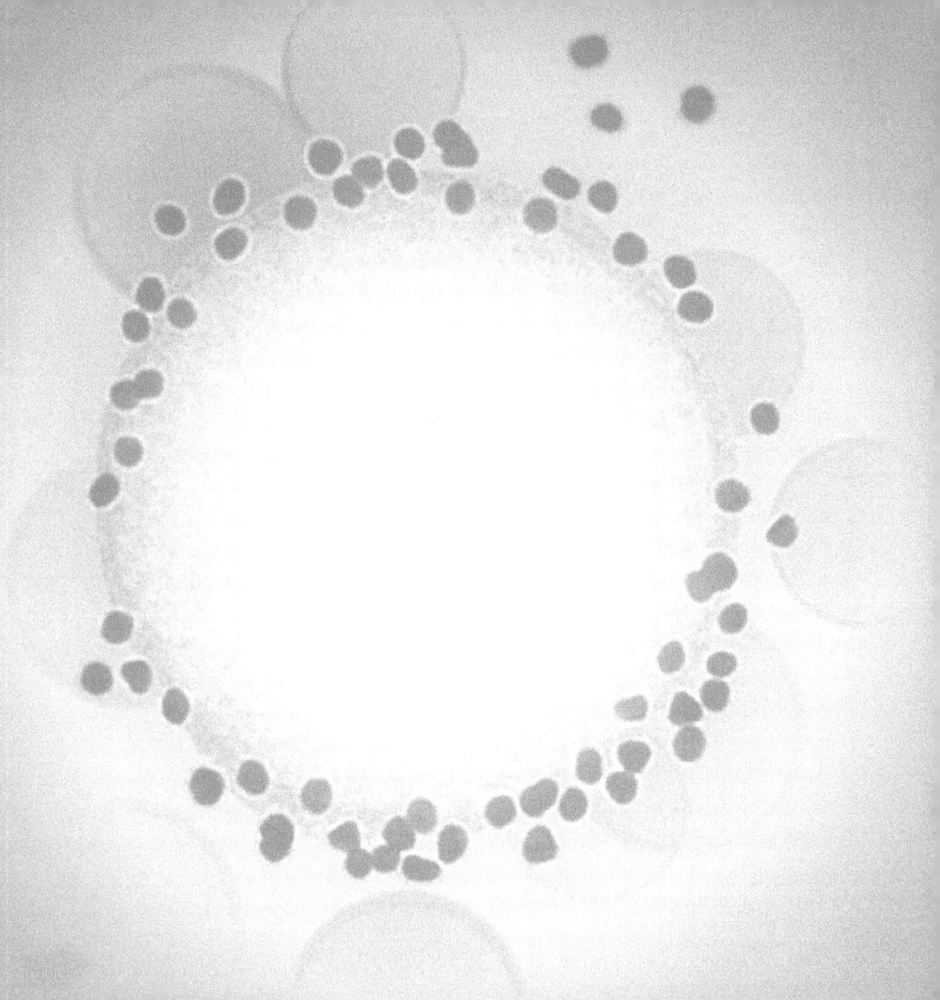

EXTRACELLULAR VESICLES IN HEALTH AND DISEASE

EXTRACELLULAR VESICLES IN HEALTH AND DISEASE

edited by Paul Harrison
Christopher Gardiner
Ian L. Sargent

Pan Stanford Publishing

Published by

Pan Stanford Publishing Pte. Ltd.
Penthouse Level, Suntec Tower 3
8 Temasek Boulevard
Singapore 038988

Email: editorial@panstanford.com
Web: www.panstanford.com

British Library Cataloguing-in-Publication Data
A catalogue record for this book is available from the British Library.

Extracellular Vesicles in Health and Disease

The cover image is a cryo-electron micrograph of a placental vesicle labeled with annexin-AV gold. Courtesy of Professor Alain Brisson, University of Bordeaux, France.

ISBN 978-981-4411-98-1 (Hardcover)
ISBN 978-981-4411-99-8 (eBook)

Printed in the USA

Contents

Preface

This book stems from the first international meeting on "Microvesicles and Nanovesicles in Health and Disease" held at Magdalen College, Oxford, on the September 23–24, 2010. Interest in the role of cellular microvesicles and nanovesicles (exosomes) is expanding rapidly. It is now apparent that far from being merely cellular debris, these vesicles play a key role in cell-to-cell communication and signaling. Moreover, they are significantly elevated in a number of diseases. This raises the question of their direct role in pathogenesis as well as their possible use as biomarkers. The purpose of the meeting was to bring together, for the first time, a range of experts from around the world to discuss the latest advances in this field. Key to the study of these vesicles is the availability of methodologies for their measurement in biological fluids. We therefore also invited a number of companies to present a range of exciting new technologies for this purpose. The presentations at this meeting form the basis of this book.

Since the Oxford meeting, the field has moved forward at a tremendous pace. A second meeting was organized in Paris by Professor Clotilde Thery in early 2011, which led to the founding of the International Society for Extracellular Vesicles (ISEV) (www.isev.org). The society's first scientific meeting was organized by Professor Jan Lotvall in Gothenburg, Sweden, in April 2012 and attracted over 400 delegates. The second meeting was organised in Boston, USA, in April 2013 by Professor Fred Hochberg with over 700 delegates, and the third meeting is being organised in April 2014 in Rotterdam, The Netherlands, by Professor Marca Wauben, with over 800 delegates expected. The society has also set up its own open access journal, *The Journal of Extracellular Vesicles*. Similar activity in the United States has seen two international meetings held in Orlando by Professor Doug Taylor and the formation of the American Society for Exosomes and Microvesicles (www.asemv.org). These meetings have provided a forum where scientists from many different disciplines with a common interest in extracellular vesicles can meet, and we look forward to many exciting new developments in this field in the future.

Paul Harrison
Christopher Gardiner
Ian L. Sargent
Oxford
Winter 2013

Chapter 1

Overview of Extracellular Vesicles in Health and Disease

Rienk Nieuwland,[a] **Edwin van der Pol,**[a,b] **and Augueste Sturk**[a]
[a]*Department of Clinical Chemistry, Academic Medical Centre of the University of Amsterdam, Meibergdreef 9, 1105AZ Amsterdam, the Netherlands*
[b]*Department of Biomedical Engineering and Physics, Academic Medical Centre of the University of Amsterdam, Meibergdreef 9, 1105AZ Amsterdam, the Netherlands*
r.nieuwland@amc.uva.nl

1.1 Introduction

1.1.1 Extracellular Vesicles

Extracellular vesicles (EVs) are spherical particles enclosed by a phospholipid bilayer and have a diameter ranging from 30 nm up to 5 μm.[1] Vesicles are present in body fluids and fractions thereof, such as serum and a conditioned medium of cultured cells. The concentration of EVs in human body fluids may be over 10^{10}/mL under normal, physiological conditions.[2,3] The classification of EVs is not straightforward, with substantial confusion throughout the literature, mainly because detection of single vesicles is cumbersome.[4,5] However, major improvements in the detection of EVs have been made recently. Because EVs contribute to health and disease, the clinical interest in EVs as noninvasive biomarkers

Extracellular Vesicles in Health and Disease
Edited by Paul Harrison, Christopher Gardiner, and Ian L. Sargent

ISBN 978-981-4411-98-1 (Hardcover), 978-981-4411-99-8 (eBook)
www.panstanford.com

for diagnosis or prognosis is emerging. Also, EVs may have several potential therapeutic applications, which are currently being explored.

1.1.2 History

In the 1940s it was shown that the clotting time of human cell-free plasma is prolonged after high-speed centrifugation. The pellet was shown to contain "the clotting factor of which the plasma is deprived," and reconstitution of the pellet in plasma again shortened the clotting time. These initial findings showed that plasma contains a subcellular factor that promotes the clotting of blood.[6] Wolf showed that the subcellular factor consists of vesicles of platelet origin ("platelet dust").[7] In parallel, exosomes were discovered when vesicles were isolated from sheep reticulocytes. These vesicles contain the plasma membrane receptor transferrin, which is absent on mature erythrocytes, suggesting that "vesicle externalization could be a mechanism for shedding of specific membrane functions, which are known to diminish during maturation of reticulocytes to erythrocytes."[8] Because the release of transferrin receptor–containing exosomes is not restricted to mammalian reticulocytes, this was postulated as a common pathway to remove redundant receptors.[9,10] Exosomes are thought to be formed within multivesicular endosomes (MVEs), also known as multivesicular bodies, and are released when membranes of MVEs fuse with the plasma membrane. This pathway of protein sorting is highly selective because the anion transporter, another common transmembrane protein, is absent in exosomes and fully retained during erythrocyte maturation.[11] Thus, exosomes contribute to the *specific* removal of redundant receptors.

1.1.3 Nomenclature

Cell-derived vesicles are called often after the cells or tissues from which they originate, for example, dexosomes (dendritic cell [DC]–derived exosomes),[12] prostasomes (prostate-derived vesicles),[13] matrix vesicles (vesicles in bone, cartilage, and atherosclerotic plaques),[14] and synaptic vesicles (vesicles from neurons).[15] However, these names do not provide any information with regard to the type of vesicles that are involved.

1.1.3.1 Recent classification of eukaryotic vesicles

In recent reviews, two to six different types of eukaryotic EVs have been distinguished.[16–19] Exosomes and microvesicles (also called shedding vesicles, shedding microvesicles, or microparticles) were acknowledged unanimous, and apoptotic vesicles (also called apoptotic blebs or apoptotic bodies) were acknowledged in three reviews.[17–19] In one review "ectosomes," "membrane particles," and "exosome-like vesicles" were also distinguished.[17]

1.1.3.2 Revised classification of eukaryotic vesicles

We propose to distinguish four different types of eukaryotic EVs, (1) exosomes, (2) microvesicles, (3) membrane particles, and (4) apoptotic vesicles, thereby omitting "ectosomes" and "exosome-like vesicles," because in our view there is insufficient evidence to support the existence of these types of vesicles. "Ectosomes" are omitted because their estimated diameter is based on the direct comparison of light scattered from beads, with light scattered from vesicles using flow cytometry, leading to an underestimation of the vesicle diameter.[4,5,20] Furthermore, "ectosomes" were observed only *in vitro*, and the cells initially reported to release "ectosomes" also release microvesicles, which were unknown when the term "ectosomes" was introduced.[21] The "exosome-like vesicles" are omitted because they are indistinguishable from exosomes, with electron microscopy pictures showing damaged and disrupted vesicles.[22–25]

1.2 Properties of Cell-Derived Vesicles

1.2.1 Exosomes

Exosomes are present in many and perhaps all biological fluids, including urine, blood, ascites, and cerebrospinal fluid,[26–29] but also, for example, in a conditioned medium of cell cultures. The reported diameter of exosomes is between 30 nm and 100 nm, the density ranges between 1.13 g/mL and 1.19 g/mL, they are usually isolated by ultracentrifugation, and their morphology is often described as "cup shaped" after negative staining and visualization by transmission electron microscopy (TEM). Figure 1.1A shows exosomes from human plasma isolated by differential centrifugation and imaged by TEM. Exosomes are surrounded by a phospholipid bilayer containing relatively high levels of cholesterol, sphingomyelin, and ceramide and containing detergent-resistant membrane domains (lipid rafts).[17,19,30,31] The transmembrane proteins of exosomes have the same outward orientation as the cell, and they have many characteristic proteins, including proteins involved in membrane transport and fusion; components of the endosomal sorting complex required for transport (ESCRT) complex such as Alix; TSG101; heat shock proteins; and tetraspanins, including CD63 and CD81.[16–19,30,32–34] Unfortunately, none of the before-mentioned properties is sufficient for specific exosome identification. Because the described properties of exosomes and other types of vesicles overlap considerably, there may be a sort of continuum of vesicles and vesicle types rather than clearly different archetypical types of vesicles, complicating the classification.

1.2.1.1 Classical formation

The "classical pathway" of exosome formation involves the generation of intraluminal vesicles within MVEs. In turn, MVEs fuse with either the plasma membrane to secrete the intraluminal vesicles, which after secretion into the cells' environment are then called exosomes, or lysosomes for degradation. There are distinct intracellular sorting pathways to direct proteins toward intraluminal vesicles predestined for either secretion or degradation[35–37] (see Chapter 2).

Ubiquitination of transmembrane receptors by the ESCRT complex directs receptors to intraluminal vesicles predestined for lysosomal degradation.[38,39] Sorting of proteins into intraluminal vesicles predestined for secretion is independent from ESCRT but is induced by ceramide, a sphingolipid.[40,41] In addition, cytosolic domains of proteins,[42] or lipid domains enriched in tetraspanins proteins CD9 or CD63, are thought to play a role in the sorting of transmembrane proteins toward intraluminal vesicles.[33,43] Finally, several small guanosine-5′-triphosphatases (GTPases) from the Rab family are involved in the sorting of MVEs toward either the plasma membrane[44–46] or lysosomes.[47]

1.2.1.2 Direct formation

T-cell and erythroleukemia cell lines release vesicles directly from their plasma membrane, both spontaneously as well as upon expression of the human immunodeficiency virus (HIV) antigens Gag or Nef or after cross-linking of surface receptors.[48–50] Because these vesicles are enriched in classical exosome markers and have a similar size and density as exosomes, they are indistinguishable from "endosomal" exosomes. To what extent this pathway contributes to the formation of exosomes by other types of cells, and whether or not such vesicles occur *in vivo,* has to be investigated.

1.2.2 Microvesicles

"Microvesicles," frequently also called microparticles to add to the confusion of nomenclature, is a term used for vesicles released from the plasma membrane during cell stress. The term "microvesicles" is also often used to describe total populations of vesicles isolated from biological fluids.[18,51,52] Microvesicles in the specific context of release from the plasma membrane are present in most if not all biological fluids, atherosclerotic plaques, and conditioned culture media.[53–57] Distinction between microvesicles and exosomes, especially when vesicles are isolated from biological fluids, is cumbersome. The size ranges of microvesicles and exosomes may overlap, for example, the reported size range of microvesicles from human plasma ranges from 20 nm to 800 nm.[3,7,58,59] The density of microvesicles is unknown. Microvesicles are usually isolated at 10,000–20,000 × g by centrifugation,[17] although the term "microvesicles" has also been used to describe the total population of vesicles isolated at 100,000 × g.[60] Vesicles larger than exosomes also often have a cup shape, as exemplified by Fig. 1.1B, showing microvesicles from human urine imaged by TEM. The usefulness of phosphatidylserine (PS) exposure to distinguish microvesicles from exosomes is unclear.[17,18,61,62] The mechanisms underlying the release of microvesicles are summarized elsewhere.[57] Taken together, distinguishing microvesicles from exosomes is difficult at present, especially when mixed populations of vesicles are studied in biological fluids.

1.2.3 Membrane Particles

Vesicles called "membrane particles" or prominosomes expose prominin-1 (CD133) and are present in, for example, the ventricular fluid of the embryonic mouse brain and human saliva.[17,63] Most of these epithelial cell–derived vesicles are small (diameter 50–80 nm) and thus are in the size range of exosomes, but large vesicles of 600 nm are also present. Figure 1.1C shows a TEM image of epithelial cell–derived vesicles, which also appear with a cup-shaped morphology. Because the smaller membrane particles have a slightly lower density than exosomes, do not expose the typical exosome marker CD63, coexist with vesicles that do expose that marker and are thus most likely exosomes, and are present in human body fluids such as as saliva and seminal fluid,[55,63] we assume that membrane particles are a distinct type of vesicle. Whether or not membrane particles are similar to exosomes released directly from the plasma membrane, and to what extent the larger type of membrane particles differ from microvesicles, will need additional studies.

1.2.4 Apoptotic Vesicles

Cells undergoing apoptosis release PS-exposing vesicles, often called apoptotic bodies or apoptotic vesicles. Because the diameter of apoptotic vesicles ranges between 1 μm and 5 μm,[58,64–66] they are relatively easy to distinguish from exosomes, microvesicles, and membrane particles but not from small cells such as platelets. Figure 1.1D shows a scanning electron microscopy (SEM) image of apoptotic vesicles being formed on an endothelial cell predestined to undergo anoikis. The density ranges between 1.16 g/mL and 1.28 g/mL, and their morphology is more heterogeneous than other cell-derived vesicles. Some but not all apoptotic vesicles contain deoxyribonucleic acid (DNA) and histones.[67]

1.3 Functions of Cell-Derived Vesicles

1.3.1 Intercellular Signaling

Because in many studies the type of vesicles being studied is unclear, the term as mentioned in the original work will be used as much as possible. These terms do not necessarily correspond to the attempted definitions of the various types of EVs mentioned so far.

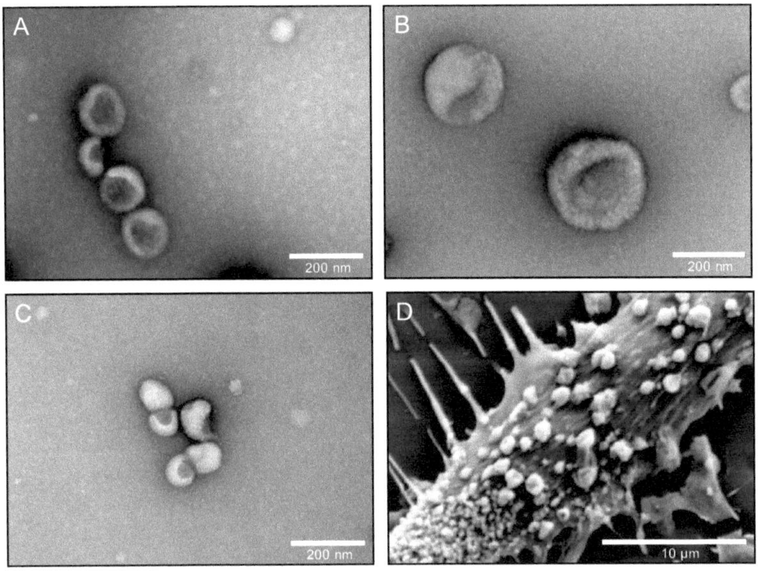

Figure 1.1 Different types of eukaryotic cell-derived vesicles. (A) Exosomes isolated from human plasma, (B) microvesicles from human urine, and (C) exosomes or membrane particles from human saliva imaged by TEM. (D) Apoptotic vesicles being formed on an endothelial cell predestined to undergo anoikis, imaged by SEM.

1.3.1.1 Immune suppression

Vesicles modulate the immune response. For example, activated T-cells and peripheral blood mononuclear cells release vesicles exposing the Fas ligand (FasL), a death receptor ligand (Fig. 1.2).[68] Intraperitoneal injection of FasL-exposing exosomes induces apoptosis of macrophages *in vivo*.[69] The ability of vesicles to induce apoptosis may play a role in growth and development. During normal pregnancy, trophoblast cells escape from the maternal immune system by releasing exosomes exposing FasL, thereby killing activated T-cells sensitized to paternal alloantigens,[70–72] and exposing ligands of the natural killer (NK) cell receptor NKG2D. The latter ligands downregulate the NKG2D receptor on NK cells, CD8+ cells, and γδ T-cells, thereby also contributing to fetal immune escape.[73]

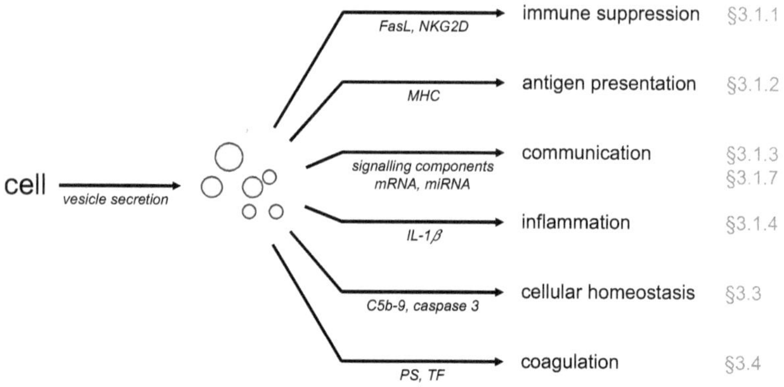

Figure 1.2 Schematic overview of the functions of vesicles from normal healthy cells. *Abbreviations:* C5b-9, complement complex 5b-9; IL-1β, interleukin-1β; MHC, major histocompatibility complex; NKG2D: natural killer receptor G2D; PAF, platelet-activating factor; TF, tissue factor.

Tumor cells use vesicles to escape from the immune system. Several types of cancer cells release FasL-exposing vesicles capable of inducing T-cell apoptosis,[74-77] and similarly, FasL-exposing microvesicles from sera of patients with oral squamous cell carcinoma induce T-cell apoptosis (Fig. 1.3). Because the tumor burden in these patients correlates with the microvesicle-associated levels of FasL, immune suppression by tumor-derived vesicles may contribute to tumor growth and development *in vivo*.[78] Exosomes from human tumor cell lines and mouse mammary tumor cells inhibit the cytotoxic response to tumor cells by blocking the interleukin-2-induced proliferation of NK cells,[79] by downregulating the expression of NKG2D,[80,81] by inducing differentiation of myeloid cells into myeloid-derived suppressor cells,[82,83] and by exposing adenosine triphosphate (ATP)- and 5′-adenosine monophosphate (5′-AMP)-phosphohydrolytic activity.[84]

Also viruses and parasites use vesicles to escape from the immune system. Vesicles from cells infected with the Epstein–Barr virus (EBV) expose latent membrane protein-1 (LMP-1), which inhibits proliferation of peripheral blood mononuclear cells and contributes to development of EBV-associated tumors (Fig. 1.4).[85] Furthermore, exosomes secreted by cells infected with the intracellular parasite *Leishmania donovani* suppress the immune response by preventing the activation of human monocytes.[86,87]

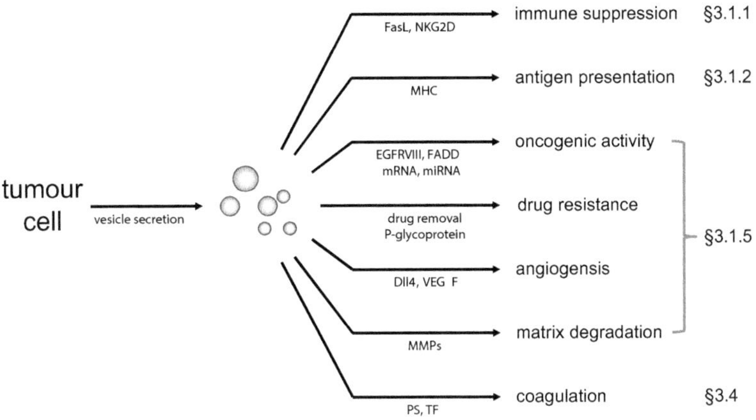

Figure 1.3 Schematic overview of the functions of vesicles from tumor cells. *Abbreviations:* Dll4, delta-like 4; EGFR, epidermal growth factor receptor; FADD, Fas-associated protein with death domain; MMP, matrix metalloprotease; VEGF, vascular endothelial growth factor.

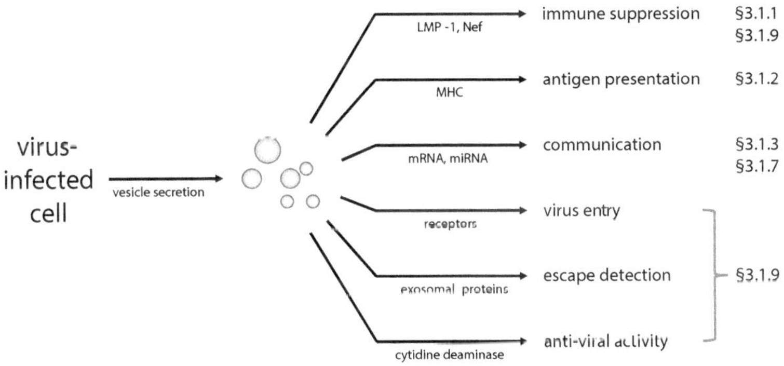

Figure 1.4 Schematic overview of the functions of vesicles from virus-infected cells. Nef: a viral protein expressed by HIV.

Vesicles can also activate the immune system. For example, exosomes from synovial fibroblasts delay T-cell activation–induced cell death, thereby contributing to apoptosis resistance.[88] Thus, the precise role(s) of EVs to affect the activity of the immune system depends on the models studied and is highly diverse and complex *in vivo*.

1.3.1.2 Antigen presentation

Vesicles often expose major histocompatibility complexes (MHCs) to present antigens (Fig. 1.2). For example, mice develop hypersensitivity reactions when exposed to certain antigens, but these reactions are suppressed when the mice receive serum from antigen-fed mice before exposure to the antigen. In other words, induction of "oral tolerance" seems possible by antigen processing in the gut. Although it has been hypothesized that exosomes from intestinal epithelial cells may play a role in the *trans*cellular transport of antigens from the lumen of the gut to immune cells,[89,90] exosomes from intestinal epithelial cells pre-exposed to pepsin/trypsin ovalbumin hydrolysate did not induce an atolerogenic immune response in mice but triggered a humoral immune response.[90] In several other models, such as allergic asthma, birch pollen allergy, and olive pollen allergy, exosomes either induce immune tolerance or enhance allergic immune responses.[23,91,92]

The initiation of T-cell-mediated antitumor immune responses requires uptake, processing, and presentation of tumor antigens by DCs (Fig. 1.3). Exosomes from mouse tumor cells bind to DCs to transfer tumor antigens *in vitro*.[93] In turn, these DCs induce CD8+ T-cell-mediated antitumor effects to established mouse tumors, illustrating that exosomes can trigger an antitumor immune response *in vivo*.[94] After uptake of antigens, DCs secrete exosomes exposing MHCs and T-cell costimulatory molecules, but efficient priming of cytotoxic T-cells by DC-derived exosomes requires the presence of mature DCs.[95]

Exosomes may also present antigens from microorganisms and allergens and can be used for immunoprophylaxis. For instance, exosomes from a mouse DC line pulsed with *Toxoplasma gondii*–derived antigens induce a systemic humoral immune response and protect against infection.[96–98] Similarly, exosomes from DCs infected with antigens from *Leishmania major*, *Mycobacterium tuberculosis*, or *Eimeriatenella* (avian cocciciosis) all induce protective immunity and immune tolerance.[99–102] Thus, the immune response can be modulated by exosome-mediated antigen presentation.

1.3.1.3 Intercellular communication

EVs may exchange signaling components between cells (Figs. 1.2–1.5). Microvesicles treated with secretory phospholipase A_2 contain lysophosphatidic acid and arachidonic acid. Lysophosphatidic acid induces platelet aggregation,[103] whereas arachidonic acid can be transferred to either platelets, where it is metabolized to thromboxane A_2, or endothelial cells, where it is metabolized to prostacylin.[104] Because the synovial fluid of inflamed joints contains high levels of secretory phospholipase A_2 as well as lysophosphatidic acid–containing microvesicles, such processes are likely to occur *in vivo*.[103] Taken together, vesicles can facilitate the intercellular exchange of lipids between cells. After exchange, these lipids can be metabolized, and this exchange may play a role in atherosclerosis and inflammation. Other examples are the transfer of Ca^{2+}-signaling tools from prostasomes to sperm cells, thereby increasing sperm motility and the fertilization rate;[105,106] suppression of the Wnt-signaling pathway by secretion of exosomes containing β-catenin;[107] and secretion of survivin, a caspase inhibitor that promotes proliferation, survival, and tumor cell invasion.[108] Thus, cells can exchange functional signaling elements via vesicles.

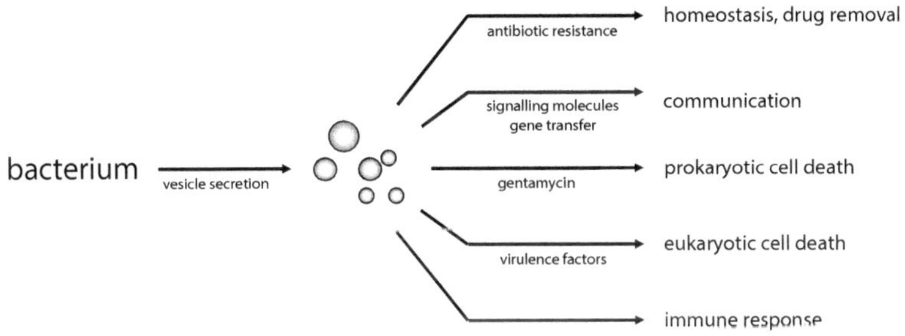

Figure 1.5 Schematic overview of the functions of bacterial (outer-membrane) vesicles.

1.3.1.4 Inflammation

EVs can modulate the inflammatory response (Fig. 1.2). For example, vesicles can stimulate cells to produce proinflammatory mediators. Blood from pre-eclamptic women contains increased concentrations of placenta-derived (syncytiotrophoblast) vesicles compared to normal pregnancy. Because these vesicles bind to monocytes and endothelial cells and thereby induce the production of proinflammatory mediators by these cells, these vesicles are thought to contribute to the increased systemic inflammatory responsiveness observed in pre-eclampsia.[109–111] In addition, vesicles from synovial fluid and microvesicles from leukocytes and platelets stimulate the production and release of interleukins, matrix metalloproteases, monocyte-chemotactic proteins, vascular endothelial growth factor (VEGF), and intercellular adhesion molecule-1 (ICAM-1) by synovial fibroblasts, indicating that these vesicles may enhance the destructive activity of synovial fibroblasts in rheumatoid arthritis.[18,112–115] Similarly, microvesicles from human airway epithelial cells stimulate the production and release of proinflammatory mediators, thereby enhancing the inflammatory airway response,[116] and microvesicles from human atherosclerotic plaques promote adhesion and transendothelial migration of monocytes by transferring ICAM-1 to endothelial cells, suggesting that such microvesicles facilitate atherosclerotic plaque progression.[117]

EVs are a major pathway for secretion of interleukin-1β,[118–120] and interleukin-1β within microvesicles from monocytes is one of the main components present in these vesicles that activates endothelial cells.[121] Not only proteins but also the lipids and enzymes involved in lipid synthesis present in microvesicles may play a role in the inflammatory response. Activation of the macrophage Toll-like receptor 4 by microvesicles is inhibited by a phospholipase D inhibitor,[122] and exosomes can contain functional enzymes for synthesis of leukotrienes, which themselves are potent lipid inflammatory mediators.[123]

Exosomes can also affect the inflammatory response more indirectly. For example, intravenous injection of exosome-like vesicles from adipose tissue induces the differentiation of monocytes into macrophages, which in turn produce and release tumor necrosis factor-α and interleukin-6.[124] Taken together, EVs can modulate the inflammatory response in various ways.

1.3.1.5 Tumor growth, metastasis, and angiogenesis

EVs can modulate tumor growth in numerous manners (Fig. 1.3). Vesicles can transfer oncogenic activity from one cell to the other by transferring oncogenic growth factor receptors such as the truncated epidermal growth factor receptor (EGFR) vIII or growth factor receptor ligands.[125,126] In addition, many cancer cells release vesicles containing the Fas-associated death domain (FADD), a key adaptor protein that transmits apoptotic signals and that becomes lost in many cancer cells.[127] Another mechanism involves the protection of tumor cells against entry or intracellular accumulation of antitumor drugs. Exosomes from breast cancer cell lines or breast cancer patients capture the humanized antibody Trastuzumab, thereby reducing the effective concentration of this anticancer drug,[128] and vesicles can exchange drug transporters such as P-glycoprotein between cells.[129] Because overexpression of P-glycoprotein by cancer cells correlates with anticancer drug failure in many types of cancer, the exchange of drug transporters by EVs is likely to contribute to drug resistance. Certain anticancer drugs can also be selectively removed from cells by secretion of exosomes, for example, cisplatin in drug-resistant ovarian carcinoma cells.[130]

EVs can promote tumor growth by facilitating vascular development, for example, by transferring the Notch ligand delta-like 4 (Dll4), from tumor cells to endothelial cells, thereby inducing a "tip cell phenotype" and increasing vessel branch formation.[131] Furthermore, tumor-derived vesicles can contain matrix-degrading enzymes such as matrix metalloproteases,[132] induce angiogenesis via VEGF,[133] or expose PS and TF,[134] mechanisms that have been excellently summarized elsewhere.[135–137] In summary, there are many different manners in which tumor-derived vesicles can promote tumor growth and development.

1.3.1.6 Morphogens

Microvesicles from T-cells or human blood induce differentiation of pluripotent erythroleukemic cells because these vesicles contain functional Hedgehog proteins.[138]

1.3.1.7 Genetic information

Bacteria and eukaryotic cells exchange genetic information via vesicles.[139,140] Exosomes from mast cells contain messenger ribonucleic acid (mRNA) and micro-RNA (miRNA)[140] (Fig. 1.2). Transfer of mouse exosomal mRNA to human mast cells induced expression of mouse proteins. Exosomes from T-cells and DCs transfer functional miRNA.[141,142] Exosomes from glioblastoma cells stimulate endothelial tubule formation and proliferation of glioma cells by transfer of mRNA, miRNA, and angiogenic proteins, and microvesicles from a colorectal cancer cell line contain mRNAs encoding cell cycle–related proteins, which induce proliferation of endothelial cells. Thus, vesicles can affect angiogenesis, tumor growth, and metastasis.[143,144] In addition, exosomes from brain tumor cells contain also mitochondrial DNA,[145] and microvesicles from several tumors and tumor cell lines contain coding as well as noncoding RNA and DNA, mutated and amplified oncogene sequences, and transposable elements (Fig. 1.3).[146] Most and perhaps all vesicles, including vesicles in biological fluids, will prove to contain genetic information, and exosomes from human saliva, plasma, and milk were shown already to contain detectable levels of mRNA.[147]

Pathogens have "hijacked" EVs to exchange genetic information between cells. Exosomes from nasopharyngeal carcinoma cells, harboring latent EBV, are enriched in several viral miRNAs when compared to the intracellular levels (Fig. 1.4). In turn, these viral miRNAs can manipulate the microenvironment of the tumor to modulate the growth of neighboring cells by downregulating target gene expression in recipient cells.[148–150]

Taken together, the role of vesicles to facilitate the exchange of functional genetic information between cells is likely to contribute to physiological and pathological processes in eukaryotes.

1.3.1.8 Prions

Exosomes contain functional, infectious prion proteins thought to contribute to the spreading of prions.[151,152] Also, β-amyloid peptides are associated with exosomes in Alzheimer disease. Because exosomal proteins accumulate in plaques of patient brains, exosomes are thought to play a role in the pathogenesis of Alzheimer disease.[153]

1.3.1.9 Viruses

Viruses utilize EVs for infection and cell survival (Fig. 1.4). Receptors essential for cell entry of HIV can be transferred to cells lacking such receptors.[154,155] Viral particles produced and released from infected cells expose exosomal (glyco)proteins, which helps viral particles to avoid recognition by the host immune system.[156,157]

One of the hallmarks of the acquired immunodeficiency syndrome is the deficiency of CD4+ T-cells. Apoptosis of these cells is induced by binding of a viral protein, Nef, to the T-cell CXCR4 receptor. Nef is exposed on leukocyte-derived exosomes.[49,158,159] LMP-1, the major oncogene of EBV, is also exposed on exosomes and inhibits T-cell proliferation, NK cytotoxicity, and proliferation of peripheral blood mononuclear cells.[160] Viruses can escape from detection by the immune system by inducing the release of empty viral particles exposing viral glycoproteins, which act as decoys to distract the immune system.[160] Moreover, the uptake of exosomes from EBV-infected cells by epithelial cells triggers the activation of growth-stimulating signaling pathways, showing that transfer of LMP-1, other signaling molecules, and growth factors is capable of manipulating the growth characteristics of neighboring cells.[160]

Because retroviruses and exosomes show many similarities, retroviruses were proposed to have adopted the pathways of exosome biogenesis to escape from immune detection, the so-called trojan exosome hypothesis.[161] Although this hypothesis is supported by the finding that an inhibitory domain of the HIV Gag protein interferes with the sorting of both viral and exosomal proteins,[162] inhibition of ceramide synthesis blocks the release of exosomes but not the release of HIV, suggesting that separate routes exist for exosome secretion and viral particle release.[160,163–166]

Exosomes also have antiviral activities. Exosomes from HIV-1-infected cells contain cytidine deaminase, which modifies the DNA sequences of the cells and may thus makes them less likely to grow properly,[167–169] and vesicles from virus-infected cells can present antigens to activate the immune system.[22,160] Thus, exosomes from virus-infected cells can have opposite effects on virus spreading.

1.3.2 Cell Adhesion

Microvesicles from platelets are thought to facilitate thrombus formation at the site of vascular injury, because these microvesicles facilitate adhesion of platelets to the endothelial cell matrix at sites of vascular injury.[170]

1.3.3 Waste Management and Protection against Stress

Although it is unknown to what extent vesicles contribute to cellular homeostasis and survival by waste management, such a general function would explain why most cells, including prokaryotes, release vesicles into their environment.[171,172] The classic example of waste management by microvesicles is the removal of the redundant transferrin receptor by exosomes from the surface of maturing reticulocytes.[8–11,173] Another well-known example is the incubation of platelets with the complement C5b-9 complex (Fig. 1.2). To escape from complement-induced lysis, platelets release vesicles enriched in the C5b-9 complex.[174] Exosomes and microvesicles from various types of viable and healthy cells contain active caspase 3, which is not detectable in the releasing cells, suggesting that caspase 3 is removed to ensure cellular survival,[175,176] and cancer cells release microvesicles enriched in anticancer drugs (Fig. 1.3).[130,177] In sum, the release of vesicles from cells is likely to contribute to cellular homeostasis and survival.

1.3.4 Coagulation

Microvesicles have been associated with the consolidation of the coagulation process (Fig. 1.2) because they can expose PS, which is an essential cofactor for coagulation, by providing the phospholipid surface to form necessary complexes of coagulation factors.[6,7] In the 1980s tumor-derived vesicles were shown to trigger the coagulation process by exposing TF.[178,179] Because tumor-derived vesicles exposing TF (Fig. 1.3) can be present within the blood of cancer patients, and because cancer patients have an increased risk of developing venous thromboembolism (VTE), these vesicles have been associated with the increased risk of developing VTE in these patients (Fig. 1.3).[180,181] The link between vesicles and coagulation is strengthened by earlier observations that increased concentrations of circulating microvesicles are associated with a hypercoagulable phenotype,[56,182–184] whereas decreased concentrations of vesicles have been associated with a bleeding tendency.[185] TF-exposing vesicles are also present in, for example, wound blood, the blood of patients with meningococcal septic shock, and disseminated intravascular coagulation;[51,186] joint fluid from inflamed joints of arthritic patients;[54] and saliva and urine.[55] TF-exposing vesicles trigger thrombus formation *in vivo*,[187] are present in human atherosclerotic plaques,[188] and are deposited at the site of vascular injury.[189–191] Vesicles present in blood collected from healthy human subjects do not expose detectable levels of TF, and these vesicles have been associated with an anticoagulant rather than a procoagulant phenotype.[192–194]

At present, there is no consensus as to whether exosomes support coagulation by exposing PS,[17,52,195] but human urine and saliva contain vesicles smaller than 100 nm that expose coagulant TF.[55] Taken together, although vesicles contribute to coagulation, the precise contribution of the various types of vesicles has to be unraveled.

1.3.5 Vascular Function and Integrity

Microvesicles isolated from plasma of pre-eclamptic women, endothelial cells, lymphocytes of diabetic or HIV patients, or T-cells, all inhibit endothelium-dependent vasorelaxation, suggesting that such microvesicles may affect vascular function *in vivo*.[196,197]

1.4 Clinical Applications

1.4.1 Therapy

1.4.1.1 Cancer

Exosomes are being tested as vaccines to suppress tumor growth, because exosomes from tumor-peptide-pulsed antigen-presenting cells (DCs) expose MHC class I and II molecules, as well as T-cell costimulatory molecules. These exosomes prime cytotoxic T-cells *in vivo*, thereby *eradicating* or *suppressing* growth of established tumors.[198,199] These exosomes also expose lactadherin, which targets exosomes to macrophages and DCs, and heat shock cognate protein 73, which triggers antitumor immune responses *in vivo*.[200]

Production and isolation protocols of DC-derived exosomes for clinical use have been optimized,[12,201–203] and several "loading strategies" have been tested to enhance the ability of exosomes as antitumor agents.[204,205] When DCs are exposed to exosomes from heat-stress-exposed tumor cells, the DCs will produce exosomes with increased immunogenicity.[206–208] Other protocols include administration of exosomes after pretreatment with cyclophosphamide,[209] priming of DCs with exosomes from genetically modified tumor cells,[210,211] or overexpression of fusion proteins consisting of tumor-associated antigens and the C1C2 domain of lactadherin.[212,213] Also, vaccination with exosomes has been tested for its efficacy to *prevent* tumor growth.[214]

The efficacy of autologous exosome immunotherapy has been tested in several clinical trials. In two phase I studies, advanced non–

small cell lung carcinoma (NSCLC) patients or melanoma patients underwent leukopheresis to produce autologous DCs from the isolated monocyte fraction. Subsequently, exosomes produced by the DCs were isolated and administered to the patient. The adjuvant administration of exosomes, however, was of limited therapeutic relevance,[215,216] which may have been due to the use of exosomes from immature (monocyte-derived) DCs. Treatment of immature DCs with interferon-γ has proven to be a suitable maturation agent for DCs, and this treatment produces exosomes that evoke a much stronger priming of the immune response *in vivo*. At present the ability of exosomes from mature DCs is tested in a phase II trial in patients with advanced NSCLC.[32]

Alternatively, tumor-derived exosomes can raise an antitumor immune response. In a phase I trial, exosomes were isolated from ascites of colorectal cancer patients. Patients received exosomes plus the granulocyte-macrophage colony-stimulating factor or exosomes only. Patients treated with the combination therapy showed a beneficial and specific antitumor cytotoxic T-lymphocyte response.[217] All patients included in aforementioned studies suffered from advanced cancers, and the patients did not respond well to prior treatments. Therefore, the aim of these studies was not primarily to cure the patients but to prolong their survival and to stabilize disease after chemotherapy or radiation therapy.[17]

1.4.1.2 Passing the blood–brain barrier

RNA drugs require efficient, tissue-specific, and nonimmunogenic delivery. Delivery of RNA to tissues at therapeutic effective doses has proven difficult thus far.[218] Recently, DC-derived exosomes were shown to be capable of delivering short interfering RNA (siRNA) in the brain of mice *in vivo* after intravenous administration, and this delivery inhibits expression and production of BACE1, a therapeutic target in Alzheimer disease.[219] This delivery of siRNA shows that exosome-mediated drug delivery can overcome the blood–brain barrier for local drug delivery.[220] In sum, the use of autologous exosomes for drug delivery is promising.

1.4.1.3 Inflammation and immune response

Exosomes can efficiently suppress inflammatory and autoimmune responses *in vivo*. For example, exosomes from genetically modified mouse DCs have immunosuppressive and anti-inflammatory properties in mouse models of experimental arthritis upon systemic or intra-articular administration.[221–224] In addition, exosomes attenuate the acute systemic inflammatory response in sepsis by facilitating the clearance of apoptotic cells and perhaps apoptotic vesicles in a lactadherin-dependent mechanism.[225]

1.4.1.4 Neovascularization

Exosomes from $CD34^+$ stem cells improve the viability of endothelial cells and induce endothelial proliferation and tube formation *in vitro*, as well as angiogenesis in various *in vivo* models, showing that exosomes promote angiogenesis *in vitro* and *in vivo*.[226,227]

1.4.2 Prognosis

The increased risk of cancer patients to develop VTE is probably explained by their increased extrinsic (TF-initiated) coagulation activation.[228] Many tumors express and produce TF. Part of this TF is released on vesicles present in the blood of cancer patients,[134,178,181,229] and tumor-derived vesicles exposing coagulant TF are associated with cancer patients who present with VTE.[180] Because the bleeding risk is too high when all cancer patients are treated prophylactically with anticoagulant therapy before the onset of VTE, a study has been initiated by our institute to predict the development of VTE in cancer patients. In this study, the ability of blood-borne and thus in part tumor-derived vesicles to trigger TF-dependent coagulation is measured in plasma from a blood sample that is collected when patients enter the hospital for a new gift of chemotherapy. On the basis of our vesicle-dependent TF-dependent plasma-clotting assay, earlier described as the fibrin generation test,[55] patients are classified as high risk or low risk for developing VTE. Patients have a 6 month follow-up for developing VTE, with at present 8 European hospitals involved and over 400 patients included. Depending on the outcome of this pilot study, the efficacy of prophylactic treatment of cancer patients identified by the vesicle-based clotting test as having a high risk for developing VTE will be studied in a large multicenter trial.

1.4.3 Biomarkers

Cell-derived vesicles are a promising source of novel clinical biomarkers. There are many examples. Exosomes in cerebrospinal fluid of Alzheimer patients contain an established biomarker of this disease, a phosphorylated form of Tau.[230] Exosomes in the urine of prostate cancer patients are associated with prostate-specific antigen and prostate-specific membrane antigen,[231] claudin-containing exosomes in peripheral blood are associated with ovarian cancer,[232] and prostasomes can be detected in the blood of prostate cancer patients.[233] To what extent these biomarkers are of clinical relevance needs further investigation.

1.4.4 Cell-Derived Vesicles In Vivo

There are still many unanswered questions with regard to the relevance of vesicles *in vivo*. We still do not know why cells release vesicles; their true biological relevance *in vivo* is to be determined. The biological relevance of vesicles will, at least in part, depend on the production and clearance of vesicles. Obviously, such a delicate balance is disturbed in, for example, cell cultures. In this part, we will briefly discuss clearance, and we will compare eukaryotic and bacterial vesicles to emphasize that vesicles and their functions may be conserved throughout evolution and therefore are certainly not unique for eukaryotic cells.

1.4.4.1 Clearance

Exosomes were shown to be internalized via phagocytosis by binding to the PS receptor TIM-4 (T-cell immunoglobulin and mucin-domain-containing molecule),[234] although it is unclear to what extent exosomes expose PS *in vivo*.[52,62,195] Phagocytosis of PS-exposing vesicles may be mediated by lactadherin because splenic macrophages in lactadherin-deficient mice have a decreased capacity to phagocytose microvesicles, resulting in increased concentrations of microvesicles and a hypercoagulable state.[235]

Injection of cancer patients with radiolabeled liposomes results in half-life times of vesicle clearance of 6 min and 5 h, suggesting a two-compartment model of distribution.[236] In other studies, half-life times of either minutes or hours were reported.[237–240] Clearly, more detailed studies will be essential to provide more insight into the clearance of vesicles and into the underlying mechanisms, especially the variation in clearance to be anticipated with vesicles of various composition.

1.4.4.2 Intercellular signaling by vesicles: An evolutionary conserved mechanism

Bacteria release vesicles (50–250 nm; Fig. 1.5), which are often called outer-membrane vesicles (OMVs) and consist of phospholipids, proteins, DNA, RNA, and lipopolysaccharides.[139,241–243] OMVs from the gram-negative bacterium *Pseudomomas aeruginosa* contain signaling molecules, which coordinate the group behavior of this bacterium.[244,245] OMVs also exchange information with eukaryotic cells and as such can be a cause of disease. OMVs of *Pseudomomas aeruginosa* contain virulence factors that kill eukarotic cells when delivered into the cytoplasm after fusion.[241,246] OMVs from *Escherichia coli* transfer heat-labile enterotoxin to recipient cells via endocytosis, thereby triggering interleukin-6 production.[247,248] Delivery of peptidoglycans of *Helicobacter pylori*, *Pseudomonas aeruginosa*, or *Neisseria gonorrhoea* by OMVs into the cytosol of host cells results in recognition by Nod1 in epithelial cells, thereby inducing both innate and adaptive immune responses.[249]

OMVs also contain death messages to kill other species of bacteria.[250,251] Gentamycin, an antibiotic used to treat infections of *Pseudonomas aeruginosa*, increases the release of OMVs. Because these OMVs contain gentamycin, the “predatory activity” of these OMVs on other bacteria becomes increased, whereas at the same time the release contributes to drug removal and homeostasis of the releasing bacterium.[171,172] In addition, OMVs can be used to deliver antimicrobials to the cytoplasm of infected host cells,[171,172] and OMVs can transfer functional genes to other bacteria, showing that OMVs are also capable of exchanging functional genetic information between cells.[139]

OMV-based vaccines have been used in vaccination protocols against meningococcal disease.[252–255] The underlying molecular mechanisms by which OMVs activate the immune system are not well known, but OMVs can offer protection against a subsequent challenge with bacteria.[242,256] Thus, OMVs and vesicles from eukaryotic cells share many common functions.

1.5 Conclusion

The research on eukaryotic vesicles is a relatively new field. Despite important developments that have been made recently, there are still many questions remaining. The mechanisms underlying the release of vesicles and the sorting of compounds into vesicles are incompletely understood, the classification of vesicles needs to be further improved, and the relevance of vesicles *in vivo* is still obscure. Nevertheless, a further understanding of the biology of vesicles will improve our understanding of health and disease and offer promising tools for monitoring and even treating disease.

References

1. Conde-Vancells J, Rodriguez-Suarez E, Embade N, Gil D, Matthiesen R, Valle M, Elortza F, Lu SC, Mato JM, Falcon-Perez JM (2008) Characterization and comprehensive proteome profiling of exosomes secreted by hepatocytes, *J. Proteome Res.*, **7**(12), 5157–5166.
2. Yuana Y, Oosterkamp TH, Bahatyrova S, Ashcroft B, Garcia RP, Bertina RM, Osanto S (2010) Atomic force microscopy: a novel approach to the detection of nanosized blood microparticles, *J. Thromb. Haemost.*, **8**(2), 315–323.
3. Dragovic RA, Gardiner C, Brooks AS, Tannetta DS, Ferguson DJ, Hole P, Carr B, Redman CW, Harris AL, Dobson PJ, Harrison P, Sargent IL (2011) Sizing and phenotyping of cellular vesicles using nanoparticle tracking analysis, *Nanomedicine*, **7**(6), 780–788.
4. van der Pol E, Hoekstra AG, Sturk A, Otto C, van Leeuwen TG, Nieuwland R (2010) Optical and non-optical methods for detection and characterization of microparticles and exosomes, *J. Thromb. Haemost.*, **8**(12), 2596–2607.
5. van der Pol E, van Gemert MJ, Sturk A, Nieuwland R, van Leeuwen TG (2012) Single versus swarm detection of microparticles and exosomes by flow cytometry, *J. Thromb. Haemost.*, **10**(5), 919–930.
6. Chargaff E, West R (1946) The biological significance of the thromboplastic protein of blood, *J. Biol. Chem.*, **166**(1), 189–197.
7. Wolf P (1967) The nature and significance of platelet products in human plasma, *BJH*, **13**(3), 269–288.
8. Johnstone RM, Adam M, Hammond JR, Orr L, Turbide C (1987) Vesicle formation during reticulocyte maturation. Association of plasma membrane activities with released vesicles (exosomes), *J. Biol. Chem.*, **262**(19), 9412–9420.
9. Johnstone RM, Mathew A, Mason AB, Teng K (1991) Exosome formation during maturation of mammalian and avian reticulocytes:

evidence that exosome release is a major route for externalization of obsolete membrane proteins, *J. Cell Physiol.*, **147**(1), 27–36.

10. Grdisa M, Mathew A, Johnstone RM (1993) Expression and loss of the transferrin receptor in growing and differentiating HD3 cells, *J. Cell Physiol.*, **155**(2), 349–357.
11. Johnstone RM (1992) The Jeanne Manery-Fisher Memorial Lecture 1991. Maturation of reticulocytes: formation of exosomes as a mechanism for shedding membrane proteins, *Biochem. Cell Biol.*, **70**(3–4), 179–190.
12. Le Pecq JB (2005) Dexosomes as a therapeutic cancer vaccine: from bench to bedside, *Blood Cells Mol. Dis.*, **35**(2), 129–135.
13. Stegmayr B, Ronquist G (1982) Promotive effect on human sperm progressive motility by prostasomes, *Urol. Res.*, **10**(5), 253–257.
14. Tanimura A, McGregor DH, Anderson HC (1983) Matrix vesicles in atherosclerotic calcification, *Proc. Soc. Exp. Biol. Med.*, **172**(2), 173–177.
15. Fischer von MG, Mignery GA, Baumert M, Perin MS, Hanson TJ, Burger PM, Jahn R, Sudhof TC (1990) rab3 is a small GTP-binding protein exclusively localized to synaptic vesicles, *Proc. Natl. Acad. Sci. U. S. A.*, **87**(5), 1988–1992.
16. Cocucci E, Racchetti G, Meldolesi J (2009) Shedding microvesicles: artefacts no more, *Trends Cell Biol.*, **19**(2), 43–51.
17. Thery C, Ostrowski M, Segura E (2009) Membrane vesicles as conveyors of immune responses, *Nat. Rev. Immunol.*, **9**(8), 581–593.
18. Beyer C, Pisetsky DS (2010) The role of microparticles in the pathogenesis of rheumatic diseases, *Nat. Rev. Rheumatol.*, **6**(1), 21–29.
19. Mathivanan S, Ji H, Simpson RJ (2010) Exosomes: extracellular organelles important in intercellular communication, *J. Proteomics*, **73**(10), 1907–1920.
20. Chandler WL, Yeung W, Tait JF (2011) A new microparticle size calibration standard for use in measuring smaller microparticles using a new flow cytometer, *J. Thromb. Haemost.*, **9**(6), 1216–1224.
21. Dalli J, Norling LV, Renshaw D, Cooper D, Leung KY, Perretti M (2008) Annexin 1 mediates the rapid anti-inflammatory effects of neutrophil-derived microparticles, *Blood*, **112**(6), 2512–2519.
22. Walker JD, Maier CL, Pober JS (2009) Cytomegalovirus-infected human endothelial cells can stimulate allogeneic CD4+ memory T cells by releasing antigenic exosomes, *J. Immunol.*, **182**(3), 1548–1559.

23. Prado N, Marazuela EG, Segura E, Fernandez-Garcia H, Villalba M, Thery C, Rodriguez R, Batanero E (2008) Exosomes from bronchoalveolar fluid of tolerized mice prevent allergic reaction, *J. Immunol.*, **181**(2), 1519–1525.
24. Kesimer M, Scull M, Brighton B, DeMaria G, Burns K, O'Neal W, Pickles RJ, Sheehan JK (2009) Characterization of exosome-like vesicles released from human tracheobronchial ciliated epithelium: a possible role in innate defense, *FASEB J.*, **23**(6), 1858–1868.
25. Admyre C, Grunewald J, Thyberg J, Gripenback S, Tornling G, Eklund A, Scheynius A, Gabrielsson S (2003) Exosomes with major histocompatibility complex class II and co-stimulatory molecules are present in human BAL fluid, *Eur. Respir. J.*, **22**(4), 578–583.
26. Pisitkun T, Shen RF, Knepper MA (2004) Identification and proteomic profiling of exosomes in human urine, *Proc. Natl. Acad. Sci. U. S. A.*, **101**(36), 13368–13373.
27. Caby MP, Lankar D, Vincendeau-Scherrer C, Raposo G, Bonnerot C (2005) Exosomal-like vesicles are present in human blood plasma, *Int. Immunol.*, **17**(7), 879–887.
28. Keller S, Sanderson MP, Stoeck A, Altevogt P (2006) Exosomes: from biogenesis and secretion to biological function, *Immunol. Lett.*, **107**(2), 102–108.
29. Vella LJ, Greenwood DL, Cappai R, Scheerlinck JP, Hill AF (2008) Enrichment of prion protein in exosomes derived from ovine cerebral spinal fluid, *Vet. Immunol. Immunopathol.*, **124**(3–4), 385–393.
30. Simons M, Raposo G (2009) Exosomes—vesicular carriers for intercellular communication, *Curr. Opin. Cell Biol.*, **21**(4), 575–581.
31. Wubbolts R, Leckie RS, Veenhuizen PT, Schwarzmann G, Mobius W, Hoernschemeyer J, Slot JW, Geuze HJ, Stoorvogel W (2003) Proteomic and biochemical analyses of human B cell-derived exosomes. Potential implications for their function and multivesicular body formation, *J. Biol. Chem.*, **278**(13), 10963–10972.
32. Chaput N, Thery C (2011) Exosomes: immune properties and potential clinical implementations, *Semin. Immunopathol.*, **33**(5), 419–440.
33. Bobrie A, Colombo M, Raposo G, Thery C (2011) Exosome secretion: molecular mechanisms and roles in immune responses, *Traffic*, **12**(12), 1659–1668.
34. Record M, Subra C, Silvente-Poirot S, Poirot M (2011) Exosomes as intercellular signalosomes and pharmacological effectors, *Biochem. Pharmacol.*, **81**(10), 1171–1182.

35. Matsuo H, Chevallier J, Mayran N, Le B, I, Ferguson C, Faure J, Blanc NS, Matile S, Dubochet J, Sadoul R, Parton RG, Vilbois F, Gruenberg J (2004) Role of LBPA and Alix in multivesicular liposome formation and endosome organization, *Science*, **303**(5657), 531–534.

36. Mobius W, Ohno-Iwashita Y, van Donselaar EG, Oorschot VM, Shimada Y, Fujimoto T, Heijnen HF, Geuze HJ, Slot JW (2002) Immunoelectron microscopic localization of cholesterol using biotinylated and non-cytolytic perfringolysin O, *J. Histochem. Cytochem.*, **50**(1), 43–55.

37. White IJ, Bailey LM, Aghakhani MR, Moss SE, Futter CE (2006) EGF stimulates annexin 1-dependent inward vesiculation in a multivesicular endosome subpopulation, *EMBO J.*, **25**(1), 1–12.

38. Babst M (2005) A protein's final ESCRT, *Traffic*, **6**(1), 2–9.

39. Michelet X, Djeddi A, Legouis R (2010) Developmental and cellular functions of the ESCRT machinery in pluricellular organisms, *Biol. Cell*, **102**(3), 191–202.

40. Trajkovic K, Hsu C, Chiantia S, Rajendran L, Wenzel D, Wieland F, Schwille P, Brugger B, Simons M (2008) Ceramide triggers budding of exosome vesicles into multivesicular endosomes, *Science*, **319**(5867), 1244–1247.

41. Marsh M, van Meer G (2008) Cell biology. No ESCRTs for exosomes, *Science*, **319**(5867), 1191–1192.

42. Theos AC, Truschel ST, Tenza D, Hurbain I, Harper DC, Berson JF, Thomas PC, Raposo G, Marks MS (2006) A lumenal domain-dependent pathway for sorting to intralumenal vesicles of multivesicular endosomes involved in organelle morphogenesis, *Dev. Cell*, **10**(3), 343–354.

43. Buschow SI, Nolte-'t Hoen EN, van NG, Pols MS, ten BT, Lauwen M, Ossendorp F, Melief CJ, Raposo G, Wubbolts R, Wauben MH, Stoorvogel W (2009) MHC II in dendritic cells is targeted to lysosomes or T cell-induced exosomes via distinct multivesicular body pathways, *Traffic*, **10**(10), 1528–1542.

44. Savina A, Fader CM, Damiani MT, Colombo MI (2005) Rab11 promotes docking and fusion of multivesicular bodies in a calcium-dependent manner, *Traffic*, **6**(2), 131–143.

45. Hsu C, Morohashi Y, Yoshimura S, Manrique-Hoyos N, Jung S, Lauterbach MA, Bakhti M, Gronborg M, Mobius W, Rhee J, Barr FA, Simons M (2010) Regulation of exosome secretion by Rab35 and its GTPase-activating proteins TBC1D10A-C, *J. Cell Biol.*, **189**(2), 223–232.

46. Ostrowski M, Carmo NB, Krumeich S, Fanget I, Raposo G, Savina A, Moita CF, Schauer K, Hume AN, Freitas RP, Goud B, Benaroch P,

Hacohen N, Fukuda M, Desnos C, Seabra MC, Darchen F, Amigorena S, Moita LF, Thery C (2010) Rab27a and Rab27b control different steps of the exosome secretion pathway, *Nat. Cell Biol.*, **12**(1), 19–30.

47. Stenmark H (2009) Rab GTPases as coordinators of vesicle traffic, *Nat. Rev. Mol. Cell Biol.*, **10**(8), 513–525.
48. Booth AM, Fang Y, Fallon JK, Yang JM, Hildreth JE, Gould SJ (2006) Exosomes and HIV Gag bud from endosome-like domains of the T cell plasma membrane, *J. Cell Biol.*, **172**(6), 923–935.
49. Lenassi M, Cagney G, Liao M, Vaupotic T, Bartholomeeusen K, Cheng Y, Krogan NJ, Plemenitas A, Peterlin BM (2010) HIV Nef is secreted in exosomes and triggers apoptosis in bystander CD4+ T cells, *Traffic*, **11**(1), 110–122.
50. Fang Y, Wu N, Gan X, Yan W, Morrell JC, Gould SJ (2007) Higher-order oligomerization targets plasma membrane proteins and HIV gag to exosomes, *PLOS Biol.*, **5**(6), e158.
51. Nieuwland R, Berckmans RJ, Rotteveel-Eijkman RC, Maquelin KN, Roozendaal KJ, Jansen PG, ten HK, Eijsman L, Hack CE, Sturk A (1997) Cell-derived microparticles generated in patients during cardiopulmonary bypass are highly procoagulant, *Circulation*, **96**(10), 3534–3541.
52. Heijnen HF, Schiel AE, Fijnheer R, Geuze HJ, Sixma JJ (1999) Activated platelets release two types of membrane vesicles: microvesicles by surface shedding and exosomes derived from exocytosis of multivesicular bodies and alpha-granules, *Blood*, **94**(11), 3791–3799.
53. Canault M, Leroyer AS, Peiretti F, Leseche G, Tedgui A, Bonardo B, Alessi MC, Boulanger CM, Nalbone G (2007) Microparticles of human atherosclerotic plaques enhance the shedding of the tumor necrosis factor-alpha converting enzyme/ADAM17 substrates, tumor necrosis factor and tumor necrosis factor receptor-1, *Am. J. Pathol.*, **171**(5), 1713–1723.
54. Berckmans RJ, Nieuwland R, Tak PP, Boing AN, Romijn FP, Kraan MC, Breedveld FC, Hack CE, Sturk A (2002) Cell-derived microparticles in synovial fluid from inflamed arthritic joints support coagulation exclusively via a factor VII-dependent mechanism, *Arthritis Rheum.*, **46**(11), 2857–2866.
55. Berckmans RJ, Sturk A, van Tienen LM, Schaap MC, Nieuwland R (2011) Cell-derived vesicles exposing coagulant tissue factor in saliva, *Blood*, **117**(11), 3172–3180.
56. Mallat Z, Benamer H, Hugel B, Benessiano J, Steg PG, Freyssinet JM, Tedgui A (2000) Elevated levels of shed membrane microparticles with

procoagulant potential in the peripheral circulating blood of patients with acute coronary syndromes, *Circulation*, **101**(8), 841–843.

57. Morel O, Jesel L, Freyssinet JM, Toti F (2011) Cellular mechanisms underlying the formation of circulating microparticles, *Arterioscler. Thromb. Vasc. Biol.*, **31**(1), 15–26.

58. Turiak L, Misjak P, Szabo TG, Aradi B, Paloczi K, Ozohanics O, Drahos L, Kittel A, Falus A, Buzas EI, Vekey K (2011) Proteomic characterization of thymocyte-derived microvesicles and apoptotic bodies in BALB/c mice, *J. Proteomics*, **74**(10), 2025–2033.

59. Gyorgy B, Modos K, Pallinger E, Paloczi K, Pasztoi M, Misjak P, Deli MA, Sipos A, Szalai A, Voszka I, Polgar A, Toth K, Csete M, Nagy G, Gay S, Falus A, Kittel A, Buzas EI (2011) Detection and isolation of cell-derived microparticles are compromised by protein complexes resulting from shared biophysical parameters, *Blood*, **117**(4), e39–e48.

60. Shet AS, Aras O, Gupta K, Hass MJ, Rausch DJ, Saba N, Koopmeiners L, Key NS, Hebbel RP (2003) Sickle blood contains tissue factor-positive microparticles derived from endothelial cells and monocytes, *Blood*, **102**(7), 2678–2683.

61. Joop K, Berckmans RJ, Nieuwland R, Berkhout J, Romijn FP, Hack CE, Sturk A (2001) Microparticles from patients with multiple organ dysfunction syndrome and sepsis support coagulation through multiple mechanisms, *Thromb. Haemost.*, **85**(5), 810–820.

62. Connor DE, Exner T, Ma DD, Joseph JE (2010) The majority of circulating platelet-derived microparticles fail to bind annexin V, lack phospholipid-dependent procoagulant activity and demonstrate greater expression of glycoprotein Ib, *Thromb. Haemost.*, **103**(5), 1044–1052.

63. Marzesco AM, Janich P, Wilsch-Brauninger M, Dubreuil V, Langenfeld K, Corbeil D, Huttner WB (2005) Release of extracellular membrane particles carrying the stem cell marker prominin-1 (CD133) from neural progenitors and other epithelial cells, *J. Cell Sci.*, **118**(Pt 13), 2849–2858.

64. Kerr JF, Wyllie AH, Currie AR (1972) Apoptosis: a basic biological phenomenon with wide-ranging implications in tissue kinetics, *Br. J. Cancer*, **26**(4), 239–257.

65. Hristov M, Erl W, Linder S, Weber PC (2004) Apoptotic bodies from endothelial cells enhance the number and initiate the differentiation of human endothelial progenitor cells *in vitro*, *Blood*, **104**(9), 2761–2766.

66. Thery C, Boussac M, Veron P, Ricciardi-Castagnoli P, Raposo G, Garin J, Amigorena S (2001) Proteomic analysis of dendritic cell-derived exosomes: a secreted subcellular compartment distinct from apoptotic vesicles, *J. Immunol.*, **166**(12), 7309–7318.

67. Bilyy RO, Shkandina T, Tomin A, Munoz LE, Franz S, Antonyuk V, Kit YY, Zirngibl M, Fuernrohr BG, Janko C, Lauber K, Schiller M, Schett G, Stoika RS, Herrmann M (2011) Macrophages discriminate glycosylation patterns of apoptotic cell-derived microparticles, *J. Biol. Chem.*, **287**(1), 496–503.

68. Martinez-Lorenzo MJ, Anel A, Gamen S, Monle n I, Lasierra P, Larrad L, Pineiro A, Alava MA, Naval J (1999) Activated human T cells release bioactive Fas ligand and APO2 ligand in microvesicles, *J. Immunol.*, **163**(3), 1274–1281.

69. Hohlbaum AM, Gregory MS, Ju ST, Marshak-Rothstein A (2001) Fas ligand engagement of resident peritoneal macrophages *in vivo* induces apoptosis and the production of neutrophil chemotactic factors, *J. Immunol.*, **167**(11), 6217–6224.

70. Abrahams VM, Straszewski-Chavez SL, Guller S, Mor G (2004) First trimester trophoblast cells secrete Fas ligand which induces immune cell apoptosis, *Mol. Hum. Reprod.*, **10**(1), 55–63.

71. Frangsmyr L, Baranov V, Nagaeva O, Stendahl U, Kjellberg L, Mincheva-Nilsson L (2005) Cytoplasmic microvesicular form of Fas ligand in human early placenta: switching the tissue immune privilege hypothesis from cellular to vesicular level, *Mol. Hum. Reprod.*, **11**(1), 35–41.

72. Taylor DD, Akyol S, Gercel-Taylor C (2006) Pregnancy-associated exosomes and their modulation of T cell signaling, *J. Immunol.*, **176**(3), 1534–1542.

73. Hedlund M, Stenqvist AC, Nagaeva O, Kjellberg L, Wulff M, Baranov V, Mincheva-Nilsson L (2009) Human placenta expresses and secretes NKG2D ligands via exosomes that down-modulate the cognate receptor expression: evidence for immunosuppressive function, *J. Immunol.*, **183**(1), 340–351.

74. Andreola G, Rivoltini L, Castelli C, Huber V, Perego P, Deho P, Squarcina P, Accornero P, Lozupone F, Lugini L, Stringaro A, Molinari A, Arancia G, Gentile M, Parmiani G, Fais S (2002) Induction of lymphocyte apoptosis by tumor cell secretion of FasL-bearing microvesicles, *J. Exp. Med.*, **195**(10), 1303–1316.

75. Martinez-Lorenzo MJ, Anel A, Alava MA, Pineiro A, Naval J, Lasierra P, Larrad L (2004) The human melanoma cell line MelJuSo secretes bioactive FasL and APO2L/TRAIL on the surface of microvesicles.

Possible contribution to tumor counterattack, *Exp. Cell Res.*, **295**(2), 315–329.

76. Abusamra AJ, Zhong Z, Zheng X, Li M, Ichim TE, Chin JL, Min WP (2005) Tumor exosomes expressing Fas ligand mediate CD8+ T-cell apoptosis, *Blood Cells Mol. Dis.*, **35**(2), 169–173.
77. Abrahams VM, Straszewski SL, Kamsteeg M, Hanczaruk B, Schwartz PE, Rutherford TJ, Mor G (2003) Epithelial ovarian cancer cells secrete functional Fas ligand, *Cancer Res.*, **63**(17), 5573–5581.
78. Kim JW, Wieckowski E, Taylor DD, Reichert TE, Watkins S, Whiteside TL (2005) Fas ligand-positive membranous vesicles isolated from sera of patients with oral cancer induce apoptosis of activated T lymphocytes, *Clin. Cancer Res.*, **11**(3), 1010–1020.
79. Liu C, Yu S, Zinn K, Wang J, Zhang L, Jia Y, Kappes JC, Barnes S, Kimberly RP, Grizzle WE, Zhang HG (2006) Murine mammary carcinoma exosomes promote tumor growth by suppression of NK cell function, *J. Immunol.*, **176**(3), 1375–1385.
80. Clayton A, Mitchell JP, Court J, Linnane S, Mason MD, Tabi Z (2008) Human tumor-derived exosomes down-modulate NKG2D expression, *J. Immunol.*, **180**(11), 7249–7258.
81. Hedlund M, Nagaeva O, Kargl D, Baranov V, Mincheva-Nilsson L (2011) Thermal- and oxidative stress causes enhanced release of NKG2D ligand-bearing immunosuppressive exosomes in leukemia/lymphoma T and B cells, *PLOS ONE*, **6**(2), e16899.
82. Liu Y, Xiang X, Zhuang X, Zhang S, Liu C, Cheng Z, Michalek S, Grizzle W, Zhang HG (2010) Contribution of MyD88 to the tumor exosome-mediated induction of myeloid derived suppressor cells, *Am. J. Pathol.*, **176**(5), 2490–2499.
83. Xiang X, Poliakov A, Liu C, Liu Y, Deng ZB, Wang J, Cheng Z, Shah SV, Wang GJ, Zhang L, Grizzle WE, Mobley J, Zhang HG (2009) Induction of myeloid-derived suppressor cells by tumor exosomes, *Int. J. Cancer*, **124**(11), 2621–2633.
84. Clayton A, Al-Taei S, Webber J, Mason MD, Tabi Z (2011) Cancer exosomes express CD39 and CD73, which suppress T cells through adenosine production, *J. Immunol.*, **187**(2), 676–683.
85. Flanagan J, Middeldorp J, Sculley T (2003) Localization of the Epstein-Barr virus protein LMP 1 to exosomes, *J. Gen. Virol.*, **84**(Pt 7), 1871–1879.
86. Silverman JM, Clos J, Horakova E, Wang AY, Wiesgigl M, Kelly I, Lynn MA, McMaster WR, Foster LJ, Levings MK, Reiner NE (2010) Leishmania exosomes modulate innate and adaptive immune responses through

effects on monocytes and dendritic cells, *J. Immunol.*, **185**(9), 5011–5022.

87. Silverman JM, Clos J, de'Oliveira CC, Shirvani O, Fang Y, Wang C, Foster LJ, Reiner NE (2010) An exosome-based secretion pathway is responsible for protein export from Leishmania and communication with macrophages, *J. Cell Sci.*, **123**(Pt 6), 842–852.
88. Zhang HG, Liu C, Su K, Yu S, Zhang L, Zhang S, Wang J, Cao X, Grizzle W, Kimberly RP (2006) A membrane form of TNF-alpha presented by exosomes delays T cell activation-induced cell death, *J. Immunol.*, **176**(12), 7385–7393.
89. van Niel G, Raposo G, Candalh C, Boussac M, Hershberg R, Cerf-Bensussan N, Heyman M (2001) Intestinal epithelial cells secrete exosome-like vesicles, *Gastroenterology*, **121**(2), 337–349.
90. van, Niel G, Mallegol J, Bevilacqua C, Candalh C, Brugiere S, Tomaskovic-Crook E, Heath JK, Cerf-Bensussan N, Heyman M (2003) Intestinal epithelial exosomes carry MHC class II/peptides able to inform the immune system in mice, *Gut*, **52**(12), 1690–1697.
91. Almqvist N, Lonnqvist A, Hultkrantz S, Rask C, Telemo E (2008) Serum-derived exosomes from antigen-fed mice prevent allergic sensitization in a model of allergic asthma, *Immunology*, **125**(1), 21–27.
92. Admyre C, Bohle B, Johansson SM, Focke-Tejkl M, Valenta R, Scheynius A, Gabrielsson S (2007) B cell-derived exosomes can present allergen peptides and activate allergen-specific T cells to proliferate and produce TH2-like cytokines, *J. Allergy Clin. Immunol.*, **120**(6), 1418–1424.
93. Segura E, Guerin C, Hogg N, Amigorena S, Thery C (2007) CD8+ dendritic cells use LFA-1 to capture MHC-peptide complexes from exosomes *in vivo*, *J. Immunol.*, **179**(3), 1489–1496.
94. Wolfers J, Lozier A, Raposo G, Regnault A, Thery C, Masurier C, Flament C, Pouzieux S, Faure F, Tursz T, Angevin E, Amigorena S, Zitvogel L (2001) Tumor-derived exosomes are a source of shared tumor rejection antigens for CTL cross-priming, *Nat. Med.*, **7**(3), 297–303.
95. Andre F, Chaput N, Schartz NE, Flament C, Aubert N, Bernard J, Lemonnier F, Raposo G, Escudier B, Hsu DH, Tursz T, Amigorena S, Angevin E, Zitvogel L (2004) Exosomes as potent cell-free peptide-based vaccine. I. Dendritic cell-derived exosomes transfer functional MHC class I/peptide complexes to dendritic cells, *J. Immunol.*, **172**(4), 2126–2136.

96. Beauvillain C, Juste MO, Dion S, Pierre J, Dimier-Poisson I (2009) Exosomes are an effective vaccine against congenital toxoplasmosis in mice, *Vaccine*, **27**(11), 1750–1757.

97. Beauvillain C, Ruiz S, Guiton R, Bout D, Dimier-Poisson I (2007) A vaccine based on exosomes secreted by a dendritic cell line confers protection against T. gondii infection in syngeneic and allogeneic mice, *Microbes Infect.*, **9**(14–15), 1614–1622.

98. Aline F, Bout D, Amigorena S, Roingeard P, Dimier-Poisson I (2004) Toxoplasma gondii antigen-pulsed-dendritic cell-derived exosomes induce a protective immune response against T. gondii infection, *Infect. Immun.*, **72**(7), 4127–4137.

99. Schnitzer JK, Berzel S, Fajardo-Moser M, Remer KA, Moll H (2010) Fragments of antigen-loaded dendritic cells (DC) and DC-derived exosomes induce protective immunity against Leishmania major, *Vaccine*, **28**(36), 5785–5793.

100. Ramachandra L, Qu Y, Wang Y, Lewis CJ, Cobb BA, Takatsu K, Boom WH, Dubyak GR, Harding CV (2010) Mycobacterium tuberculosis synergizes with ATP to induce release of microvesicles and exosomes containing major histocompatibility complex class II molecules capable of antigen presentation, *Infect. Immun.*, **78**(12), 5116–5125.

101. Anand PK, Anand E, Bleck CK, Anes E, Griffiths G (2010) Exosomal Hsp70 induces a pro-inflammatory response to foreign particles including mycobacteria, *PLOS ONE*, **5**(4), e10136.

102. Del Cacho E, Gallego M, Lee SH, Lillehoj HS, Quilez J, Lillehoj EP, Sanchez-Acedo C (2011) Induction of protective immunity against Eimeria tenella infection using antigen-loaded dendritic cells (DC) and DC-derived exosomes, *Vaccine*, **29**(21), 3818–3825.

103. Fourcade O, Simon MF, Viode C, Rugani N, Leballe F, Ragab A, Fournie B, Sarda L, Chap H (1995) Secretory phospholipase A2 generates the novel lipid mediator lysophosphatidic acid in membrane microvesicles shed from activated cells, *Cell*, **80**(6), 919–927.

104. Barry OP, Pratico D, Savani RC, FitzGerald GA (1998) Modulation of monocyte-endothelial cell interactions by platelet microparticles, *J. Clin. Invest*, **102**(1), 136–144.

105. Park KH, Kim BJ, Kang J, Nam TS, Lim JM, Kim HT, Park JK, Kim YG, Chae SW, Kim UH (2011) Ca2+ signaling tools acquired from prostasomes are required for progesterone-induced sperm motility, *Sci. Signal*, **4**(173), ra31.

106. Ren D (2011) Calcium signaling in sperm: help from prostasomes, *Sci. Signal*, **4**(173), e27.

107. Chairoungdua A, Smith DL, Pochard P, Hull M, Caplan MJ (2010) Exosome release of beta-catenin: a novel mechanism that antagonizes Wnt signaling, *J. Cell Biol.*, **190**(6), 1079–1091.

108. Khan S, Jutzy JM, Aspe JR, McGregor DW, Neidigh JW, Wall NR (2011) Survivin is released from cancer cells via exosomes, *Apoptosis*, **16**(1), 1–12.

109. Germain SJ, Sacks GP, Sooranna SR, Sargent IL, Redman CW (2007) Systemic inflammatory priming in normal pregnancy and preeclampsia: the role of circulating syncytiotrophoblast microparticles, *J. Immunol.*, **178**(9), 5949–5956.

110. Messerli M, May K, Hansson SR, Schneider H, Holzgreve W, Hahn S, Rusterholz C (2010) Feto-maternal interactions in pregnancies: placental microparticles activate peripheral blood monocytes, *Placenta*, **31**(2), 106–112.

111. Southcombe J, Tannetta D, Redman C, Sargent I (2011) The immunomodulatory role of syncytiotrophoblast microvesicles, *PLOS ONE*, **6**(5), e20245.

112. Berckmans RJ, Nieuwland R, Kraan MC, Schaap MC, Pots D, Smeets TJ, Sturk A, Tak PP (2005) Synovial microparticles from arthritic patients modulate chemokine and cytokine release by synoviocytes, *Arthritis Res. Ther.*, **7**(3), R536–R544.

113. Distler JH, Jungel A, Huber LC, Seemayer CA, Reich CF, III, Gay RE, Michel BA, Fontana A, Gay S, Pisetsky DS, Distler O (2005) The induction of matrix metalloproteinase and cytokine expression in synovial fibroblasts stimulated with immune cell microparticles, *Proc. Natl. Acad. Sci. U. S. A.*, **102**(8), 2892–2897.

114. Boilard E, Nigrovic PA, Larabee K, Watts GF, Coblyn JS, Weinblatt ME, Massarotti EM, Remold-O'Donnell E, Farndale RW, Ware J, Lee DM (2010) Platelets amplify inflammation in arthritis via collagen-dependent microparticle production, *Science*, **327**(5965), 580–583.

115. Distler JH, Distler O (2010) Inflammation: microparticles and their roles in inflammatory arthritides, *Nat. Rev. Rheumatol.*, **6**(7), 385–386.

116. Cerri C, Chimenti D, Conti I, Neri T, Paggiaro P, Celi A (2006) Monocyte/macrophage-derived microparticles up-regulate inflammatory mediator synthesis by human airway epithelial cells, *J. Immunol.*, **177**(3), 1975–1980.

117. Rautou PE, Leroyer AS, Ramkhelawon B, Devue C, Duflaut D, Vion AC, Nalbone G, Castier Y, Leseche G, Lehoux S, Tedgui A, Boulanger

CM (2011) Microparticles from human atherosclerotic plaques promote endothelial ICAM-1-dependent monocyte adhesion and transendothelial migration, *Circ. Res.*, **108**(3), 335–343.

118. MacKenzie A, Wilson HL, Kiss-Toth E, Dower SK, North RA, Surprenant A (2001) Rapid secretion of interleukin-1beta by microvesicle shedding, *Immunity*, **15**(5), 825–835.

119. Wilson HL, Francis SE, Dower SK, Crossman DC (2004) Secretion of intracellular IL-1 receptor antagonist (type 1) is dependent on P2X7 receptor activation, *J. Immunol.*, **173**(2), 1202–1208.

120. Bianco F, Pravettoni E, Colombo A, Schenk U, Moller T, Matteoli M, Verderio C (2005) Astrocyte-derived ATP induces vesicle shedding and IL-1 beta release from microglia, *J. Immunol.*, **174**(11), 7268–7277.

121. Wang JG, Williams JC, Davis BK, Jacobson K, Doerschuk CM, Ting JP, Mackman N (2011) Monocytic microparticles activate endothelial cells in an IL-1beta-dependent manner, *Blood*, **118**(8), 2366–2374.

122. Thomas LM, Salter RD (2010) Activation of macrophages by P2X7-induced microvesicles from myeloid cells is mediated by phospholipids and is partially dependent on TLR4, *J. Immunol.*, **185**(6), 3740–3749.

123. Esser J, Gehrmann U, D'Alexandri FL, Hidalgo-Estevez AM, Wheelock CE, Scheynius A, Gabrielsson S, Radmark O (2010) Exosomes from human macrophages and dendritic cells contain enzymes for leukotriene biosynthesis and promote granulocyte migration, *J. Allergy Clin. Immunol.*, **126**(5), 1032–1040.

124. Deng ZB, Poliakov A, Hardy RW, Clements R, Liu C, Liu Y, Wang J, Xiang X, Zhang S, Zhuang X, Shah SV, Sun D, Michalek S, Grizzle WE, Garvey T, Mobley J, Zhang HG (2009) Adipose tissue exosome-like vesicles mediate activation of macrophage-induced insulin resistance, *Diabetes*, **58**(11), 2498–2505.

125. Al-Nedawi K, Meehan B, Micallef J, Lhotak V, May L, Guha A, Rak J (2008) Intercellular transfer of the oncogenic receptor EGFRvIII by microvesicles derived from tumour cells, *Nat. Cell Biol.*, **10**(5), 619–624.

126. Higginbotham JN, Demory BM, Gephart JD, Franklin JL, Bogatcheva G, Kremers GJ, Piston DW, Ayers GD, McConnell RE, Tyska MJ, Coffey RJ (2011) Amphiregulin exosomes increase cancer cell invasion, *Curr. Biol.*, **21**(9), 779–786.

127. Tourneur L, Mistou S, Schmitt A, Chiocchia G (2008) Adenosine receptors control a new pathway of Fas-associated death domain protein expression regulation by secretion, *J. Biol. Chem.*, **283**(26), 17929–17938.

128. Ciravolo V, Huber V, Ghedini GC, Venturelli E, Bianchi F, Campiglio M, Morelli D, Villa A, Mina PD, Menard S, Filipazzi P, Rivoltini L, Tagliabue E, Pupa SM (2012) Potential role of HER2-overexpressing exosomes in countering trastuzumab-based therapy, *J. Cell Physiol.*, **227**(2), 658–667.

129. Gong J, Jaiswal R, Mathys JM, Combes V, Grau GE, Bebawy M (2011) Microparticles and their emerging role in cancer multidrug resistance, *Cancer Treat. Rev.*, **38**(3), 226–234.

130. Safaei R, Larson BJ, Cheng TC, Gibson MA, Otani S, Naerdemann W, Howell SB (2005) Abnormal lysosomal trafficking and enhanced exosomal export of cisplatin in drug-resistant human ovarian carcinoma cells, *Mol. Cancer Ther.*, **4**(10), 1595–1604.

131. Sheldon H, Heikamp E, Turley H, Dragovic R, Thomas P, Oon CE, Leek R, Edelmann M, Kessler B, Sainson RC, Sargent I, Li JL, Harris AL (2010) New mechanism for Notch signaling to endothelium at a distance by delta-like 4 incorporation into exosomes, *Blood*, **116**(13), 2385–2394.

132. Hakulinen J, Sankkila L, Sugiyama N, Lehti K, Keski-Oja J (2008) Secretion of active membrane type 1 matrix metalloproteinase (MMP-14) into extracellular space in microvesicular exosomes, *J. Cell Biochem.*, **105**(5), 1211–1218.

133. Taverna S, Flugy A, Saieva L, Kohn EC, Santoro A, Meraviglia S, De LG, Alessandro R (2011) Role of exosomes released by chronic myelogenous leukemia cells in angiogenesis, *Int. J. Cancer*, **130**(9), 2033–2043.

134. Uno K, Homma S, Satoh T, Nakanishi K, Abe D, Matsumoto K, Oki A, Tsunoda H, Yamaguchi I, Nagasawa T, Yoshikawa H, Aonuma K (2007) Tissue factor expression as a possible determinant of thromboembolism in ovarian cancer, *Br. J. Cancer*, **96**(2), 290–295.

135. Zoller M (2009) Tetraspanins: push and pull in suppressing and promoting metastasis, *Nat. Rev. Cancer*, **9**(1), 40–55.

136. Peinado H, Lavotshkin S, Lyden D (2011) The secreted factors responsible for pre-metastatic niche formation: old sayings and new thoughts, *Semin. Cancer Biol.*, **21**(2), 139–146.

137. Rak J (2010) Microparticles in cancer, *Semin. Thromb. Haemost.*, **36**(8), 888–906.

138. Martinez MC, Larbret F, Zobairi F, Coulombe J, Debili N, Vainchenker W, Ruat M, Freyssinet JM (2006) Transfer of differentiation signal by membrane microvesicles harboring hedgehog morphogens, *Blood*, **108**(9), 3012–3020.

139. Yaron S, Kolling GL, Simon L, Matthews KR (2000) Vesicle-mediated transfer of virulence genes from Escherichia coli O157:H7 to other enteric bacteria, *Appl. Environ. Microbiol.*, **66**(10), 4414–4420.

140. Valadi H, Ekstrom K, Bossios A, Sjostrand M, Lee JJ, Lotvall JO (2007) Exosome-mediated transfer of mRNAs and microRNAs is a novel mechanism of genetic exchange between cells, *Nat. Cell Biol.*, **9**(6), 654–659.

141. Mittelbrunn M, Gutierrez-Vazquez C, Villarroya-Beltri C, Gonzalez S, Sanchez-Cabo F, Gonzalez MA, Bernad A, Sanchez-Madrid F (2011) Unidirectional transfer of microRNA-loaded exosomes from T cells to antigen-presenting cells, *Nat. Commun.*, **2**282.

142. Montecalvo A, Larregina AT, Shufesky WJ, Beer SD, Sullivan ML, Karlsson JM, Baty CJ, Gibson GA, Erdos G, Wang Z, Milosevic J, Tkacheva OA, Divito SJ, Jordan R, Lyons-Weiler J, Watkins SC, Morelli AE (2011) Mechanism of transfer of functional microRNAs between mouse dendritic cells via exosomes, *Blood*, **119**(3), 756–766.

143. Skog J, Wurdinger T, van RS, Meijer DH, Gainche L, Sena-Esteves M, Curry WT, Jr., Carter BS, Krichevsky AM, Breakefield XO (2008) Glioblastoma microvesicles transport RNA and proteins that promote tumour growth and provide diagnostic biomarkers, *Nat. Cell Biol.*, **10**(12), 1470–1476.

144. Hong BS, Cho JH, Kim H, Choi EJ, Rho S, Kim J, Kim JH, Choi DS, Kim YK, Hwang D, Gho YS (2009) Colorectal cancer cell-derived microvesicles are enriched in cell cycle-related mRNAs that promote proliferation of endothelial cells, *BMC Genomics*, **10**556.

145. Guescini M, Genedani S, Stocchi V, Agnati LF (2010) Astrocytes and glioblastoma cells release exosomes carrying mtDNA, *J. Neural Transm.*, **117**(1), 1–4.

146. Balaj L, Lessard R, Dai L, Cho YJ, Pomeroy SL, Breakefield XO, Skog J (2011) Tumour microvesicles contain retrotransposon elements and amplified oncogene sequences, *Nat. Commun.*, **2**180.

147. Lasser C, Alikhani VS, Ekstrom K, Eldh M, Paredes PT, Bossios A, Sjostrand M, Gabrielsson S, Lotvall J, Valadi H (2011) Human saliva, plasma and breast milk exosomes contain RNA: uptake by macrophages, *J. Transl. Med.*, **9**, 9.

148. Meckes DG, Jr., Shair KH, Marquitz AR, Kung CP, Edwards RH, Raab-Traub N (2010) Human tumor virus utilizes exosomes for intercellular communication, *Proc. Natl. Acad. Sci. U. S. A.*, **107**(47), 20370–20375.

149. Pegtel DM, Cosmopoulos K, Thorley-Lawson DA, van Eijndhoven MA, Hopmans ES, Lindenberg JL, de Gruijl TD, Wurdinger T, Middeldorp JM

(2010) Functional delivery of viral miRNAs via exosomes, *Proc. Natl. Acad. Sci. U. S. A.*, **107**(14), 6328–6333.

150. Gourzones C, Gelin A, Bombik I, Klibi J, Verillaud B, Guigay J, Lang P, Temam S, Schneider V, Amiel C, Baconnais S, Jimenez AS, Busson P (2010) Extra-cellular release and blood diffusion of BART viral micro-RNAs produced by EBV-infected nasopharyngeal carcinoma cells, *Virol. J.*, **7**, 271.

151. Fevrier B, Vilette D, Archer F, Loew D, Faigle W, Vidal M, Laude H, Raposo G (2004) Cells release prions in association with exosomes, *Proc. Natl. Acad. Sci. U. S. A.*, **101**(26), 9683–9688.

152. Vella LJ, Sharples RA, Lawson VA, Masters CL, Cappai R, Hill AF (2007) Packaging of prions into exosomes is associated with a novel pathway of PrP processing, *J. Pathol.*, **211**(5), 582–590.

153. Rajendran L, Honsho M, Zahn TR, Keller P, Geiger KD, Verkade P, Simons K (2006) Alzheimer's disease beta-amyloid peptides are released in association with exosomes, *Proc. Natl. Acad. Sci. U. S. A.*, **103**(30), 11172–11177.

154. Mack M, Kleinschmidt A, Bruhl H, Klier C, Nelson PJ, Cihak J, Plachy J, Stangassinger M, Erfle V, Schlondorff D (2000) Transfer of the chemokine receptor CCR5 between cells by membrane-derived microparticles: a mechanism for cellular human immunodeficiency virus 1 infection, *Nat. Med.*, **6**(7), 769–775.

155. Rozmyslowicz T, Majka M, Kijowski J, Murphy SL, Conover DO, Poncz M, Ratajczak J, Gaulton GN, Ratajczak MZ (2003) Platelet- and megakaryocyte-derived microparticles transfer CXCR4 receptor to CXCR4-null cells and make them susceptible to infection by X4-HIV, *AIDS*, **17**(1), 33–42.

156. Wiley RD, Gummuluru S (2006) Immature dendritic cell-derived exosomes can mediate HIV-1 trans infection, *Proc. Natl. Acad. Sci. U. S. A.*, **103**(3), 738–743.

157. Krishnamoorthy L, Bess JW, Jr., Preston AB, Nagashima K, Mahal LK (2009) HIV-1 and microvesicles from T cells share a common glycome, arguing for a common origin, *Nat. Chem. Biol.*, **5**(4), 244–250.

158. Shelton MN, Huang MB, Ali SA, Powell MD, Bond VC (2012) SMR-derived peptide disrupts HIV-1 Nef's interaction with mortalin and blocks virus and Nef exosome release, *J. Virol.*, **86**(1), 406–419.

159. Raymond AD, Campbell-Sims TC, Khan M, Lang M, Huang MB, Bond VC, Powell MD (2011) HIV Type 1 Nef is released from infected cells in CD45(+) microvesicles and is present in the plasma of HIV-infected individuals, *AIDS Res. Hum. Retroviruses*, **27**(2), 167–178.

160. Meckes DG, Jr., Raab-Traub N (2011) Microvesicles and viral infection, *J. Virol.*, **85**(24), 12844–12854.

161. Gould SJ, Booth AM, Hildreth JE (2003) The Trojan exosome hypothesis, *Proc. Natl. Acad. Sci. U. S. A.*, **100**(19), 10592–10597.

162. Gan X, Gould SJ (2011) Identification of an inhibitory budding signal that blocks the release of HIV particles and exosome/microvesicle proteins, *Mol. Biol. Cell*, **22**(6), 817–830.

163. Brugger B, Glass B, Haberkant P, Leibrecht I, Wieland FT, Krausslich HG (2006) The HIV lipidome: a raft with an unusual composition, *Proc. Natl. Acad. Sci. U. S. A.*, **103**(8), 2641–2646.

164. Izquierdo-Useros N, Naranjo-Gomez M, Archer J, Hatch SC, Erkizia I, Blanco J, Borras FE, Puertas MC, Connor JH, Fernandez-Figueras MT, Moore L, Clotet B, Gummuluru S, Martinez-Picado J (2009) Capture and transfer of HIV-1 particles by mature dendritic cells converges with the exosome-dissemination pathway, *Blood*, **113**(12), 2732–2741.

165. Sato K, Aoki J, Misawa N, Daikoku E, Sano K, Tanaka Y, Koyanagi Y (2008) Modulation of human immunodeficiency virus type 1 infectivity through incorporation of tetraspanin proteins, *J. Virol.*, **82**(2), 1021–1033.

166. Jolly C, Sattentau QJ (2007) Human immunodeficiency virus type 1 assembly, budding, cell-cell spread in T cells take place in tetraspanin-enriched plasma membrane domains, *J. Virol.*, **81**(15), 7873–7884.

167. Tumne A, Prasad VS, Chen Y, Stolz DB, Saha K, Ratner DM, Ding M, Watkins SC, Gupta P (2009) Noncytotoxic suppression of human immunodeficiency virus type 1 transcription by exosomes secreted from CD8+ T cells, *J. Virol.*, **83**(9), 4354–4364.

168. Khatua AK, Taylor HE, Hildreth JE, Popik W (2010) Inhibition of LINE-1 and Alu retrotransposition by exosomes encapsidating APOBEC3G and APOBEC3F, *Virology*, **400**(1), 68–75.

169. Khatua AK, Taylor HE, Hildreth JE, Popik W (2009) Exosomes packaging APOBEC3G confer human immunodeficiency virus resistance to recipient cells, *J. Virol.*, **83**(2), 512–521.

170. Merten M, Pakala R, Thiagarajan P, Benedict CR (1999) Platelet microparticles promote platelet interaction with subendothelial matrix in a glycoprotein IIb/IIIa-dependent mechanism, *Circulation*, **99**(19), 2577–2582.

171. Kadurugamuwa JL, Beveridge TJ (1999) Membrane vesicles derived from Pseudomonas aeruginosa and Shigella flexneri can be integrated into the surfaces of other gram-negative bacteria, *Microbiology*, **145**(Pt 8), 2051–2060.

172. Kadurugamuwa JL, Beveridge TJ (1998) Delivery of the non-membrane-permeative antibiotic gentamicin into mammalian cells by using Shigella flexneri membrane vesicles, *Antimicrob. Agents Chemother.*, **42**(6), 1476–1483.

173. Johnstone RM, Bianchini A, Teng K (1989) Reticulocyte maturation and exosome release: transferrin receptor containing exosomes shows multiple plasma membrane functions, *Blood*, **74**(5), 1844–1851.

174. Sims PJ, Faioni EM, Wiedmer T, Shattil SJ (1988) Complement proteins C5b-9 cause release of membrane vesicles from the platelet surface that are enriched in the membrane receptor for coagulation factor Va and express prothrombinase activity, *J. Biol. Chem.*, **263**(34), 18205–18212.

175. de Gassart A, Geminard C, Fevrier B, Raposo G, Vidal M (2003) Lipid raft-associated protein sorting in exosomes, *Blood*, **102**(13), 4336–4344.

176. Abid Hussein MN, Nieuwland R, Hau CM, Evers LM, Meesters EW, Sturk A (2005) Cell-derived microparticles contain caspase 3 *in vitro* and *in vivo*, *J. Thromb. Haemost.*, **3**(5), 888–896.

177. Shedden K, Xie XT, Chandaroy P, Chang YT, Rosania GR (2003) Expulsion of small molecules in vesicles shed by cancer cells: association with gene expression and chemosensitivity profiles, *Cancer Res.*, **63**(15), 4331–4337.

178. Dvorak HF, Quay SC, Orenstein NS, Dvorak AM, Hahn P, Bitzer AM, Carvalho AC (1981) Tumor shedding and coagulation, *Science*, **212**(4497), 923–924.

179. Bastida E, Ordinas A, Escolar G, Jamieson GA (1984) Tissue factor in microvesicles shed from U87MG human glioblastoma cells induces coagulation, platelet aggregation, thrombogenesis, *Blood*, **64**(1), 177–184.

180. Tesselaar ME, Romijn FP, van der Linden IK, Prins FA, Bertina RM, Osanto S (2007) Microparticle-associated tissue factor activity: a link between cancer and thrombosis? *J. Thromb. Haemost.*, **5**(3), 520–527.

181. Zwicker JI, Liebman HA, Neuberg D, Lacroix R, Bauer KA, Furie BC, Furie B (2009) Tumor-derived tissue factor-bearing microparticles are associated with venous thromboembolic events in malignancy, *Clin. Cancer Res.*, **15**(22), 6830–6840.

182. Hugel B, Socie G, Vu T, Toti F, Gluckman E, Freyssinet JM, Scrobohaci ML (1999) Elevated levels of circulating procoagulant microparticles in patients with paroxysmal nocturnal hemoglobinuria and aplastic anemia, *Blood*, **93**(10), 3451–3456.

183. Wiedmer T, Hall SE, Ortel TL, Kane WH, Rosse WF, Sims PJ (1993) Complement-induced vesiculation and exposure of membrane prothrombinase sites in platelets of paroxysmal nocturnal hemoglobinuria, *Blood*, **82**(4), 1192–1196.

184. Babiker AA, Ronquist G, Nilsson UR, Nilsson B (2002) Transfer of prostasomal CD59 to CD59-deficient red blood cells results in protection against complement-mediated hemolysis, *Am. J. Reprod. Immunol.*, **47**(3), 183–192.

185. Dachary-Prigent J, Pasquet JM, Fressinaud E, Toti F, Freyssinet JM, Nurden AT (1997) Aminophospholipid exposure, microvesiculation and abnormal protein tyrosine phosphorylation in the platelets of a patient with Scott syndrome: a study using physiologic agonists and local anaesthetics, *BJH*, **99**(4), 959–967.

186. Nieuwland R, Berckmans RJ, McGregor S, Boing AN, Romijn FP, Westendorp RG, Hack CE, Sturk A (2000) Cellular origin and procoagulant properties of microparticles in meningococcal sepsis, *Blood*, **95**(3), 930–935.

187. Biro E, Sturk-Maquelin KN, Vogel GM, Meuleman DG, Smit MJ, Hack CE, Sturk A, Nieuwland R (2003) Human cell-derived microparticles promote thrombus formation *in vivo* in a tissue factor-dependent manner, *J. Thromb. Haemost.*, **1**(12), 2561–2568.

188. Mallat Z, Hugel B, Ohan J, Leseche G, Freyssinet JM, Tedgui A (1999) Shed membrane microparticles with procoagulant potential in human atherosclerotic plaques: a role for apoptosis in plaque thrombogenicity, *Circulation*, **99**(3), 348–353.

189. Del Conde I, Shrimpton CN, Thiagarajan P, Lopez JA (2005) Tissue-factor-bearing microvesicles arise from lipid rafts and fuse with activated platelets to initiate coagulation, *Blood*, **106**(5), 1604–1611.

190. Furie B, Furie BC (2004) Role of platelet P-selectin and microparticle PSGL-1 in thrombus formation, *Trends Mol. Med.*, **10**(4), 171–178.

191. Furie B, Furie BC (2008) Mechanisms of thrombus formation, *N. Engl. J. Med.*, **359**(9), 938–949.

192. Berckmans RJ, Nieuwland R, Boing AN, Romijn FP, Hack CE, Sturk A (2001) Cell-derived microparticles circulate in healthy humans and support low grade thrombin generation, *Thromb. Haemost.*, **85**(4), 639–646.

193. Tans G, Rosing J, Thomassen MC, Heeb MJ, Zwaal RF, Griffin JH (1991) Comparison of anticoagulant and procoagulant activities of stimulated platelets and platelet-derived microparticles, *Blood*, **77**(12), 2641–2648.

194. Dahlback B, Wiedmer T, Sims PJ (1992) Binding of anticoagulant vitamin K-dependent protein S to platelet-derived microparticles, *Biochemistry*, **31**(51), 12769–12777.

195. Laulagnier K, Motta C, Hamdi S, Roy S, Fauvelle F, Pageaux JF, Kobayashi T, Salles JP, Perret B, Bonnerot C, Record M (2004) Mast cell- and dendritic cell-derived exosomes display a specific lipid composition and an unusual membrane organization, *Biochem. J.*, **380**(Pt 1), 161–171.

196. VanWijk MJ, Svedas E, Boer K, Nieuwland R, Vanbavel E, Kublickiene KR (2002) Isolated microparticles, but not whole plasma, from women with preeclampsia impair endothelium-dependent relaxation in isolated myometrial arteries from healthy pregnant women, *Am. J. Obstet. Gynecol.*, **187**(6), 1686–1693.

197. Martin S, Tesse A, Hugel B, Martinez MC, Morel O, Freyssinet JM, Andriantsitohaina R (2004) Shed membrane particles from T lymphocytes impair endothelial function and regulate endothelial protein expression, *Circulation*, **109**(13), 1653–1659.

198. Andre F, Schartz NE, Movassagh M, Flament C, Pautier P, Morice P, Pomel C, Lhomme C, Escudier B, Le CT, Tursz T, Amigorena S, Raposo G, Angevin E, Zitvogel L (2002) Malignant effusions and immunogenic tumour-derived exosomes, *Lancet*, **360**(9329), 295–305.

199. Zitvogel L, Regnault A, Lozier A, Wolfers J, Flament C, Tenza D, Ricciardi-Castagnoli P, Raposo G, Amigorena S (1998) Eradication of established murine tumors using a novel cell-free vaccine: dendritic cell-derived exosomes, *Nat. Med.*, **4**(5), 594–600.

200. Thery C, Regnault A, Garin J, Wolfers J, Zitvogel L, Ricciardi-Castagnoli P, Raposo G, Amigorena S (1999) Molecular characterization of dendritic cell-derived exosomes. Selective accumulation of the heat shock protein hsc73, *J. Cell Biol.*, **147**(3), 599–610.

201. Johansson SM, Admyre C, Scheynius A, Gabrielsson S (2008) Different types of *in vitro* generated human monocyte-derived dendritic cells release exosomes with distinct phenotypes, *Immunology*, **123**(4), 491–499.

202. Lamparski HG, Metha-Damani A, Yao JY, Patel S, Hsu DH, Ruegg C, Le Pecq JB (2002) Production and characterization of clinical grade exosomes derived from dendritic cells, *J. Immunol. Methods*, **270**(2), 211–226.

203. Viaud S, Ploix S, Lapierre V, Thery C, Commere PH, Tramalloni D, Gorrichon K, Virault-Rocroy P, Tursz T, Lantz O, Zitvogel L, Chaput N (2011) Updated technology to produce highly immunogenic dendritic

cell-derived exosomes of clinical grade: a critical role of interferon-gamma, *J. Immunother.*, **34**(1), 65–75.

204. Chaput N, Schartz NE, Andre F, Taieb J, Novault S, Bonnaventure P, Aubert N, Bernard J, Lemonnier F, Merad M, Adema G, Adams M, Ferrantini M, Carpentier AF, Escudier B, Tursz T, Angevin E, Zitvogel L (2004) Exosomes as potent cell-free peptide-based vaccine. II. Exosomes in CpG adjuvants efficiently prime naive Tc1 lymphocytes leading to tumor rejection, *J. Immunol.*, **172**(4), 2137–2146.

205. Adams M, Navabi H, Croston D, Coleman S, Tabi Z, Clayton A, Jasani B, Mason MD (2005) The rationale for combined chemo/immunotherapy using a Toll-like receptor 3 (TLR3) agonist and tumour-derived exosomes in advanced ovarian cancer, *Vaccine*, **23**(17–18), 2374–2378.

206. Dai S, Wan T, Wang B, Zhou X, Xiu F, Chen T, Wu Y, Cao X (2005) More efficient induction of HLA-A*0201-restricted and carcinoembryonic antigen (CEA)-specific CTL response by immunization with exosomes prepared from heat-stressed CEA-positive tumor cells, *Clin. Cancer Res.*, **11**(20), 7554–7563.

207. Chen W, Wang J, Shao C, Liu S, Yu Y, Wang Q, Cao X (2006) Efficient induction of antitumor T cell immunity by exosomes derived from heat-shocked lymphoma cells, *Eur. J. Immunol.*, **36**(6), 1598–1607.

208. Chen T, Guo J, Yang M, Zhu X, Cao X (2011) Chemokine-containing exosomes are released from heat-stressed tumor cells via lipid raft-dependent pathway and act as efficient tumor vaccine, *J. Immunol.*, **186**(4), 2219–2228.

209. Taieb J, Chaput N, Schartz N, Roux S, Novault S, Menard C, Ghiringhelli F, Terme M, Carpentier AF, Darrasse-Jeze G, Lemonnier F, Zitvogel L (2006) Chemoimmunotherapy of tumors: cyclophosphamide synergizes with exosome based vaccines, *J. Immunol.*, **176**(5), 2722–2729.

210. Yang Y, Xiu F, Cai Z, Wang J, Wang Q, Fu Y, Cao X (2007) Increased induction of antitumor response by exosomes derived from interleukin-2 gene-modified tumor cells, *J. Cancer Res. Clin. Oncol.*, **133**(6), 389–399.

211. Zhang Y, Luo CL, He BC, Zhang JM, Cheng G, Wu XH (2010) Exosomes derived from IL-12-anchored renal cancer cells increase induction of specific antitumor response *in vitro*: a novel vaccine for renal cell carcinoma, *Int. J. Oncol.*, **36**(1), 133–140.

212. Hartman ZC, Wei J, Glass OK, Guo H, Lei G, Yang XY, Osada T, Hobeika A, Delcayre A, Le Pecq JB, Morse MA, Clay TM, Lyerly HK (2011) Increasing

vaccine potency through exosome antigen targeting, *Vaccine*, **29**(50), 9361–9367.

213. Rountree RB, Mandl SJ, Nachtwey JM, Dalpozzo K, Do L, Lombardo JR, Schoonmaker PL, Brinkmann K, Dirmeier U, Laus R, Delcayre A (2011) Exosome targeting of tumor antigens expressed by cancer vaccines can improve antigen immunogenicity and therapeutic efficacy, *Cancer Res.*, **71**(15), 5235–5244.

214. Altieri SL, Khan AN, Tomasi TB (2004) Exosomes from plasmacytoma cells as a tumor vaccine, *J. Immunother.*, **27**(4), 282–288.

215. Morse MA, Garst J, Osada T, Khan S, Hobeika A, Clay TM, Valente N, Shreeniwas R, Sutton MA, Delcayre A, Hsu DH, Le Pecq JB, Lyerly HK (2005) A phase I study of dexosome immunotherapy in patients with advanced non-small cell lung cancer, *J. Transl. Med.*, **3**(1), 9.

216. Escudier B, Dorval T, Chaput N, Andre F, Caby MP, Novault S, Flament C, Leboulaire C, Borg C, Amigorena S, Boccaccio C, Bonnerot C, Dhellin O, Movassagh M, Piperno S, Robert C, Serra V, Valente N, Le Pecq JB, Spatz A, Lantz O, Tursz T, Angevin E, Zitvogel L (2005) Vaccination of metastatic melanoma patients with autologous dendritic cell (DC) derived-exosomes: results of thefirst phase I clinical trial, *J. Transl. Med.*, **3**(1), 10.

217. Dai S, Wei D, Wu Z, Zhou X, Wei X, Huang H, Li G (2008) Phase I clinical trial of autologous ascites-derived exosomes combined with GM-CSF for colorectal cancer, *Mol. Ther.*, **16**(4), 782–790.

218. van den Boorn JG, Schlee M, Coch C, Hartmann G (2011) SiRNA delivery with exosome nanoparticles, *Nat. Biotechnol.*, **29**(4), 325–326.

219. Alvarez-Erviti L, Seow Y, Yin H, Betts C, Lakhal S, Wood MJ (2011) Delivery of siRNA to the mouse brain by systemic injection of targeted exosomes, *Nat. Biotechnol.*, **29**(4), 341–345.

220. Lakhal S, Wood MJ (2011) Exosome nanotechnology: an emerging paradigm shift in drug delivery: exploitation of exosome nanovesicles for systemic *in vivo* delivery of RNAi heralds new horizons for drug delivery across biological barriers, *Bioessays*, **33**(10), 737–741.

221. Kim SH, Bianco NR, Shufesky WJ, Morelli AE, Robbins PD (2007) Effective treatment of inflammatory disease models with exosomes derived from dendritic cells genetically modified to express IL-4, *J. Immunol.*, **179**(4), 2242–2249.

222. Kim SH, Bianco N, Menon R, Lechman ER, Shufesky WJ, Morelli AE, Robbins PD (2006) Exosomes derived from genetically modified DC expressing FasL are anti-inflammatory and immunosuppressive, *Mol. Ther.*, **13**(2), 289–300.

223. Bianco NR, Kim SH, Ruffner MA, Robbins PD (2009) Therapeutic effect of exosomes from indoleamine 2,3-dioxygenase-positive dendritic cells in collagen-induced arthritis and delayed-type hypersensitivity disease models, *Arthritis Rheum.*, **60**(2), 380–389.

224. Kim SH, Lechman ER, Bianco N, Menon R, Keravala A, Nash J, Mi Z, Watkins SC, Gambotto A, Robbins PD (2005) Exosomes derived from IL-10-treated dendritic cells can suppress inflammation and collagen-induced arthritis, *J. Immunol.*, **174**(10), 6440–6448.

225. Miksa M, Wu R, Dong W, Komura H, Amin D, Ji Y, Wang Z, Wang H, Ravikumar TS, Tracey KJ, Wang P (2009) Immature dendritic cell-derived exosomes rescue septic animals via milk fat globule epidermal growth factor-factor VIII [corrected], *J. Immunol.*, **183**(9), 5983–5990.

226. Sahoo S, Klychko E, Thorne T, Misener S, Schultz KM, Millay M, Ito A, Liu T, Kamide C, Agrawal H, Perlman H, Qin G, Kishore R, Losordo DW (2011) Exosomes from human CD34(+) stem cells mediate their proangiogenic paracrine activity, *Circ. Res.*, **109**(7), 724–728.

227. Lai RC, Arslan F, Lee MM, Sze NS, Choo A, Chen TS, Salto-Tellez M, Timmers L, Lee CN, El Oakley RM, Pasterkamp G, de Kleijn DP, Lim SK (2010) Exosome secreted by MSC reduces myocardial ischemia/reperfusion injury, *Stem Cell Res.*, **4**(3), 214–222.

228. Kakkar AK, DeRuvo N, Chinswangwatanakul V, Tebbutt S, Williamson RC (1995) Extrinsic-pathway activation in cancer with high factor VIIa and tissue factor, *Lancet*, **346**(8981), 1004–1005.

229. Carr JM, Dvorak AM, Dvorak HF (1985) Circulating membrane vesicles in leukemic blood, *Cancer Res.*, **45**(11 Pt 2), 5944–5951.

230. Saman S, Kim W, Raya M, Visnick Y, Miro S, Saman S, Jackson B, McKee AC, Alvarez VE, Lee NC, Hall GF (2012) Exosome-associated tau is secreted in tauopathy models and is selectively phosphorylated in cerebrospinal fluid (CSF) in early Alzheimer's Disease, *J. Biol. Chem.*, **287**(6), 3842–3849.

231. Mitchell PJ, Welton J, Staffurth J, Court J, Mason MD, Tabi Z, Clayton A (2009) Can urinary exosomes act as treatment response markers in prostate cancer? *J. Transl. Med.*, **7**, 4.

232. Li J, Sherman-Baust CA, Tsai-Turton M, Bristow RE, Roden RB, Morin PJ (2009) Claudin-containing exosomes in the peripheral circulation of women with ovarian cancer, *BMC Cancer*, **9**, 244.

233. Tavoosidana G, Ronquist G, Darmanis S, Yan J, Carlsson L, Wu D, Conze T, Ek P, Semjonow A, Eltze E, Larsson A, Landegren UD, Kamali-Moghaddam M (2011) Multiple recognition assay reveals prostasomes as promising plasma biomarkers for prostate cancer, *Proc. Natl. Acad. Sci. U. S. A.*, **108**(21), 8809–8814.

234. Feng D, Zhao WL, Ye YY, Bai XC, Liu RQ, Chang LF, Zhou Q, Sui SF (2010) Cellular internalization of exosomes occurs through phagocytosis, *Traffic*, **11**(5), 675–687.

235. Dasgupta SK, Abdel-Monem H, Niravath P, Le A, Bellera RV, Langlois K, Nagata S, Rumbaut RE, Thiagarajan P (2009) Lactadherin and clearance of platelet-derived microvesicles, *Blood*, **113**(6), 1332–1339.

236. Lopez-Berestein G, Kasi L, Rosenblum MG, Haynie T, Jahns M, Glenn H, Mehta R, Mavligit GM, Hersh EM (1984) Clinical pharmacology of 99mTc-labeled liposomes in patients with cancer, *Cancer Res.*, **44**(1), 375–378.

237. Rand ML, Wang H, Bang KW, Packham MA, Freedman J (2006) Rapid clearance of procoagulant platelet-derived microparticles from the circulation of rabbits, *J. Thromb. Haemost.*, **4**(7), 1621–1623.

238. Rank A, Nieuwland R, Crispin A, Grutzner S, Iberer M, Toth B, Pihusch R (2011) Clearance of platelet microparticles *in vivo*, *Platelets.*, **22**(2), 111–116.

239. Luo SS, Ishibashi O, Ishikawa G, Ishikawa T, Katayama A, Mishima T, Takizawa T, Shigihara T, Goto T, Izumi A, Ohkuchi A, Matsubara S, Takeshita T, Takizawa T (2009) Human villous trophoblasts express and secrete placenta-specific microRNAs into maternal circulation via exosomes, *Biol. Reprod.*, **81**(4), 717–729.

240. van den Goor JM, Nieuwland R, Rutten PM, Tijssen JG, Hau C, Sturk A, Eijsman L, de Mol BA (2007) Retransfusion of pericardial blood does not trigger systemic coagulation during cardiopulmonary bypass, *Eur. J. Cardiothorac. Surg.*, **31**(6), 1029–1036.

241. Kadurugamuwa JL, Beveridge TJ (1995) Virulence factors are released from Pseudomonas aeruginosa in association with membrane vesicles during normal growth and exposure to gentamicin: a novel mechanism of enzyme secretion, *J. Bacteriol.*, **177**(14), 3998–4008.

242. Rivera J, Cordero RJ, Nakouzi AS, Frases S, Nicola A, Casadevall A (2010) Bacillus anthracis produces membrane-derived vesicles containing biologically active toxins, *Proc. Natl. Acad. Sci. U. S. A.*, **107**(44), 19002–19007.

243. Mashburn-Warren L, McLean RJ, Whiteley M (2008) Gram-negative outer membrane vesicles: beyond the cell surface, *Geobiology*, **6**(3), 214–219.

244. Mashburn LM, Whiteley M (2005) Membrane vesicles traffic signals and facilitate group activities in a prokaryote, *Nature*, **437**(7057), 422–425.

245. Winans SC (2005) Microbiology: bacterial speech bubbles, *Nature*, **437**(7057), 330.

246. Bomberger JM, Maceachran DP, Coutermarsh BA, Ye S, O'Toole GA, Stanton BA (2009) Long-distance delivery of bacterial virulence factors by Pseudomonas aeruginosa outer membrane vesicles, *PLOS Pathog.*, **5**(4), e1000382.

247. Chen DJ, Osterrieder N, Metzger SM, Buckles E, Doody AM, DeLisa MP, Putnam D (2010) Delivery of foreign antigens by engineered outer membrane vesicle vaccines, *Proc. Natl. Acad. Sci. U. S. A.*, **107**(7), 3099–3104.

248. Chutkan H, Kuehn MJ (2011) Context-dependent activation kinetics elicited by soluble versus outer membrane vesicle-associated heat-labile enterotoxin, *Infect. Immun.*, **79**(9), 3760–3769.

249. Kaparakis M, Turnbull L, Carneiro L, Firth S, Coleman HA, Parkington HC, Le BL, Karrar A, Viala J, Mak J, Hutton ML, Davies JK, Crack PJ, Hertzog PJ, Philpott DJ, Girardin SE, Whitchurch CB, Ferrero RL (2010) Bacterial membrane vesicles deliver peptidoglycan to NOD1 in epithelial cells, *Cell Microbiol.*, **12**(3), 372–385.

250. Kadurugamuwa JL, Beveridge TJ (1996) Bacteriolytic effect of membrane vesicles from Pseudomonas aeruginosa on other bacteria including pathogens: conceptually new antibiotics, *J. Bacteriol.*, **178**(10), 2767–2774.

251. Kadurugamuwa JL, Mayer A, Messner P, Sara M, Sleytr UB, Beveridge TJ (1998) S-layered Aneurinibacillus and Bacillus spp. are susceptible to the lytic action of Pseudomonas aeruginosa membrane vesicles, *J. Bacteriol.*, **180**(9), 2306–2311.

252. Bjune G, Hoiby EA, Gronnesby JK, Arnesen O, Fredriksen JH, Halstensen A, Holten E, Lindbak AK, Nokleby H, Rosenqvist E, (1991) Effect of outer membrane vesicle vaccine against group B meningococcal disease in Norway, *Lancet*, **338**(8775), 1093–1096.

253. Sadarangani M, Pollard AJ (2010) Serogroup B meningococcal vaccines—an unfinished story, *Lancet Infect. Dis.*, **10**(2), 112–124.

254. Caron F, du Chatelet IP, Leroy JP, Ruckly C, Blanchard M, Bohic N, Massy N, Morer I, Floret D, Delbos V, Hong E, Revillion M, Berthelot G, Lemee L, Deghmane AE, Benichou J, Levy-Bruhl D, Taha MK (2011) From tailor-made to ready-to-wear meningococcal B vaccines: longitudinal study of a clonal meningococcal B outbreak, *Lancet Infect. Dis.*, **11**(6), 455–463.

255. van der Ley P, van den Dobbelsteen G (2011) Next-generation outer membrane vesicle vaccines against Neisseria meningitidis based on nontoxic LPS mutants, *Hum. Vaccin.*, **7**(8), 886–890.

256. Alaniz RC, Deatherage BL, Lara JC, Cookson BT (2007) Membrane vesicles are immunogenic facsimiles of Salmonella typhimurium that potently activate dendritic cells, prime B and T cell responses, and stimulate protective immunity *in vivo*, *J. Immunol.*, **179**(11), 7692–7701.

Chapter 2

The Biogenesis of Exosomes

D. Michiel Pegtel
Department of Pathology, VU University Medical Center, Cancer Center Amsterdam (CCA), Room 3.08, De Boelelaan 1117, 1081HV Amsterdam, the Netherlands
d.pegtel@vumc.nl

2.1 Exosome Biogenesis and Physiological Function(s)

2.1.1 A Historical Perspective

Exosomes as secreted vesicles were first reported in 1983 by Johnstone et al. While studying maturing reticulocytes (immature erythrocytes) by electron microscopy (EM), small 100 nm vesicles were observed on the outside of the cells. Immature erythrocytes are the precursors of red blood cells and contain cytoplasmic ribosomes and remnants of organelles such as mitochondria, the endoplasmic reticulum (ER), the Golgi apparatus, and an intact endosomal system.[1] During their differentiation into mature erythrocytes, the ribosomes and other organelles are lost and several plasma membrane proteins are cleared, resulting in a decrease in size and loss of some specific

Extracellular Vesicles in Health and Disease
Edited by Paul Harrison, Christopher Gardiner, and Ian L. Sargent

ISBN 978-981-4411-98-1 (Hardcover), 978-981-4411-99-8 (eBook)
www.panstanford.com

plasma membrane activities. The secretion of these vesicles termed "exosomes" was conceived as an unconventional mechanism to eliminate unwanted materials like proteins and lipids that facilitated maturation and loss in size. While the maturing reticulocytes clearly seem to benefit from exosome secretion, the exact role and fate of the expelled exosomes remained unknown.[2-4] Similarly, the non-nucleated platelets (thrombocytes) also secrete large amounts of extracellular vesicles (EVs), including exosomes,[5] as do their immature progenitors, the megakaryocytes.[6]

In the last decade it has become clear that most, if not all, cell types have the ability to secrete exosome-like vesicles *in vitro*,[7] and widespread dispersal of exosomes *in vivo* has been established.[8] Importantly, more recent, detailed molecular and cell biological studies yielded important new clues into the molecular mechanism(s) behind their biogenesis, sorting, secretion,[8-14] and the finding that exosomes can transfer active signaling molecules to recipient cells.[15,16] The interest became even higher as exosomes from mammalian cells also seem to influence cell–cell communication rather unexpectedly through their ability to carry and transfer functional genetic material.[17-22] These findings are supported by findings that human (tumor) viruses exploit this pathway for their benefit[21-23] and that micro–ribonucleic acids (miRNAs) can cross a functional immunological synapse *in vitro*[24] and possibly *in vivo*,[25,26] but for the most part, physiological evidence for the role of exosome secretion *in vivo* has not been established. An exception to this rule are immune cell–derived exosomes that have recognized physiological properties in mice and humans.[8] This may be partly for historic reasons since functional exosomes were shown to be secreted by immune cells first, but also due to the detailed understanding of the molecular mechanisms behind antigen processing, loading, and sorting in the endocytic pathway.

2.1.2 Physiological Function of Exosomes

The clearest physiological evidence for exosomes as an intracellular signaling device arose from studies looking into the mechanisms of antigen loading and trafficking of class II major histocompatibility complex (MHC) molecules in immune cells. The role of the endocytic pathway and in particular multivesicular bodies (MVBs), which are abundant in activated antigen-presenting cells (APCs) and

T-cells, led to the discovery that exosomes contain functional MHC II molecules,[27] reviewed by Denser et al. in 2001.[3] In particular activated B-cell blasts as well as (mature) dendritic cells (DCs) seem to secrete large amounts of MHC-carrying exosomes upon physiological stimuli.[28–31] The exosomes secreted are rich in peptide-MHC II and can thus in principle be considered as an extracellular source of antigen for activation of naive T-cells by DCs *in vivo*.[19,32–36] Interestingly, MHC II–bearing exosomes have also been identified on the surface of follicular DCs (fDCs) in human tonsils,[37] and the immune-stimulatory capacity of exosomes is being exploited for use in cell-free antitumor vaccines.[38–40] While these studies indirectly suggested that exosomes have a function *in vivo*, the precise physiological role of APC-derived exosomes in immunity and the mechanisms that regulate their release from APCs are elusive. Nevertheless it has been calculated that over 12% of MHC II complexes are secreted by activated B-cells via exosomes per day.[31] This secretory pathway is probably activated through nuclear factor kappa B (NFκB) activation as proteins that stimulate NFκB signaling, such as the constitutively active Epstein–Barr virus latent membrane protein 1 (LMP1), dramatically affect the endo-exosomal pathway.[12]

Raposo et al. were the first to show clear physiological relevance for MVB-engineered exosomes outside the cells in which they were produced.[27] A subsequent *in vivo* study showed that even though fDCs themselves do not express MHC II on their surface they can be decorated with MHC II–positive exosomes *in vivo*.[37] This may have physiological relevance as fDCs have an important role in Ag presentation and B-cell development in germinal centers.[41] These studies raised many fundamental questions that, for a large part, have remained unanswered. It is still not clear what specialized endosomal (sub) compartments (if any) produce MHC II–containing exosomes and what makes these distinct from the MVBs that fuse with lysosomes. Secondly the kinetics of exosome secretion remains ill-described; estimates have been made that some cells secrete up to ~10,000 exosomes a day,[42] but this still needs further confirmation. Moreover the exact contribution of exosomes to immune responses *in vivo* has not been determined, and finally the molecular mechanisms controlling their release and subsequent fate *in vivo* are far from clear.

What is clear, however, is that APCs, and in particular DCs and activated B-cell blasts, contain MHC II–rich storage compartments known as MIICs.[43,44] Early cell biological studies showed that these compartments contain both endocytosed and endogenous antigens (Ag), which are loaded into MHC II complexes, a process aided by HLA-DM, as shown by fluorescence resonance energy transfer (FRET) studies.[45] Although MIICs have endosomal characteristics they seem to share the ability with secretory lysosomes (SLs) to fuse with the plasma membrane—a phenomenon first proposed in other cell types in particular cytotoxic T-cells (CTLs).[46] MIICs in B-cell blasts contain the invariant chain (li) and CLIP,[47] and more details of this process are reviewed elsewhere.[48] We found by quantitative proteomic analysis that these molecules are present in sucrose-purified exosomes (Pegtel et al. unpublished data), supporting the idea that MIICs or closely related intracellular compartments seem responsible for exosome production. Despite this, MIIC fusion with the plasma membrane has not unequivocally been demonstrated such that exosome budding directly from the plasma membrane cannot be ruled out completely in having a contribution to exosome secretion. Indeed a proportion of the invariant chain (Ii) is transported to the plasma membrane, but since Ii is rapidly internalized (with an estimated half-life of 3–4 min at the surface)[49] it is more reasonable to presume that the major pool of li found in exosomes is, in fact, derived from intracellular compartments such as MIICs and not derived from the plasma membrane.

Recent data from Buschow et al. confirm that recycling of MHC II molecules via specialized endocytic compartments is relevant for secretion via exosomes. Whether peptide-loaded MHC II (pMHC II) is stably expressed at the plasma membrane or degraded in lysosomes is in fact determined by the activation status of the APC.[10] Immature DCs have low cell surface levels of pMHCII because these molecules are ubiquitinated, endocytosed, and sorted into the interior of MVBs destined to fuse with lysosomes, thus leading to their degradation. In activated DCs, however, pMHCII molecules are preferentially sorted into MVBs that produce exosomes independently of ubiquitilation.[10] Thus lysosomal targeting of MHC II is dependent on ubiquitination, while exosomal targeting seems dependent on the association within incorporation into tetraspanin (CD9)-enriched domains. These findings are in agreement with other studies in nonimmune cell types, demonstrating that ubiquitination does not affect MHC II secretion via exosomes.[50]

Despite the large body of evidence that exosomes derived from immune cells originate from intracellular endocytic compartments, it must be noted that although many subtypes of MVBs exist, in particular in APCs, such as activated blasts and DCs,[43] only few of these compartments are likely to produce exosomes. In fact some MVBs with clearly alternate physical forms such as multilaminar bodies (MLBs) that do not contain intraluminal vesicles (ILVs) may have similar functions in Ag loading as classic MVBs but are unlikely to produce exosomes and may be resistant to fusion with the plasma membrane. It would be of importance to identify in these cell types what molecular requirements favor exosome production and secretion. Since Raposo et al. proposed a first clear purpose for exosomes, these initial observations in activated B-cells culminated into the current defining pathway for exosome biogenesis.[33,36,51] On a cautionary note, alternative pathways for exosome biogenesis—besides the one described here—may exist.

Recent elegant studies focusing on enterocytes (small-intestinal epithelial cells) indicate that microvilli (specialized membrane protrusions on the apical side of these cells) are vesicle-producing organelles that resemble exosomes.[52] In the gut lumen, the microvilli continuously "pinch off" nanometer-size vesicles that are present in large numbers. These vesicles function by detoxifying bacterial products, thus controlling the epithelial–microbial interactions.[53] In addition to surface-produced exosome-like vesicles, exosomes may also be produced in endosome-like compartments that also lack the requirement for MVB fusion with the plasma membrane for their secretion. It appears that exosomes from macrophages can derive from particular endosomal compartments that are in direct contact with the exterior via deep-membrane invaginations.[54] Although it is unclear whether these observations in macrophages can be translated into a more general phenomenon, these observations highlight the notion that not all cell types generate exosomes via the "classic" model as described by Johnstone et al. in reticulocytes and by Raposo et al. in activated B-cells. Despite these important nuances, a large body of evidence derived from proteomic, cell biological, and functional genetic studies indicates that exosome biogenesis and secretion are strongly correlated with the endosomal pathway and in particular MVB and ILV formation.

The examples from enterocytes and immune cells underscore the broadly accepted assumption that the mechanisms underlying exosome biogenesis are related to the cellular background, activation, and differentiation state, as well as to the local microenvironment. Autophagy, for instance, could also play a significant role in the exosomes' composition, as discussed recently.[55] Despite the recent advances into exosome biogenesis and physiology, crucial molecular details, as pointed out previously by van Niel et al.,[7] need to be addressed. Although progress has been made into a more universal understanding of exosome biogenesis within the last decade[8,32] some long-term lingering questions remain. A particular concern is how to prove the physiological function of exosomes *in vivo*, be it derived from neuronal, epithelial, and endothelial origin or from stem cells. The reason why this is such a difficult task lies in the incomplete knowledge on their biogenesis and secretion, thus precluding specific blockage or induction of exosome secretion in specific cell types or conditions *in vivo*.

2.1.3 Exosomes as an Alternative End of the Endosomal Pathway

Exosomes are classically defined as vesicles that are derived from late endosomes (LEs). LEs are recognized as highly dynamic and heterogeneous structural compartments in eukaryotic cells. The endosomal membranes serve as major sorting platforms for molecules that have been taken up and internalized by invaginations at the plasma membrane. Collectively this process is known as endocytosis, and the endosomal system thus represents a key direct contact of the cell with its exterior apart from the plasma membrane.

Endocytosis starts with the internalization of fluid, solutes, macromolecules, plasma membrane components, and particles through various pathways of endocytic trafficking.[56] Not surprisingly, the endocytic pathway is heavily exploited by viruses and other intracellular parasites that wish to gain entry into specific host cells to support their life cycle. A functioning endosomal system is essential for normal development and physiology, and impaired endosomal function contributes to many human diseases, including cancer,[57] neurological diseases, and immune disorders.[58-61] Endocytosed material from the plasma membrane is deposited via small transport vesicles into the early endosomes, also known as sorting endosomes. The early endosomes receive their cargo via multiple mechanisms, the clathrin- or non-clathrin-mediated pathways. The non-clathrin-dependent pathways are caveolar, GEEC- (GPI-enriched endocytic compartments) and ARF6-mediated.[62]

Genetic studies in yeast identified a crucial, evolutionary, conserved molecular constituent of the endosomal compartments, most notably, the endosomal sorting complex required for transport (ESCRT) machinery.[63] The ESCRT complexes (0, I, II, and III) contain multiple conserved proteins (typically named "Vps" for vacuolar protein sorting–associated protein) that act similarly in yeast as in mammalian cells. Nevertheless in mammalian ESCRT proteins also have other functions, besides protein recycling and endocytosis, such as controlling cytokinesis and maintaining cell polarity.[64] Detailed biochemical analysis and *in vitro* reconstitution experiments have elucidated and confirmed the important aspects of molecular trafficking through this compartment in many cell types and mammalian animal models.[65]

Directly after internalization from the plasma membrane, transport vesicles fuse with each other or with an existing sorting endosome. These fusion events are the main driving force behind the formation of larger sorting endosomes with tubular structures[66] that are typically identified by the early endosomal marker early endosome antigen 1 (EEA1). Sorting endosomes select the proteins for recycling back to the plasma membrane or direct those that are subject to degradation to lysosomes, a process that relies heavily on ubiquitination and the formation of ILVs.[63] The current dogma in metazoan cells dictates, however, that proteins selected for secretion via exosomes not require ubiquitination[67] and that the ESCRT machinery be not required for MVB biogenesis and ILV production. Although sorting of molecules into ILVs appears affected in ESCRT-depleted cells, the effect on exosomes was not reported.[68] In agreement with this, results of our own studies and those by several other groups show no evidence for preferential sorting of proteins into exosomes upon ubiquitination.[10,12,50,69] The structure of the sorting endosome is composed of both vacuolar and tubular domains, which are considered functionally important. Tubular membranes are involved in sorting of cargo targeted for recycling, whereas cargo sorted into the vacuolar domains is typically intended for degradation.[66] Broadly put, the endocytic system recycles proteins, lipids, and, as it seems, even genetic material,[70] and while some of this cargo will remain at the limiting membranes (LMs) and/or in the cytoplasm, other molecules are selectively sorted into the interior for degradation in lysosomes by acidic hydrolases.[71] An exception to this are molecules that are neither recycled nor degraded but instead sorted for secretion into specialized LEs/MVBs[10] or lysosomes.[72] These MVBs are suspected to preferentially fuse with the plasma membrane, as proposed by Simons and Raposo,[73] possibly by analogy with SLs, but whether this occurs and how this is regulated are just a few of the major outstanding questions that need to be addressed.

2.2 ESCRT Machinery in ILV Formation

Arguably the most crucial part of exosome physiology is ILV formation and the mechanisms that control the selective incorporation of cargo. ILV biogenesis starts by deformation of the LM of MVBs through budding away from the cytoplasm and their subsequent detachment in single vesicular entities. Multiple conflicting models have been proposed on the basis of contradictory observations, possibly because these models are cell-type and organism dependent.

A generalized model predicts that the ESCRT pathway represents the major protein-sorting and membrane fission events required for release of cargo-selected vesicles but also viruses. The presumed molecular mechanism of ILV formation during MVB biogenesis has been extensively reviewed elsewhere and can be summed as follows: ESCRT-0 consisting of Hrs (hepatocyte growth factor [HGF]-regulated tyrosine kinase substrate) (Vps27 in yeast) and signal-transducing adaptor molecule (STAM) are recruited to the endosomal membrane by the interaction with phosphatidylinositol-3-phosphate (PI3P), a phosphoinositide enriched in endosomal membranes. Both subunits of ESCRT-0 bind ubiquitin moieties on their cargo clustered in clathrin-coated pits. In addition to the interaction with PI3P and ubiquitin, HRS also recruits and interacts with tumor susceptibility gene 101 (Tsg101) in mammalian cells (Vps23 in yeast) of ESCRT-I, starting a cascade of multiple ESCRT complex interactions. ESCRT-I interacts with ESCRT-II, which recruits charged multivesicular body protein 6 (CHMP6) (Vps20) to the neck and binds Snf7 (VpsS32) to form the main engine for neck scission that promotes cargo internalization into ILVs. ESCRT-III will remain largely on the outside of the endosomal LM and is recycled by mammalian Vps4.[74–78]

What can we learn from this molecular network, and how does this relate to exosomes? Firstly, extensive proteomic analyses by multiple independent groups have indicated that several key endosomal proteins are prominent constituents of exosomes and thus indicative of their endocytic origin.[79] The dominant endosomal proteins present in exosomes that are associated with ESCRT-mediated ILV biogenesis and sorting are TSG101 and ALIX. TSG101 functions as a subunit of the heterotetrameric ESCRT-I complex, together with several other MVB-related proteins, including VPS28, VPS37, and MVB12,[80–84] but these are less frequently described in exosomes. ALIX or PDCD6IP also functions in the MVB pathway, where it interacts with multiple endosomal proteins, including TSG101 or ESCRT-I.[85–88] Despite this, depletion of Tsg101 and/or ALIX at least in mammalian cells does not seem to impair MVB biogenesis[68] and certainly does not seem to control protein sorting,[10,12] casting doubt on the relevance of these proteins for exosome physiology. These molecules have one thing in common that they bind ubiquitinated cargo for selection into ILVs, yet current studies in mammalian systems find little evidence that ubiquitin has a role in sorting of proteins into exosomes. Interestingly, however, both TSG101 and ALIX bind directly to retroviral Gag proteins, including human immunodeficiency virus-1 (HIV-1) Gag, facilitating late stages of virus budding.[89–91] This highlights the striking similarities between virus budding and exosome biogenesis and may be more related to the overall physiology of exosomes rather than having a role in exosomal sorting. Intriguing similarities between exosomes and several types of viruses have been found[92,93] and are reviewed elsewhere.[94] Alternatively it is possible that ESCRT-dependent budding events at the plasma membrane may contaminate endosome-derived exosomal preparations.[95]

Many *in vivo* observations related to changes in membrane morphology, including tubulation, vesiculation, fusion, and fission, can be mimicked *in vitro* in the complete absence of proteins using liposomes with the appropriate lipid composition.[96] In such a reductionist approach proteins appear inessential in inducing membrane topology changes. It is thus impossible to fully discern the contributions of a multitude of proteins in membrane-based cellular functions in such *in vitro* models. Nevertheless these model systems are in line with *in vivo* studies, indicating that ESCRT, at least in metazoan cells, does not drive MVB vesicle formation but rather is involved in cargo sorting, in addition to coordinating the overall maturation of endosomes. In support of this model, in *Drosophila* the absence of the entire ESCRT machinery did not seem to block the formation of MVB vesicles but resulted in impaired cargo sorting into ILVs and variations in the ILV number and size.[68,97] Moreover protein modifications that do not seem to be associated with ESCRT function may be responsible for sorting into exosomes, besides their lipid domain affiliation.[98]

2.3 MVB and ILV Biogenesis Multiple Pathways for Differential Sorting

The earliest models of ESCRT function are primarily based on pioneering studies in yeast, where these complexes were first discovered and since yeast ESCRT complexes have been analyzed extensively. However, the straightforward paradigm from yeast—in which each component of the Hrs-STAM and ESCRTI, ESCRT-II, ESCRT-III complexes acts primarily in MVB sorting and biogenesis—has been challenged by evidence from studies in *Drosophila*, pointing to more diversified roles, for instance, by genetic screens in metazoans.[97] In addition mammalian HEP-2 cells depleted for individual members of all ESCRT components by short interfering RNAs (siRNAs) still maintained the ability to form MVBs and ILVs. Although morphology was altered and the sorting of the (ubiquitinated) epidermal growth factor (EGF) receptor impaired,[68] these studies indicate that other mechanisms for exosome biogenesis and sorting exist.

Seemingly unrelated observations over the years appear relevant for the mechanisms behind exosomal sorting. Indeed specific exosomal proteins that have been studied extensively indicate that ubiquitination is not a requirement for their selective sorting into exosomes (PMEL, HLA, LMP1, CD63), confirming the idea that "classic ESCRT function," as defined in yeast,[63] does not play an essential role in exosome sorting and biogenesis. The question emerging is, What are the molecular differences that distinguish the primordial MVBs in yeast from "sophisticated" exosome-producing MVBs described in metazoans? On the basis of the assumption that such differences indeed exist one could hypothesize that the ESCRT system in yeast is essential for MVB biogenesis and sorting, while during evolution, additional ESCRT-independent pathways may have evolved to form ILVs that specifically regulate sorting into exosomes as observed in multicellular organisms. Besides eukaryotes gram-negative bacteria also produce outer membrane vesicles (OMVs) that play a role in the delivery of virulence factors to host cells, and comparisons with the molecular machineries that regulate OMV biogenesis in bacteria (or plants) may be of use.[99] In fact, the human oral pathogen *Porphyromonas gingivalis* selectively sorts virulence factors into OMVs, while abundant outer-membrane proteins were excluded from the protein cargo.[100]

The most studied example on an ESCRT-independent mechanism is the recently identified "ceramide pathway," which is based on the synthesis of sphingolipid ceramide, which is produced by neutral sphingomyelinase-1 (nSMase1).[67] nSMase1 hydrolyzes the membrane lipid sphingomyelin into ceramide, and Trajkovic et al.[9] measured a major reduction in exosome production by Oli-neu cells when nSMase1 was chemically inhibited or depleted by RNA interference. Importantly in this respect they also observed impaired ILV formation *in vivo* and *in vitro*, suggesting that in wt Oli-neu cell membrane invagination could be spontaneously triggered through local enrichment of ceramide, possibly due to its "cone shaped" structure. Exosomal cargo, in this case proteolipid protein (PLP), was abrogated from sorting into ILVs and exosomes upon nSnMase1 inhibition, leading to PLP accumulation in the cell. It is tempting to speculate that ESCRT-independent ILV production pathways evolved specifically to promote vesicle-mediated intercellular communication by promoting exosome biogenesis.

In conclusion, "unconventional" ILV pathways seem to be driven largely by the presence of lipids, including lysobisphosphatidic acid (LBPA) and ceramides, although tetraspanins may be an exception.[12,69] These molecules have one thing in common that they are able to organize into specialized membrane domains simply because a local lipid composition promotes inward budding to form vesicles. Indeed observations *in vitro* using so-called giant unilamellar vesicles (GUVs) confirm the idea that lipids alone are sufficient for ILV budding and that proteins seem to facilitate this lipid-driven process. Thus the contribution of ESCRT to MVB biogenesis and ILV formation may be restricted to the actual sorting of ubiquitinated proteins for lysosomal degradation and seems irrelevant for sorting of proteins into exosomes.

2.3.1 Tetraspanins

ILVs and exosomes are enriched in multiple tetraspanins: CD9, CD37, CD53, CD63, CD81, and CD82.[101,102] Duffield et al. show that CD63 interacts with the β-subunit of hydrogen/potassium (HK)–adenosine triphosphatase (ATPase) in gastric parietal cells, resulting in the internalization of HKβ into ILVs.[103] They indicate that CD63 might be involved in chaperoning its interaction partners to the machinery of the endosomal system. Another study performed by Källquist et al.[104] determined an interaction between CD63 and neutrophil elastase (NE). Coexpression of proNE and CD63 in COS cells leads to an interaction resulting in internalization of NE. Overall, tetraspanin CD63 is probably involved in sorting of specific cargo into ILVs of MVB-related organelles such as melanosomes. The importance of CD63 is further illustrated by our own studies showing that the viral protein LMP1 recruits CD63 in low-cholesterol MVBs to escape degradation from lysosomes. Indeed LMP1 is abundantly and rapidly secreted via exosomes, which seems to attenuate downstream NFκB signaling by this molecule.[12]

2.4 MVB Formation, Maturation, and Cellular Movement

The crucial compartments for exosome biogenesis and secretion are MVBs, sometimes also referred to as multivesicular endosomes (MVEs). MVBs were first observed in EM studies decades ago as membrane-enclosed structures with ILVs. Subsequently the "inward" invagination of the LM of MVBs was noticed, indicating that these structures might mediate some form of autophagy.[105] Indeed it was proposed that MVBs are in fact part of the endocytic pathway and are crucially involved in internalization and recycling of the EGF receptor (reviewed by Katzmann and Emr). In fact it was recently suggested that the MVB pathway is closely associated with autophagy.[106] In exosome-producing B-cells that we have studied we found by EM hybrid structures, presumably endo-lysosomes that appear to have incorporated vesicles with features of autophagosomes.

MVBs are part of the complex and highly dynamic maturation program of endosomes. One could wonder why such an elaborate network is required. It has been proposed that the heterogeneity in endosomal structures is needed to separate endosomes that have "closed down" their sorting machinery from those endosomes that are primed for fusion with the lysosomes where the degradative program is active and irreversible.[56] One could speculate that in between the sorting and degradation programs MVB subtypes originate and are triggered to fuse with the plasma membrane. In fact, a gene network was recently indentified that regulates lysosomal biogenesis[107] that increases the pool of SLs in the proximity of the plasma membrane, promoting fusion by the simultaneous elevation of intracellular calcium levels.[108] In addition, some but not all endosome populations are able to acquire secretory properties by fusing with the plasma membrane, a process that is in part mediated by Rab27.[109] These findings are in agreement with the notion that secretion of the MHC II molecules via exosomes is controlled by Rab27[110] but raises the question whether SLs and exosomes produced in the MVB are in fact the same phenomenon.[111] From these studies it seems clear that distinctions, if possible, should be made between exosome-producing MVBs and SLs. While SLs have functions such as controlling membrane repair, exosomes from MVBs are more likely to have specific effects on recipient cells. What is currently lacking is unequivocal experimental evidence that MVBs or specialized subsets preferentially move to and fuse with the plasma membrane, leading to exosome secretion. Even though Rab27 is a likely candidate regulating exosome secretion,[110,111] the involvement of this molecule in exocytosis from lysosomal compartments reduces its attractiveness as a selective Rab required for exosome secretion. Additional molecules on the LM of MVBs are likely to provide such selectivity and may include soluble *N*-ethylmaleimide-sensitive factor attachment protein receptors (SNAREs).

Before docking with the plasma membrane and releasing their content as exosomes, the MVBs must move to the cellular periphery. This direction is diametrically opposite the movement of maturing MVBs that have a tendency to move to the perinuclear region (Woodman–Futter). Early endosomes that have initiated ILV formation mature via homotypic fusion into LEs by Rab conversion via the exchange of Rab5 for Rab7, also known as the "Rab switch."[112] Endocytic vesicles undergo active movement that is mediated by microtubule (MT)- or actin-dependent motors.[113] Early endosomes containing the protein marker EEA1 tend to distribute peripherally compared to late endocytic organelles (i.e., LEs and lysosomes) that contain the lysosome-associated membrane protein 1 (LAMP1). Indeed lysosomes and MVBs normally accumulate in a juxtanuclear region around the MT-organizing center (MTOC).[114]

An understudied aspect of exosome secretion is the involvement of molecular motors that are required to direct the MVBs to their presumed docking sites at the plasma membrane. These motors include dynein and the dynactin-associated complex, both of which regulate movement toward the minus end of MTs (toward the nucleus), whereas the kinesin superfamily members KIF2β and KIF1Bβ3 orchestrate MT-dependent movement toward the cell periphery. Myosin Iα and other members of the myosin family are involved in endosomal movement via actin fibers.[115] The physiological significance of the movement and steady-state localization of LEs and MVBs in relationship to exosomes is unknown, although their positioning clearly affects the overall efficacy of the processes of organelle biogenesis and protein sorting. It is plausible that movements of MVBs toward the cellular periphery will positively affect exosome secretion, while stimulating their repositioning toward the perinuclear region would lead to a decrease in exosome secretion. One important factor in endosomal positioning within the cell is the level of cholesterol. Indeed, blocking cholesterol efflux from endosomes leads to cholesterol accumulation in LEs, which in turn restricts their movement to the cellular periphery.[116,117] However, not all endosomes are rich in cholesterol[118,119], and those that exclude cholesterol are resistant to the cholesterol-sequestering drug U18666A and may still move to the cell's periphery.[12] It thus appears that multiple subtypes of MVBs exist that correspond with their movement within the cell. This preferential movement will ultimately affect their ability to dock and fuse with the plasma membrane. Indeed the cholesterol sensor ORP1L is critically involved in LE positioning. Rocha et al. proposed a model in which cholesterol depletion causes LE scattering and high cholesterol levels induce LE clustering, as observed in Niemann–Pick type C disease.[120]

Besides the role of cholesterol in endosome positioning, cholesterol is also a major constituent of exosomes themselves. Indeed, Mobius showed clear cholesterol presence in the ILVs of some MVB subtypes, corresponding to the presence of cholesterol in exosomes.[119] Although I mentioned earlier that both lipids and ESCRT proteins have been implicated in ILV formation, since most budding mechanisms rely on membrane microdomains, it seems plausible that the ESCRT-mediated budding involves lipid-enriched domains that initiate the budding.[78] However, limited knowledge is available on how ESCRT may regulate the formation of such microdomains and whether these microdomains indeed form spontaneously, as expected, and if cholesterol is involved.[121] Recent studies, however, indicated that Hrs/Vps27 has a crucial role in intracellular (endosomal) cholesterol transport*. Hrs may be able to initiate the ESCRT pathway by orchestrating the formation of putative cholesterol-rich microdomains that help drive the budding away from cytosol. The authors hypothesize that Hrs may selectively remove cholesterol from specific regions of the LM of MVBs, thereby generating cholesterol-rich neighboring regions that could form the basis of membrane deformation, which in turn initiates the ESCRT pathway. Thus in mammalian cells, different subcellular organelles have distinct cholesterol concentrations, which is thought to be critical for biological functions. Oxysterol-binding protein-related proteins (ORPs) have been assumed to mediate nonvesicular cholesterol trafficking in cells; however, their *in vivo* functions and therefore the biological significance of cholesterol in each organelle were not fully understood. Kobuna et al. studied cholesterol restriction in *Caenorhabditis elegans* (*C. elegans*) and found that this induced the formation of enlarged LEs/lysosomes that increased embryonic lethality similarly to knockdown of MVB-related genes.[122] This study showed that knockdown of ORP1L, a mammalian ORP family member, induces the formation of enlarged MVBs in mammalian (HeLa) cells. *In vivo* findings suggest that the proper cholesterol level of LEs/lysosomes generated by ORPs is required for normal MVB formation and MVB-mediated membrane protein degradation.[122]

The above implication that cholesterol is important for membrane domain formation is related to the idea that the local enrichment of cone-shaped ceramide leads to microdomains in the LM of MVBs that promote deformation, budding, and ultimately scission into ILVs.[9] Moreover, ceramide has been shown to induce local loss of cholesterol in caveolin-enriched membranes (CEMs), suggesting ceramide and cholesterol compete for microdomain formation.[123] How the distinct roles of cholesterol in ILV budding,

*Du et al. (2012) *Dev. Cell*, in press.

MVB biogenesis, and endosome movement can be reconciled must be further investigated.

2.5 Exosome Secretion: Rabs and SNAREs

It is presumed that the final step of exosome biogenesis is secretion upon fusion of the MVB with the plasma membrane, releasing the ILVs as "exosomes." Ostrowski et al. (2010) were first to show the involvement of Rab27a and Rab27b in the movement of the MVB toward the plasma membrane, which is important for the subsequent fusion step.[110] First the authors identified five Rab proteins (Rab2b, Rab9a, Rab5a, Rab27a, and Rab27b) that appeared to be involved in exosome secretion as measured by a bead-capture technique coupled to fluorescence-activated cell sorting (FACS) analysis. A decrease in the level of exosome secretion, without major alterations in the regular secretory pathway, was seen when Rab27 proteins were silenced using short hairpin RNA. Silencing of Rab27a and Rab27b lead to an overall decrease in exosome secretion, but no major changes in protein content were observed, thus suggesting that Rab27a and Rab27b do not participate in the sorting of cargo into exosomes. Further investigation led to the discovery of three Rab27 effectors: Slac2b, Slp4, and possibly Munc 13-4. Silencing of the associated genes, EXPH5, SYTL4, and UNC13D, respectively, induced decreased exosome secretion. Silencing of Slp4 induced a similar phenotype as with silencing Rab27a, and a similar phenotype was also observed when silencing both together, thus indicating that Slp4 mediates the function of Rab27a.

These findings raise multiple questions: for instance, is the RAB27-mediated secretion pathway of exosomes ESCRT- or ceramide dependent? Since sorting was not effected one may presume that the ceramide pathway functions through Rab27. Although the secretion of exosomes was uncoupled from classic protein secretion, it is unclear if inhibition of SLs was measured. As mentioned previously, discriminating between MVB fusion and lysosome fusion with the plasma membrane may be difficult. In fact it has long been recognized that SLs rely on Rab27 for their movement to the plasma membrane.[124] One way to exclude lysosome involvement would be to measure the secretion of lysosomal enzymes (hydrolases) in the conditioned media. Despite these open questions it is safe to conclude that Rab27a and Rab27b are clearly involved in the movement of endosomal and/or lysosomal subpopulations to the plasma membrane, which is in agreement with recent studies by Laulagnier and Gruenberg.[109] Knockdown of Rab27a or Rab27b also resulted in a decreased number of endosomes and may reduce docking and fusion with the plasma membrane. Nevertheless, Rab27 is not likely to have a role in the membrane fusion process itself, as it seems to act more as a bridge between vesicles and the cytoskeleton controlling movement and perhaps docking rather than fusion.[125] The finding of Rab27 to have a role in exosome secretion is in line with previous findings that this molecule is important for secretion of compounds in other cell types but raises the question how molecules and MVBs are selected for secretion instead of degradation. Indeed molecules that can specifically be targeted to reduce exosome secretion alone, while leaving other secretory mechanisms intact, still need to be discovered.[126]

In a separate study performed by Hsu et al.[14] Rab35 was identified as one of the most prevalent Rab proteins in exosomes secreted by an oligodendrocyte cell line. Additionally, they discovered that the Rab guanosine-5′-triphosphatase (GTPase)-activating protein TBC1D10C controls PLP release via exosomes. TBC1D10C interacts with Rab35, and knockdown of Rab35 by siRNA reduces PLP secretion through exosomes. Furthermore, Rab35 and its Rab GTPase-activating protein TBC1D10C were shown to localize to the plasma membrane, suggesting Rab35 may be involved in the docking and fusion process. Indeed, no decrease in motility was measured introducing an inactive mutant of Rab35, compared to an active mutant, indicating that Rab35 is in fact likely to be involved in fusion with the plasma membrane rather than regulating the transport to the plasma membrane.[14] Savina et al., on the other hand, identified a high amount of the Rab11 protein in exosomes secreted by K562 cells (erythroleukemic cells). These exosomes were considered similar to reticulocyte exosomes, and decreased levels of secreted exosomes were observed using a Rab11 mutant. Moreover Rab11 overexpressing K562 cells leads to an enlargement of MVBs and localization of these MVBs near the plasma membrane, where upon the addition of calcium fusion takes place.[127,128] In turn, the involvement of a specific combination of SNARE proteins in the fusion of MVBs with the plasma membrane has been proposed by Bobrie et al. (2011).[129] For a schematic overview of the molecules involved in exosome biogenesis and secretion see Fig. 2.1.

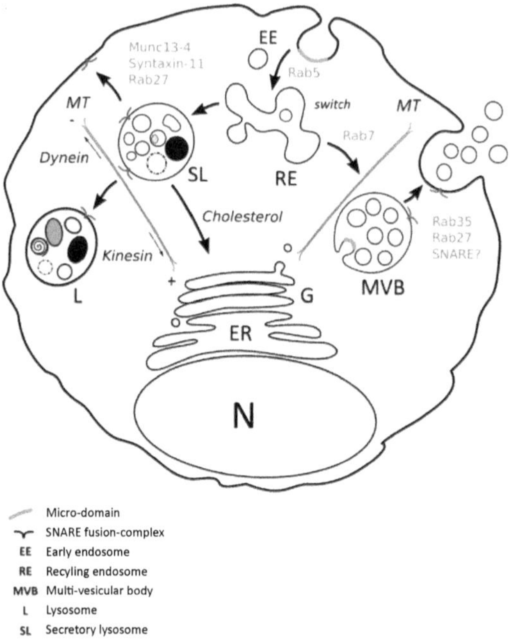

Figure 2.1 Exosomes are tightly linked with the endosomal pathway. The endosomes and MVBs regulate sorting, processing, recycling, storing, activating, silencing, and degradation of external material as well as surface receptors. By doing so, endosomes are responsible for regulation and fine-tuning of several key pathways in the cell.[56] Res, also called sorting endosomes, receive material in the cellular periphery from early endosomes that originate at the plasma membrane. In addition, material can be delivered directly via the biosynthetic/secretory transport pathway from the ER and Golgi apparatus. The REs mature into LEs such as MVBs where the rate of inward budding of the LM is increased. Under "normal conditions," the maturing endosomes are coupled to motor proteins (kinesin) and moved along MTs toward the perinuclear region—a process controlled at least in part by cholesterol. However, under polarizing conditions, SLs move into the opposite direction (dynein) toward the plasma membrane, where docking and fusion of the SL are mediated by specific Rabs (Rab27) and SNAREs.[141] Exosome secretion from MVBs may follow a similar path, and although Rab27 (as well as Rab35 depending on the cell type) seems to have an important role in the movement of MVBs to and docking at the plasma membrane, specific SNAREs that mediate fusion have yet to be determined. *Abbreviation*: RE, recycling endosome.

SNAREs are evolutionary conserved membrane proteins involved in all fusion events of the secretory pathway and are characterized by a conserved coiled-coil segment termed as the "SNARE motif." During fusion, the SNARE motifs of interacting SNARE proteins assemble into a parallel four-helix bundle that is initiated by the N-terminal tips and zips up toward the C-terminal membrane anchors. As a result, the membranes are pulled into close apposition and fusion is initiated. On the basis of a highly conserved set of amino acid side chains, SNARE motifs are classified into Qa-, Qb-, Qc-, and R-SNAREs, and every functional SNARE complex contains one SNARE motif of each subfamily.[130] The lysosome-to-plasma membrane transport pathway is proposed to play a role in plasma membrane repair, entry of parasites, and tumor cell invasiveness.[131–135] In addition, at least some cell types, MVEs—or some specialized subpopulation—may acquire the capacity to fuse with plasma membrane and thereby release their intralumenal vesicles in the extracellular milieu as exosomes. Although SNARE involvement in exosome secretion has not been identified, some clues can be obtained from nonspecialized fibroblast-like cells where a transient rise in intracellular calcium is believed to initiate the fusion of lysosomes with the plasma membrane.[72,131] This process is regulated by synaptotagmin VII[136] and is dependent on SNAREs (soluble *N*-ethylmaleimide-sensitive factor [NSF] attachment protein [SNAP] receptor) VAMP7, syntaxin 4, and SNAP23.[137] Inhibition of the formation of this complex induced inhibition of lysosome exocytosis. Interestingly SytVII associates directly with CD63[138], a well-established component of the endosomal–exosomal pathway.[12,69,101] Moreover, TI-VAMP/VAMP7 is the vesicle SNARE that mediates the lysosome exocytosis of ATP and cathepsin B in glia cells in a calcium-dependent manner.[139] In immune cells other SNARE molecules appear to have a role in SLs.[140] With total internal reflection fluorescence (TIRF) imaging Elstak et al. showed that munc13-4 Rab27(a) restricts the motility of SLs in the cytoplasm and proposed that Rab27 and munc13-4 tether lysosomes to the plasma membrane, which is a requirement for the formation of a cognate SNARE complex for fusion.[141] When compared to the results by Ostrowski et al. it is clear that the mechanisms for the secretion for exosomes share many similarities.[110]

2.6 Conclusions and Perspectives

An overwhelming body of evidence indicates that most cell types are able to secrete exosomes, either constitutively or upon activation, and thus it may not be surprising that these nanometer-size vesicles are present in virtually all body fluids examined. Despite this we have only just begun to understand the complex physiological properties of these vesicles. Unfortunately in many published reports clear evidence is lacking that the extracellular vesicles (EVs) studied are indeed produced by MVBs and hold typical exosomal characteristics such as a defined density, a size, and the presence of exosomal markers, including tetraspanins such as CD63, members of the ESCRT pathway (Alix), annexins (annexin II and VI), and heat shock proteins (HSP70). In fact, exosome-sized vesicles and larger microvesicles can also be generated by many cell types at the plasma membrane by activation of specific pathways such as ARF6.[142] Although the role of such microvesicles *in vivo* is less well understood and more difficult to distinguish from cellular debris,[8] they could also impact the physiology of target cells, in particular in cancer.[15] To make things even more complex, microvesicles derived from the plasma membrane often share biochemical and biological characteristics with exosomes, including ESCRT complex involvement for their budding[95], although, as discussed previously, not all "exosomes" seem to require ESCRT.[9,68] These complicating factors must be resolved before exosome research will reach full maturity, which may only happen when the molecular details that control EV biogenesis are fully elucidated and confirmed using *in vivo* models. Despite this, exosome secretion by immune cells has proven its physiological relevance for the immune response *in vivo*,[8] and similar physiological evidence for plasma membrane-derived vesicles may be provided soon, considering the rising interest in this field.

What should be the focus of future exosome research? One of the most intriguing and still outstanding questions is which MVBs are able to escape fusion with lysosomes and show that this may be regulated.[111] Multiple studies have now clearly showed that subtypes of MVBs exist and that some but not all have the capability of producing exosomes.[73] Studies by Mobius et al. clearly showed distinct MVBs at the lipid level, whereas subsequently functional evidence was provided that two distinct MVB pathways, one for lysosomal targeting and the other for exosome secretion, exist.[10] Indeed detailed cell biological studies by Ostrowski showed that only a small proportion of MVEs dock at the plasma membrane.[110] Moreover studies in our own laboratory point to a diversification of exosome-producing MVBs within one cell type,[12] as hypothesized previously.[73] Future detailed cell biological studies such as these should be able to characterize the possible molecular distinctions between subpopulations of MVBs in order to identify the presumed exosome donor population. Such a finding would also be highly relevant for the identification of the molecular mechanisms by which luminal cargo, be it protein, lipids, or RNA molecules, is sorted into exosome-producing MVBs. It is clear that besides heterogeneity in MVBs, the individual ILVs within MVBs are often heterogeneous not only in size and shape but also in composition. For instance, the unusual lipid bis(monoacylglycero)phosphate (lysobisphosphatidic acid) is present in some MVBs, yet exosomes seem to lack this lipid.[96] Besides the sorting of proteins and lipids into exosomes an exciting study by Ratajczak et al. discovered that embryonic stem cell–derived EVs reprogram hematopoietic progenitors through transfer of messenger RNA (mRNA) and protein delivery.[19] Subsequent studies indicated that mast cell exosomes contain functional mRNA and a class of miRNAs.[17] Importantly EVs secreted by tumor cells are present in biofluids and can be used for diagnostic purposes.[18] How such nucleic acids are loaded into exosomes remains unclear and their potential role *in vivo* unsubstantiated. On the basis of findings using virus-infected B-cell blasts we proposed that exosomes derived from MVBs are likely to contain the class of small RNAs,[22] in part because exosomes are physically incapable of carrying (many) large mRNAs. Indeed bioanalyzer studies on exosomal RNA usually indicate that small RNAs are abundantly present compared to larger-size RNAs [21] which agrees with studies in *Drosophila* and mammalian cells that MVBs are closely linked with miRNA function.[70,143] Some of these nucleic acids, in particular the class of miRNAs, can be delivered in a functional manner to recipient cells, and evidence of transfer has been provided in humans with exploiting viral miRNAs.[21] Not surprisingly, the findings mentioned here stirred the interest in exosomes and secreted membrane vesicles even more because of their biological, diagnostic, and therapeutic implications. The role of nucleic acids in the biogenesis of exosomes is still a "black box." It is presumed that miRNAs from intracellular granules can be delivered into ILVs of MVBs, possibly via a process of autophagy,[30,144] although mechanistic studies have not yet been performed. Undoubtedly, with the rapid progression of this field, the answer may come sooner as expected. In fact, a link between autophagy and exosomes may have been provided in yeast.[145]

Acknowledgments

The author would like to thank Dr. Thomas Wurdinger for inspiring discussions and valuable suggestions and Frederik Verweij for assistance with the artwork and critical reading of this chaper.

References

1. Pan BT, Johnstone RM (1983) Fate of the transferrin receptor during maturation of sheep reticulocytes *in vitro*: selective externalization of the receptor, *Cell*, **33**(3), 967–978.
2. Carayon K, et al. (2011) Proteolipidic composition of exosomes changes during reticulocyte maturation, *J. Biol. Chem.*, **286**(39), 34426–34439.
3. Denzer K, Kleijmeer MJ, Heijnen HF, Stoorvogel W, Geuze HJ (2000) Exosome: from internal vesicle of the multivesicular body to intercellular signaling device, *J. Cell Sci.*, **113**(Pt 19), 3365–3374.
4. Johnstone RM, Adam M, Hammond JR, Orr L, Turbide C (1987) Vesicle formation during reticulocyte maturation. Association of plasma membrane activities with released vesicles (exosomes), *J. Biol. Chem.*, **262**(19), 9412–9420.
5. Heijnen HF, Schiel AE, Fijnheer R, Geuze HJ, Sixma JJ (1999) Activated platelets release two types of membrane vesicles: microvesicles by surface shedding and exosomes derived from exocytosis of multivesicular bodies and alpha-granules, *Blood*, **94**(11), 3791–3799.
6. Flaumenhaft R, et al. (2009) Megakaryocyte-derived microparticles: direct visualization and distinction from platelet-derived microparticles, *Blood*, **113**(5), 1112–1121.
7. van NG, Porto-Carreiro I, Simoes S, Raposo G (2006) Exosomes: a common pathway for a specialized function, *J. Biochem.*, **140**(1), 13–21.
8. Thery C, Ostrowski M, Segura E (2009) Membrane vesicles as conveyors of immune responses, *Nat. Rev. Immunol.*, **9**(8), 581–493.
9. Trajkovic K, et al. (2008) Ceramide triggers budding of exosome vesicles into multivesicular endosomes, *Science*, **319**(5867), 1244–1247.

10. Buschow SI, et al. (2009) MHC II in dendritic cells is targeted to lysosomes or T cell-induced exosomes via distinct multivesicular body pathways, *Traffic*, **10**(10), 1528–1542.

11. Fang Y, Wu N, Gan X, Yan W, Morrell JC, Gould SJ (2007) Higher-order oligomerization targets plasma membrane proteins and HIV gag to exosomes, *PLOS Biol.*, **5**(6), e158.

12. Verweij FJ, et al. (2011) LMP1 association with CD63 in endosomes and secretion via exosomes limits constitutive NF-kappaB activation, *EMBO J.*, **30**(11), 2115–2129.

13. Chairoungdua A, Smith DL, Pochard P, Hull M, Caplan MJ (2010) Exosome release of beta-catenin: a novel mechanism that antagonizes Wnt signaling, *J. Cell Biol.*, **190**(6), 1079–1091.

14. Hsu C, et al. (2010) Regulation of exosome secretion by Rab35 and its GTPase-activating proteins TBC1D10A-C, *J. Cell Biol.*, **189**(2), 223–232.

15. Al-Nedawi K, et al. (2008) Intercellular transfer of the oncogenic receptor EGFRvIII by microvesicles derived from tumour cells, *Nat. Cell Biol.*, **10**(5), 619–624.

16. Meckes DG, Jr., Shair KH, Marquitz AR, Kung CP, Edwards RH, Raab-Traub N (2010) Human tumor virus utilizes exosomes for intercellular communication, *Proc. Natl. Acad. Sci. U. S. A.*, **107**(47), 20370–20375.

17. Valadi H, Ekstrom K, Bossios A, Sjostrand M, Lee JJ, Lotvall JO (2007) Exosome-mediated transfer of mRNAs and microRNAs is a novel mechanism of genetic exchange between cells, *Nat. Cell. Biol.*, **9**(6), 654–659.

18. Skog J, et al. (2008) Glioblastoma microvesicles transport RNA and proteins that promote tumour growth and provide diagnostic biomarkers, *Nat. Cell. Biol.*, **10**(12), 1470–1476.

19. Ratajczak J, et al. (2006) Embryonic stem cell-derived microvesicles reprogram hematopoietic progenitors: evidence for horizontal transfer of mRNA and protein delivery, *Leukemia*, **20**(5), 847–856.

20. Zhang Y, et al. (2010) Secreted monocytic miR-150 enhances targeted endothelial cell migration 4, *Mol. Cell*, **39**(1), 133–144.

21. Pegtel DM, et al. (2010) Functional delivery of viral miRNAs via exosomes, *Proc. Natl. Acad. Sci. U. S. A.*, **107**(14), 6328–6333.

22. Zomer A, Vendrig T, Hopmans ES, van EM, Middeldorp JM, Pegtel DM (2010) Exosomes: fit to deliver small RNA, *Commun. Integr. Biol.*, **3**(5), 447–450.

23. Jochum S, Ruiss R, Moosmann A, Hammerschmidt W, Zeidler R (2012) RNAs in Epstein-Barr virions control early steps of infection, *Proc. Natl. Acad. Sci. U. S. A.*, **109**(21), E1396–E1404.

24. Mittelbrunn M, et al. (2011) Unidirectional transfer of microRNA-loaded exosomes from T cells to antigen-presenting cells, *Nat. Commun.*, **2**, 282.

25. Montecalvo A, et al. (2008) Exosomes as a short-range mechanism to spread alloantigen between dendritic cells during T cell allorecognition, *J. Immunol.*, **180**(5), 3081–3090.

26. Montecalvo A, et al. (2012) Mechanism of transfer of functional microRNAs between mouse dendritic cells via exosomes, *Blood*, **119**(3), 756–766.

27. Raposo G, et al. (1996) B lymphocytes secrete antigen-presenting vesicles, *J. Exp. Med.*, **183**(3), 1161–1172.

28. Saunderson SC, et al. (2008) Induction of exosome release in primary B cells stimulated via CD40 and the IL-4 receptor, *J. Immunol.*, **180**(12), 8146–8152.

29. Arita S, et al. (2008) B cell activation regulates exosomal HLA production, *Eur. J. Immunol.*, **38**(5), 1423–1434.

30. Gibbings D, Voinnet O (2010) Control of RNA silencing and localization by endolysosomes, *Trends Cell Biol.*, **20**(8), 491–501.

31. Muntasell A, Berger AC, Roche PA (2007) T cell-induced secretion of MHC class II-peptide complexes on B cell exosomes, *EMBO J.*, **26**(19), 4263–4272.

32. Thery C, Zitvogel L, Amigorena S (2002) Exosomes: composition, biogenesis and function, *Nat. Rev. Immunol.*, **2**(8), 569–579.

33. Zitvogel L, et al. (1998) Eradication of established murine tumors using a novel cell-free vaccine: dendritic cell-derived exosomes, *Nat. Med.*, **4**(5), 594–600.

34. Segura E, et al. (2005) ICAM-1 on exosomes from mature dendritic cells is critical for efficient naive T-cell priming, *Blood*, **106**(1), 216–223.

35. Segura E, Amigorena S, Thery C (2005) Mature dendritic cells secrete exosomes with strong ability to induce antigen-specific effector immune responses, *Blood Cells Mol. Dis.*, **35**(2), 89–93.

36. Morelli AE, et al. (2004) Endocytosis, intracellular sorting, and processing of exosomes by dendritic cells, *Blood*, **104**(10), 3257–3266.

37. Denzer K, van EM, Kleijmeer MJ, Jakobson E, de GC, Geuze HJ (2000) Follicular dendritic cells carry MHC class II-expressing microvesicles at their surface, *J. Immunol.*, **165**(3), 1259–1265.

38. Taieb J, Chaput N, Zitvogel L (2005) Dendritic cell-derived exosomes as cell-free peptide-based vaccines, *Crit. Rev. Immunol.*, **25**(3), 215–223.

39. Mignot G, Roux S, Thery C, Segura E, Zitvogel L (2006) Prospects for exosomes in immunotherapy of cancer, *J. Cell Mol. Med.*, **10**(2), 376–388.

40. Zeelenberg IS, et al. (2011) Antigen localization controls T cell-mediated tumor immunity, *J. Immunol.*, **187**(3), 1281–1288.

41. Allen CD, Cyster JG (2008) Follicular dendritic cell networks of primary follicles and germinal centers: phenotype and function, *Semin. Immunol.*, **20**(1), 14–25.

42. Balaj L, et al. (2011) Tumour microvesicles contain retrotransposon elements and amplified oncogene sequences, *Nat. Commun.*, **2**, 180.

43. Kleijmeer MJ, Morkowski S, Griffith JM, Rudensky AY, Geuze HJ (1997) Major histocompatibility complex class II compartments in human and mouse B lymphoblasts represent conventional endocytic compartments, *J. Cell Biol.*, **139**(3), 639–649.

44. Peters PJ, et al. (1995) Major histocompatibility complex class II compartments in human B lymphoblastoid cells are distinct from early endosomes, *J. Exp. Med.*, **182**(2), 325–334.

45. Zwart W, et al. (2005) Spatial separation of HLA-DM/HLA-DR interactions within MIIC and phagosome-induced immune escape, *Immunity*, **22**(2), 221–233.

46. Peters PJ, Geuze HJ, van der Donk HA, Borst J (1990) A new model for lethal hit delivery by cytotoxic T lymphocytes, *Immunol. Today*, **11**(1), 28–32.

47. Stang E, Guerra CB, Amaya M, Paterson Y, Bakke O, Mellins ED (1998) DR/CLIP (class II-associated invariant chain peptides) and DR/peptide complexes colocalize in prelysosomes in human B lymphoblastoid cells, *J. Immunol.*, **160**(10), 4696–4707.

48. Rocha N, Neefjes J (2008) MHC class II molecules on the move for successful antigen presentation, *EMBO J.*, **27**(1), 1–5.

49. Henne C, Schwenk F, Koch N, Moller P (1995) Surface expression of the invariant chain (CD74) is independent of concomitant expression of major histocompatibility complex class II antigens, *Immunology*, **84**(2), 177–182.

50. Gauvreau ME, et al. (2009) Sorting of MHC class II molecules into exosomes through a ubiquitin-independent pathway, *Traffic*, **10**(10), 1518–1527.
51. Nolte-'t Hoen EN, Buschow SI, Anderton SM, Stoorvogel W, Wauben MH (2009) Activated T cells recruit exosomes secreted by dendritic cells via LFA-1, *Blood*, **113**(9), 1977–1981.
52. McConnell RE, Higginbotham JN, Shifrin DA, Jr., Tabb DL, Coffey RJ, Tyska MJ (2009) The enterocyte microvillus is a vesicle-generating organelle, *J. Cell Biol.*, **185**(7), 1285–1298.
53. Shifrin DA, Jr., McConnell RE, Nambiar R, Higginbotham JN, Coffey RJ, Tyska MJ (2012) Enterocyte microvillus-derived vesicles detoxify bacterial products and regulate epithelial-microbial interactions, *Curr. Biol.*, **22**(7), 627–631.
54. Welsch S, Keppler OT, Habermann A, Allespach I, Krijnse-Locker J, Krausslich HG (2007) HIV-1 buds predominantly at the plasma membrane of primary human macrophages, *PLOS Pathog.*, **3**(3), e36.
55. Fader CM, Colombo MI (2009) Autophagy and multivesicular bodies: two closely related partners, *Cell Death Differ.*, **16**(1), 70–78.
56. Huotari J, Helenius A (2011) Endosome maturation, *EMBO J.*, **30**(17), 3481–3500.
57. Stenmark H (2009) Seeing is believing, *Nat.* Rev *Mol. Cell Biol.*, **10**(9), 582.
58. Jeyakumar M, Dwek RA, Butters TD, Platt FM (2005) Storage solutions: treating lysosomal disorders of the brain, *Nat. Rev. Neurosci.*, **6**(9), 713–725.
59. Huizing M, Helip-Wooley A, Westbroek W, Gunay-Aygun M, Gahl WA (2008) Disorders of lysosome-related organelle biogenesis: clinical and molecular genetics, *Annu. Rev. Genomics Hum. Genet.*, **9**, 359–386.
60. Mosesson Y, Mills GB, Yarden Y (2008) Derailed endocytosis: an emerging feature of cancer, *Nat. Rev. Cancer*, **8**(11), 835–850.
61. Fevrier B, et al. (2004) Cells release prions in association with exosomes, *Proc. Natl. Acad. Sci. U. S. A.*, **101**(26), 9683–9688.
62. Mayor S, Pagano RE (2007) Pathways of clathrin-independent endocytosis, *Nat. Rev. Mol. Cell Biol.*, **8**(8), 603–612.
63. Katzmann DJ, Babst M, Emr SD (2001) Ubiquitin-dependent sorting into the multivesicular body pathway requires the function of a conserved endosomal protein sorting complex, ESCRT-I, *Cell*, **106**(2), 145–155.

64. Hurley JH, Emr SD (2006) The ESCRT complexes: structure and mechanism of a membrane-trafficking network, *Annu. Rev. Biophys. Biomol. Struct.*, **35**, 277–298.
65. Henne WM, Buchkovich NJ, Emr SD (2011) The ESCRT pathway, *Dev. Cell*, **21**(1), 77–91.
66. Mellman I (1996) Endocytosis and molecular sorting, *Annu. Rev. Cell Dev. Biol.*, **12**, 575–625.
67. Marsh M, van MG (2008) Cell biology. No ESCRTs for exosomes, *Science*, **319**(5867), 1191–1192.
68. Stuffers S, Sem WC, Stenmark H, Brech A (2009) Multivesicular endosome biogenesis in the absence of ESCRTs, *Traffic*, **10**(7), 925–937.
69. van NG, et al. (2011) The tetraspanin CD63 regulates ESCRT-independent and -dependent endosomal sorting during melanogenesis, *Dev. Cell*, **21**(4), 708–721.
70. Gibbings DJ, Ciaudo C, Erhardt M, Voinnet O (2009) Multivesicular bodies associate with components of miRNA effector complexes and modulate miRNA activity, *Nat. Cell. Biol.*, **11**(9), 1143–1149.
71. Piper RC, Luzio JP (2007) Ubiquitin-dependent sorting of integral membrane proteins for degradation in lysosomes, *Curr. Opin. Cell Biol.*, **19**(4), 459–465.
72. Andrews NW (2000) Regulated secretion of conventional lysosomes, *Trends Cell Biol.*, **10**(8), 316–321.
73. Simons M, Raposo G (2009) Exosomes—vesicular carriers for intercellular communication, *Curr. Opin. Cell Biol.*, **21**, 575–581.
74. Babst M, Katzmann DJ, Snyder WB, Wendland B, Emr SD (2002) Endosome-associated complex, ESCRT-II, recruits transport machinery for protein sorting at the multivesicular body, *Dev. Cell*, **3**(2), 283–289.
75. Lin Y, Kimpler LA, Naismith TV, Lauer JM, Hanson PI (2005) Interaction of the mammalian endosomal sorting complex required for transport (ESCRT) III protein hSnf7-1 with itself, membranes, and the AAA+ ATPase SKD1, *J. Biol. Chem.*, **280**(13), 12799–12809.
76. Babst M, Wendland B, Estepa EJ, Emr SD (1998) The Vps4p AAA ATPase regulates membrane association of a Vps protein complex required for normal endosome function, *EMBO J.*, **17**(11), 2982–2993.
77. Wollert T, Hurley JH (2010) Molecular mechanism of multivesicular body biogenesis by ESCRT complexes, *Nature*, **464**(7290), 864–869.

78. Hurley JH, Hanson PI (2010) Membrane budding and scission by the ESCRT machinery: it's all in the neck, *Nat.* Rev *Mol. Cell Biol.*, **11**(8), 556–566.

79. Simpson RJ, Jensen SS, Lim JW (2008) Proteomic profiling of exosomes: current perspectives, *Proteomics*, **8**(19), 4083–4099.

80. Chu T, Sun J, Saksena S, Emr SD (2006) New component of ESCRT-I regulates endosomal sorting complex assembly, *J. Cell Biol.*, **175**(5), 815–823.

81. Kostelansky MS, et al. (2007) Molecular architecture and functional model of the complete yeast ESCRT-I heterotetramer, *Cell*, **129**(3), 485–498.

82. Morita E, Sandrin V, Alam SL, Eckert DM, Gygi SP, Sundquist WI (2007) Identification of human MVB12 proteins as ESCRT-I subunits that function in HIV budding, *Cell Host Microbe*, **2**(1), 41–53.

83. Oestreich AJ, Davies BA, Payne JA, Katzmann DJ (2007) Mvb12 is a novel member of ESCRT-I involved in cargo selection by the multivesicular body pathway, *Mol. Biol. Cell*, **18**(2), 646–657.

84. Curtiss M, Jones C, Babst M (2007 Efficient cargo sorting by ESCRT-I and the subsequent release of ESCRT-I from multivesicular bodies requires the subunit Mvb12, *Mol. Biol. Cell*, **18**(2), 636–645.

85. Martin-Serrano J, Zang T, Bieniasz PD (2003) Role of ESCRT-I in retroviral budding, *J. Virol.*, **77**(8), 4794–4804.

86. von Schwedler UK, et al. (2003) The protein network of HIV budding, *Cell*, **114**(6), 701–713.

87. Odorizzi G (2006) The multiple personalities of Alix, *J. Cell Sci.*, **119**(Pt 15), 3025–3032.

88. Strack B, Calistri A, Craig S, Popova E, Gottlinger HG (2003) AIP1/ALIX is a binding partner for HIV-1 p6 and EIAV p9 functioning in virus budding, *Cell*, **114**(6), 689–699.

89. Demirov DG, Freed EO (2004) Retrovirus budding, *Virus Res.*, **106**(2), 87–102.

90. Morita E, Sundquist WI (2004) Retrovirus budding, *Annu. Rev. Cell Dev. Biol.*, **20**, 395–425.

91. Bieniasz PD (2006) Late budding domains and host proteins in enveloped virus release, *Virology*, **344**(1), 55–63.

92. Izquierdo-Useros N, et al. (2009) Capture and transfer of HIV-1 particles by mature dendritic cells converges with the exosome-dissemination pathway, *Blood*, **113**(12), 2732–2741.

93. Booth AM, Fang Y, Fallon JK, Yang JM, Hildreth JE, Gould SJ (2006) Exosomes and HIV Gag bud from endosome-like domains of the T cell plasma membrane, *J. Cell Biol.*, **172**(6), 923–935.

94. Izquierdo-Useros N, et al. (2010) HIV and mature dendritic cells: trojan exosomes riding the Trojan horse? *PLOS Pathog.*, **6**(3), e1000740.

95. Nabhan JF, Hu R, Oh RS, Cohen SN, Lu Q (2012) Formation and release of arrestin domain-containing protein 1-mediated microvesicles (ARMMs) at plasma membrane by recruitment of TSG101 protein, *Proc. Natl. Acad. Sci. U. S. A.*, **109**(11), 4146–4151.

96. Matsuo H, et al. (2004) Role of LBPA and Alix in multivesicular liposome formation and endosome organization, *Science*, **303**(5657), 531–534.

97. Vaccari T, et al. (2009) Comparative analysis of ESCRT-I, ESCRT-II and ESCRT-III function in Drosophila by efficient isolation of ESCRT mutants, *J. Cell Sci.*, **122**(Pt 14), 2413–2423.

98. Shen B, Wu N, Yang JM, Gould SJ (2011) Protein targeting to exosomes/microvesicles by plasma membrane anchors, *J. Biol. Chem.*, **286**(16), 14383–14395.

99. Kuehn MJ, Kesty NC (2005) Bacterial outer membrane vesicles and the host-pathogen interaction, *Genes Dev.*, **19**(22), 2645–2655.

100. Haurat MF, et al. (2011) Selective sorting of cargo proteins into bacterial membrane vesicles, *J. Biol. Chem.*, **286**(2), 1269–1276.

101. Escola JM, Kleijmeer MJ, Stoorvogel W, Griffith JM, Yoshie O, Geuze HJ (1998) Selective enrichment of tetraspan proteins on the internal vesicles of multivesicular endosomes and on exosomes secreted by human B-lymphocytes, *J. Biol. Chem.*, **273**(32), 20121–20127.

102. Wubbolts R, et al. (2003) Proteomic and biochemical analyses of human B cell-derived exosomes. Potential implications for their function and multivesicular body formation, *J. Biol. Chem.*, **278**(13), 10963–10972.

103. Duffield A, Kamsteeg EJ, Brown AN, Pagel P, Caplan MJ (2003) The tetraspanin CD63 enhances the internalization of the H,K-ATPase beta-subunit, *Proc. Natl. Acad. Sci. U. S. A.*, **100**(26), 15560–15565.

104. Kallquist L, et al. (2008) The tetraspanin CD63 is involved in granule targeting of neutrophil elastase, *Blood*, **112**(8), 3444–3454.

105. Piper RC, Katzmann DJ (2007) Biogenesis and function of multivesicular bodies, *Annu. Rev. Cell Dev. Biol.*, **23**, 519–547.

106. Fader CM, Sanchez DG, Mestre MB, Colombo MI (2009) TI-VAMP/VAMP7 and VAMP3/cellubrevin: two v-SNARE proteins involved

in specific steps of the autophagy/multivesicular body pathways, *Biochim. Biophys. Acta*, **1793**(12), 1901–1916.

107. Sardiello M, et al. (2009) A gene network regulating lysosomal biogenesis and function, *Science*, **325**(5939), 473–477.
108. Medina DL, et al. (2011) Transcriptional activation of lysosomal exocytosis promotes cellular clearance, *Dev. Cell*, **21**(3), 421–430.
109. Laulagnier K, Schieber NL, Maritzen T, Haucke V, Parton RG, Gruenberg J (2011) Role of AP1 and Gadkin in the traffic of secretory endo-lysosomes, *Mol. Biol. Cell*, **22**(12), 2068–2082.
110. Ostrowski M, et al. (2009) Rab27a and Rab27b control different steps of the exosome secretion pathway, *Nat. Cell. Biol.*, **12**(1), 19–30.
111. Pfeffer SR (2010) Two Rabs for exosome release, *Nat. Cell. Biol.*, **12**(1), 3–4.
112. Rink J, Ghigo E, Kalaidzidis Y, Zerial M (2005) Rab conversion as a mechanism of progression from early to late endosomes, *Cell*, **122**(5), 735–749.
113. Schroer TA (2000) Motors, clutches and brakes for membrane traffic: a commemorative review in honor of Thomas Kreis, *Traffic*, **1**(1), 3–10.
114. Matteoni R, Kreis TE (1987) Translocation and clustering of endosomes and lysosomes depends on microtubules, *J. Cell Biol.*, **105**(3), 1253–1265.
115. Raposo G, et al. (1999) Association of myosin I alpha with endosomes and lysosomes in mammalian cells, *Mol. Biol. Cell*, **10**(5), 1477–1494.
116. Kobayashi T, Gu F, Gruenberg J (1998) Lipids, lipid domains and lipid-protein interactions in endocytic membrane traffic, *Semin. Cell Dev. Biol.*, **9**(5), 517–526.
117. Gruenberg J (2001) The endocytic pathway: a mosaic of domains, *Nat. Rev. Mol. Cell Biol.*, **2**(10), 721–730.
118. Mobius W, et al. (2002) Immunoelectron microscopic localization of cholesterol using biotinylated and non-cytolytic perfringolysin O, *J. Histochem. Cytochem.*, **50**(1), 43–55.
119. Mobius W, et al. (2003) Recycling compartments and the internal vesicles of multivesicular bodies harbor most of the cholesterol found in the endocytic pathway, *Traffic*, **4**(4), 222–231.
120. Rocha N, et al. (2009) Cholesterol sensor ORP1L contacts the ER protein VAP to control Rab7-RILP-p150 Glued and late endosome positioning, *J. Cell Biol.*, **185**(7), 1209–1225.

121. Babst M (2011) MVB vesicle formation: ESCRT-dependent, ESCRT-independent and everything in between, *Curr. Opin. Cell Biol.*, **23**(4), 452–457.

122. Kobuna H, et al. (2010) Multivesicular body formation requires OSBP-related proteins and cholesterol, *PLOS Genet.*, **6**(8).

123. Yu C, Alterman M, Dobrowsky RT (2005) Ceramide displaces cholesterol from lipid rafts and decreases the association of the cholesterol binding protein caveolin-1, *J. Lipid Res.*, **46**(8), 1678–1691.

124. Griffiths G (2002) What's special about secretory lysosomes? *Semin. Cell Dev. Biol.*, **13**(4), 279–284.

125. Goud B (2002) How Rab proteins link motors to membranes, *Nat. Cell. Biol.*, **4**(4), E77–E78.

126. Tolmachova T, et al. (2004) A general role for Rab27a in secretory cells, *Mol. Biol. Cell*, **15**(1), 332–344.

127. Savina A, Furlan M, Vidal M, Colombo MI (2003) Exosome release is regulated by a calcium-dependent mechanism in K562 cells, *J. Biol. Chem.*, **278**(22), 20083–20090.

128. Savina A, Fader CM, Damiani MT, Colombo MI (2005) Rab11 promotes docking and fusion of multivesicular bodies in a calcium-dependent manner, *Traffic*, **6**(2), 131–143.

129. Bobrie A, Colombo M, Raposo G, Thery C (2011) Exosome secretion: molecular mechanisms and roles in immune responses, *Traffic*, **12**(12), 1659–1668.

130. Jahn R, Scheller RH (2006) SNAREs—engines for membrane fusion, *Nat. Rev. Mol. Cell Biol.*, **7**(9), 631–643.

131. Rodriguez A, Webster P, Ortego J, Andrews NW (1997) Lysosomes behave as Ca2+-regulated exocytic vesicles in fibroblasts and epithelial cells, *J. Cell Biol.*, **137**(1), 93–104.

132. Reddy A, Caler EV, Andrews NW (2001) Plasma membrane repair is mediated by Ca(2+)-regulated exocytosis of lysosomes, *Cell*, **106**(2), 157–169.

133. Andrews NW (2002) Lysosomes and the plasma membrane: trypanosomes reveal a secret relationship, *J. Cell Biol.*, **158**(3), 389–394.

134. Kroemer G, Jaattela M (2005) Lysosomes and autophagy in cell death control, *Nat. Rev. Cancer*, **5**(11), 886–897.

135. Eder C (2009) Mechanisms of interleukin-1beta release, *Immunobiology*, **214**(7), 543–553.

136. Czibener C, et al. (2006) Ca2+ and synaptotagmin VII-dependent delivery of lysosomal membrane to nascent phagosomes, *J. Cell Biol.*, **174**(7), 997–1007.

137. Rao SK, Huynh C, Proux-Gillardeaux V, Galli T, Andrews NW (2004) Identification of SNAREs involved in synaptotagmin VII-regulated lysosomal exocytosis, *J. Biol. Chem.*, **279**(19), 20471–20479.

138. Flannery AR, Czibener C, Andrews NW (2010) Palmitoylation-dependent association with CD63 targets the Ca2+ sensor synaptotagmin VII to lysosomes, *J. Cell Biol.*, **191**(3), 599–613.

139. Verderio C, et al. (2012) TI-VAMP/VAMP7 is the SNARE of secretory lysosomes contributing to ATP secretion from astrocytes, *Biol. Cell*, **104**(4), 213–228.

140. Elstak ED, et al. (2011) The munc13–4-rab27 complex is specifically required for tethering secretory lysosomes at the plasma membrane, *Blood*, **118**(6), 1570–1578.

141. Elstak ED, Neeft M, Nehme NT, Callebaut I, de Saint BG, van der SP (2012) Munc13–4*rab27 complex tethers secretory lysosomes at the plasma membrane, *Commun. Integr. Biol.*, **5**(1), 64–67.

142. Muralidharan-Chari V, et al. (2009) ARF6-regulated shedding of tumor cell-derived plasma membrane microvesicles, *Curr. Biol.*, **19**(22), 1875–1885.

143. Lee YS, et al. (2009) Silencing by small RNAs is linked to endosomal trafficking, *Nat. Cell Biol.*, **11**(9), 1150–1156.

144. Pegtel DM, van dG, Middeldorp JM (2011) Viral miRNAs exploiting the endosomal-exosomal pathway for intercellular cross-talk and immune evasion, *Biochim. Biophys. Acta.*, **1809**(11-12), 715–721.

145. Dupont N, Jiang S, Pilli M, Ornatowski W, Bhattacharya D, Deretic V (2011) Autophagy-based unconventional secretory pathway for extracellular delivery of IL-1beta, *EMBO J.*, **30**(23), 4701–4711.

Chapter 3

Extracellular Vesicle–Mediated Epigenetic Reprogramming of Cells

Giovanni Camussi, Federica Collino, and Maria C. Deregibus
Department of Medical Sciences, Molecular Biotechnology Center (MBC) and Centre for Research in Experimental Medicine (CeRMS), Torino, Italy
giovanni.camussi@unito.it

3.1 Introduction

The cell differentiation fate and phenotype are tightly controlled by transcription factors that operate in a defined microenvironment. Within this microenvironment the exchange of information between cells is critical and influences the final cell phenotype. Cells may communicate with each other, not only through soluble factors such as cytokines, chemokines, and hormones,[1] but also by direct cell-to-cell contact through cytonemes that connect adjacent cells, allowing transfer of surface-associated molecules, as well as tunneling nanotubules, which also enable transfer of cytoplasmic components.[2] Moreover, cells secrete vesicles into the extracellular space membrane that act as carriers for exchange of information between cells. The extracellular vesicles (EVs) released from cells

Extracellular Vesicles in Health and Disease
Edited by Paul Harrison, Christopher Gardiner, and Ian L. Sargent

ISBN 978-981-4411-98-1 (Hardcover), 978-981-4411-99-8 (eBook)
www.panstanford.com

both *in vitro* and *in vivo* are a mix of different vesicles with a spherical shape constituted of a lipid bilayer and hydrophilic proteins. Extracellular membrane vesicles include shedding vesicles, also known as ectosomes, microparticles, exovesicles, or microvesicles (MVs), formed by budding of the cell plasma membrane, along with exosomes derived from the endosomal membrane compartment by exocytosis.[3,4] Shedding vesicles include a heterogeneous population with sizes ranging from 100 nm to 1,000 nm, whereas exosomes are a more homogeneous population of vesicles with sizes ranging from 30 nm to 120 nm.

Formation of shedding vesicles involves calcium influx and cytoskeleton reorganization with redistribution of membrane components, leading to formation of membrane nanodomains[5] (Fig. 3.1a). These vesicles express high levels of cholesterol and phosphatidylserine (PS),[4] as well as other molecules generally ascribed to lipid rafts, including tissue factor (TF) and flotillin-1.[6] Plasma membrane budding is associated with changes in membrane lipid asymmetry with exposure of PS by a process that involves the increase of cytosolic Ca^{2+}, leading to alteration of the transmembrane enzymatic balance of several enzymes such as calpain, scramblase, floppase, flippase, and gelsolin.[7] Vesicle shedding from the cell surface then follows after Ca^{2+}-dependent proteolysis.[8] Therefore, the release of shedding vesicles can be triggered by all stimuli that increase intracellular Ca^{2+} levels. During formation, shedding vesicles collect specific transmembrane and cytosolic proteins[4] and therefore carry specific molecules of the cell of origin. For instance, shedding vesicles may contain cytokines and chemokines (e.g., interleukin-1β and interleukin-8), vascular endothelial growth factor (VEGF), and fibroblast growth factor 2 (FGF2) produced by the cell of origin[9] or oncogenic growth factor receptors (e.g., epidermal growth factor receptor variant III [EGFRvIII]) as in the case of shedding vesicles derived from gliomas.[10] Different groups have observed the instrumental role of sphingomyelinases (SMases) in the mechanism of EV release. Bianco et al. recently demonstrated a causal relationship between acidic SMase (A-SMase) activity and MV shedding and showed that A-SMase is both sufficient and necessary for the MV-shedding process in glial cells.[11]

The biogenesis of exosomes and the precise intracellular compartment from which they derive is still unidentified. Exosomes are thought to originate from the endosomal compartment following inward invaginations of the cell membrane under the control of the endosomal sorting complex required for transport (ESCRT) (Fig. 3.1b). The involvement of the ESCRT has been suggested by the presence of Alix, a component of the ESCRT, within exosomes.[12] After compartmentalization into late endosome/multivesicular bodies (MVBs), exosomes may be released from the cell after fusion with the cell membrane.[13] Other studies suggest a critical role for ceramide, which induces exosome release after invagination of the cell membrane.[14] This mechanism is catalyzed by neutral SMAse (N-SMAse),[14] rather than A-SMAse,[11] and exosome release is reduced following inhibition of neutral sphingomyelinases. This and other studies indicate that different members of the SMase family specifically control the budding and release of distinct populations of EVs.[11,14,15]

Exosomes, unlike shedding vesicles, express molecules such as TSG101, Alix, HSC70, CD81, CD63, CD9, and a low content of PS.[16] Other classes of vesicles include gesicles with a size of 100 nm and exosome-like vesicles with sizes ranging from 20 nm to 50nm, expressing the full-length 55 kDa tumor necrosis factor (TNF) receptor-1.[17]

Cells simultaneously produce several types of EVs bearing a specific cargo and potentially capable of different biological activities.[13] Moreover, membrane composition and also the size of vesicles may differ depending on the cell of origin; it is therefore difficult to exactly categorize different EVs. Irrespectively of their origin, EVs contain cytosol delimited by a plasma membrane expressing proteins and receptors from the cell of origin; nucleic acids are also present in both exosomes and shedding vesicles.

Even though the exact physiological role of EVs remains to be elucidated, it is becoming clear that they may transfer proteins, receptors, bioactive lipids, messenger ribonucleic acid (mRNA), and micro-RNA (miRNA) from the cell of origin to the recipient cell, which may modify their phenotype and functions.[18,19] Interestingly, high-density lipoprotein (HDL) and low-density lipoprotein (LDL), which are the other major lipid-based particles in the plasma, also contain miRNAs. Highly pure HDL fractions were found to contain distinct miRNA signatures, which are altered in cardiovascular diseases. Furthermore, HDL was demonstrated to deliver miRNAs to Huh7 hepatocellular carcinoma cells.[20]

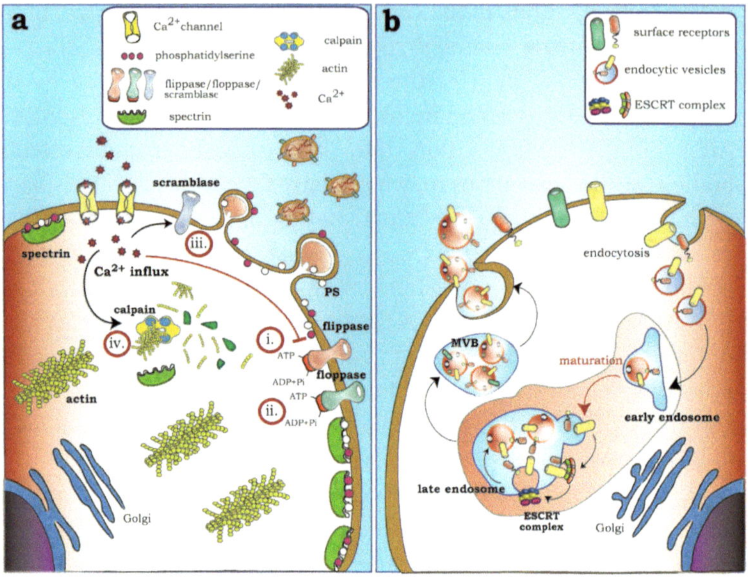

Figure 3.1 Biogenesis of extracellular vesicles. (a) Changes in membrane lipid asymmetry induces blebbing of the cell plasma membrane with generation of shedding vesicles. Formation of shedding vesicles is dependent on the increase of the intracellular calcium concentration that leads to (i) inactivation of ATP-dependent aminophospholipid translocase (flippase) with consequent translocation of PS to the external leaflet of the cell membrane bilayer; (ii) inactivation of floppase, an ATP-dependent enzyme that induces the transfer of lipids from the inner leaflet of the cell membrane to the outer together with flippase; (iii) activation of scramblase, which determines back-and-forth movements of phospholipids between the two leaflets of the cell membrane; and (iv) activated calpain, which cleaves the long actin filaments of the cytoskeleton, allowing membrane budding and shedding of vesicles. (b) Exosomes derive from the endosomal compartment as a result of maturation of early endosomes that progress into MVBs under the control of the ESCRT. Exosomes are released into the extracellular space after fusion of MVBs with the cell plasma membrane.

Increasing evidence indicates that exosome-/MV-dependent transfer of transcriptional regulators may induce epigenetic and functional changes that at least under certain conditions can be stable. Therefore the intercellular communication mediated by EVs uncovers an unexpected high plasticity of the cellular system.

3.2 Microvesicles and Exosomes as Haulers of Information

It has been suggested that EVs represent a very early and evolutionary conserved mechanism of cell-to-cell communication that served as a template for the development of subsequent, more sophisticated mechanisms of cell communication.[18] EVs can bind selectively and be incorporated by specific recipient cells;[21] they therefore only interact with target cells that can recognize them, as opposed to just any cell present in the microenvironment.[22] Binding to the recipient cell through ligand–receptor interactions or surface-expressed lipids is followed by internalization of EVs into endocytic compartments or direct fusion with the plasma membrane, leading to transfer of membrane components and delivery of exosome/MV contents to the cytoplasm of the recipient cell.[23] Alternatively, they may remain segregated within endosomes and be subsequently dismissed by the cells following fusion with the plasma membrane, thus leading to the process of transcytosis.[23] As the molecular composition of EVs released from diverse cell types is different, functional consequences to recipient cells also vary. Therefore, surface and intravesicular molecules are the determinant of functional and phenotype changes occurring in cells that interact with EVs.

3.2.1 Receptor–Ligand Signaling Complexes

EVs may influence the behavior of target cells in a variety of ways.

EVs may ***act as signaling complexes and directly stimulate target cells.*** This is the case with monocyte-derived MVs that express P-selectin glycoprotein ligand-1, instrumental in the monocytes' binding to platelets or MV release from platelets that are enriched with integrins and P-selectin.[18,23] Through these surface molecules they may directly activate endothelial, inflammatory, and malignant human hematopoietic cells.[22] MVs expressing TF in their membrane, such as those derived from platelets, monocytes, or tumor cells, also act by direct surface interaction. In addition, MVs released from neutrophils that express activated Mac-1 are able to induce platelet activation.[22] Moreover, the anionic aminophospholipid PS present on the MV surface may act as a platform for the assembly of coagulation factors.

3.2.2 Shedding Vesicles as Protein Carriers

3.2.2.1 Transfer of functional receptors

Another mechanism of action of EVs is related to fusion with the plasma membrane of the recipient cell with the ***transfer of receptors*** that are expressed by EVs. This mechanism has been shown for bystander B-cells that may acquire antigen receptors from activated B-cells by membrane transfer,[24] leading to expansion of cells able to present a specific antigen to CD4 T-cells. Other receptors known to be transferred by MVs include the Fas ligand that, once transferred from tumor cells, induces apoptosis of activated T-cells, promoting tumor immune escape.[25] In addition, there is the adhesion molecule CD41 that, once transferred via MVs from platelets to endothelial cells[26] or tumor cells,[18] induces proadhesive properties to these cells. The MV transfer of CXCR4 and CCR5 chemokine coreceptors for the human immunodeficiency virus-1 (HIV-1) virus to non-lympho-hemopoietic cells may favor the entry of the virus into these cells.[27,28] Another example of surface protein delivery by EVs is the transfer of TF and certain oncogenic receptors (i.e., EGFRvIII) from tumor cells to adjacent endothelial cells, a process that could clearly influence tumor angiogenesis.[29,30] Interestingly, it has been described that MV production can be triggered by the activity of some of the same oncogenic receptors that subsequently become incorporated into the MV cargo.[22]

3.2.2.2 Transfer of signaling proteins

EVs may also ***deliver proteins within the cytoplasm of target cells***, as seen for MVs derived from endotoxin-stimulated monocytes, which are able to transfer a cell death message via encapsulated caspase-1.[31] Sheldon et al. demonstrated that tumor- and endothelial cell–derived exosomes can transfer Delta-like 4 (Dll4), an important regulator of the Notch pathway, in recipient cells, both *in vitro* and *in vivo*. Exososomal Dll4, may affect target endothelial cells through the inhibition of Notch signaling, loss of Notch receptors, and change of the endothelial phenotype toward tip cells.[32] It has been also suggested that MVs may contribute to propagation of several infective agents, such as prions or HIV.[22] The so called "trojan exosome hypothesis" contemplates the direct delivery of HIV-1 by MVs.[33]

3.2.3 Extracellular Vesicles as Carriers of Genetic Information

The discovery that exosomes and MVs also contain nucleic acids makes them an attractive vehicle for intercellular trafficking of RNA. The presence of extracellular RNA described by Stroun et al. in 1978 suggests that, in addition to proteins, secretory RNA may also participate in cell-to-cell communication.[34] However, the presence of RNA-degradative enzymes in the extracellular space rendered it difficult to fully comprehend the biological relevance of secretory RNA. A breakthrough in extracellular RNA biology followed the revelation that EVs derived from different cell sources contain mRNA and miRNA. Indeed the encapsulated RNA within EVs is protected from extracellular ribonuclease (RNase), facilitating the extracellular transport of RNA. Ratajczak et al.[35] demonstrated that MVs derived from murine embryonic stem (ES) cells carry specific mRNA. Baj-Krzyworzeka et al.[36] demonstrated that MVs derived from tumor cells carried mRNA for growth factors that could be transferred to monocytes. In addition, Valadi et al.[37] demonstrated that the exosome-transferred RNA was also translated into proteins in the recipient cells. We found that MVs derived from human endothelial progenitor cells (EPCs) contained selected patterns of mRNA that, once transferred into quiescent endothelial cells, activate an angiogenic program.[30] Moreover, recipient cells incubated with MVs derived from green fluorescent protein (GFP)-transfected EPCs start to express GFP, demonstrating the transfer of functional mRNA. MV-mediated delivery of human mRNA to mouse cells and its translation into proteins has not only been shown *in vitro*[37,38] but also *in vivo*.[39,40] Aliotta et al.[41] recently showed that MV-treated bone marrow cells may undergo phenotypic changes due to MV delivery of mRNA or indirect translation of tissue-specific mRNA. Besides mRNA, Valadi et al. demonstrated that exosomes released from mouse and human mast cells contain and transfer miRNAs,[37] a class of post-transcriptional gene expression regulators, suggesting that they can alter gene expression in recipient cells. The miRNAs seem to be compartmentalized within EVs as their concentration is relatively more abundant than within the whole cell. Evidence of *in vitro* transfer of specific subsets of miRNAs to mouse embryonic fibroblasts has been also provided by Yuan et al.[42] We also found that miRNAs are enriched in MVs released by adult human mesenchymal stem cells (MSCs). Interestingly, the miRNA profile of cell-secreted MVs reflects those of the donor cells, but some miRNAs are selectively accumulated within MVs, suggesting a regulated process of RNA compartmentalization.[43] The gene ontology analysis of the miRNAs targets, highly expressed inside MVs, suggests that they may be involved in immune system regulation, cell survival, differentiation, and multiorgan development. The observation that target proteins of miRNA in recipient cells are affected by MV administration suggests that the transferred miRNAs are indeed functional.[43]

Several other studies have reported the transfer of secretory miRNAs through EVs. Zhang et al.[44] demonstrated the presence of miR-150 in exosomes secreted from cultured human monocyte/macrophage cell lines and its subsequent transfer to endothelial cells. The increased concentration of miR-150 into endothelial cells resulted in functional alterations, including c-Myb expression and enhanced migration. Kosaka et al. demonstrated the transfer via exosomes of the tumor-suppressive miR-143 from normal to cancerous prostate cells, resulting in the suppression of its target gene and inhibition of tumor cell growth.[45] Pegtel et al. showed that exosomes released from Epstein–Barr virus (EBV)-infected cells contain mature EBV-encoded miRNAs.[46] Recently, miRNAs have been also detected in body fluids such as blood plasma, urine, saliva, and sperm. The generation of miRNA profiles in biological fluids has supported the role of extracellular miRNAs as putative biomarkers of pathological states. The relatively high stability of circulatory miRNAs in physiological fluids has been attributed to their associations with RNA-binding proteins or their encapsulation within EVs.[17] The miRNAs contained in EVs may function both as passive indicators of disease and as real effectors of EV function.

Some reports have recently described the possible shuttle by EVs of parental deoxyribonucleic acid (DNA). Waldenstrom et al. identified the presence of 343 different chromosomal DNA sequences in EVs from mouse cardiomyocytes. DNA shuttled by EVs can be incorporated and localized both in the cytoplasm and in the nuclei of recipient cells.[47] This is in line with previous studies that identified the presence of chromosomal DNA in human prostasomes isolated from PC-3-cells.[48] Experiments with both nuclease treatment and different DNA staining indicated an internal location of the prostasomal DNA. Other studies have also suggested that mitochondrial DNA is present in exosomes isolated from astrocytes and glioblastoma cells, suggesting that MVs released from cancer cells may contain traces of mitochondrial DNA.[49]

The mechanism of accumulation and compartmentalization of ribonucleoproteins within EVs still remains to be defined. However, recent studies suggest a selective mechanism of RNA packaging.[50] Some information on the mechanisms of RNA accumulation within EVs may derive from studies on the ribonucleoproteins involved in the intracellular trafficking of RNAs and on the comparison of the miRNA species contained in EVs and in the cells of origin. We found that MVs released from stem cells contained stress granule–specific proteins, including ribonucleoproteins involved in RNA storage. Among these proteins TIA, TIAR, and HuR, together with Stau 1 and 2 proteins expressed in nuclei and stress granules but not in P-bodies, have been detected within MVs.[43] The absence of P-body-specific proteins and the presence of proteins involved in miRNA organization, such as Ago2 and GW182, were also observed in exosomes released from cultured monocytes.[50] We observed the presence of Ago2,[43] a protein of the argonaute family, which is involved in miRNA transport and processing, in MVs derived from human MSCs. These studies suggest a role of ribonucleoproteins in RNA transport, storage, and stability within EVs.[43]

The important role of lipids in the selective release of miRNAs inside exosomes has recently been unveiled.[20] Ceramide has been shown to play a role in intraluminal vesicle formation as well as in their sorting into lysosomes for cargo degradation or into exosomes for secretion into the extracellular space.[14] Accordingly, experiments of the knockdown of neutral nSMase2, the key enzyme in ceramide synthesis, induced a relevant decrease in the cellular export of miR-16 and miR-146a.[51] Furthermore, chemical inhibition of nSMase2 attenuated the release of exosomes and in turn miR-16 shuttled by exosomes.[20] Other groups observed that inhibition of nSMase2 with GW4869 resulted in a significant increase of miRNA release and transport through the HDL.[20]

3.3 Modulation of Extracellular Composition and Secretion of Vesicles by Different Stimuli

EVs contain RNA and proteins that reflect those of donor cells. Their content may vary depending on cell stimulation. It has been suggested that EVs protect target cells from cellular stress via transfer of RNA and proteins. De Jong et al. recently demonstrated that endothelial cell–derived exosomes mediate communication of stress-related signals.[52] They observed that exposure of endothelial cells to hypoxia or TNF-α modulates both protein and mRNA content of exosomes derived from these cells. In pathological states such as cancer, the microenvironmental stimuli not only affect the quantity of EVs released but also affect their composition.[13] Recently, Hedlund et al. demonstrated that leukemia/lymphoma T- and B-cell lines, subjected to thermal and oxidative stress, enhance the release of immunosuppressive exosomes in models of hematopoietic malignancies.[53] They demonstrated that cellular stress significantly enhances the secretion of NKG2D-ligand-carrying exosomes by tumor cells. Functional studies demonstrated that NKG2DL transferred by exosomes abrogates NKG2D-mediated NK cell cytotoxicity, and therefore it might contribute to the immune evasion of leukemia cells.[53]

In addition, Lv et al. showed that treatment with a chemotherapeutic agent enhances immunogenicity of exosomes released from cells resistant to anticancer drugs. The mechanism is related to the induction of HSP-specific NK cell responses, ascribed to an enhancement of exosome-carried HSPs (HSP60, HSP70, and HSP90).[54]

In agreement with these findings, Chen et al.[55] previously found that exosomes derived from heat-stressed tumor cells (HS-TEX) could also chemo-attract and activate dendritic cells (DCs) and T-cells more potently than activation by conventional tumor-derived exosomes. Accordingly, intratumoral injection of HS-TEX enhances the antitumor immune response, inhibiting tumor growth.[55]

3.4 Extracellular Vesicle–Induced Changes in the Cellular Phenotype

Epigenetic changes induced by EVs have been particularly studied in the context of immunology and cancer and stem cell biology.

3.4.1 Role in Immune Response

Numerous recent studies have focused on the role of EVs as regulators of immune responses.[17] Firstly, the role of EVs in immune system cell cross-talk was discovered by studies demonstrating that EBV-transformed B-lymphocytes were able to secrete exosomes through the fusion of MVBs with the plasma membrane.[56] Lymphocyte B–released exosomes were able to express major histocompatibility complex (MHC) class II dimers bound to antigenic peptides on their surface, presentable to specific T-cells, supporting the active function of exosomes in adaptive immune responses.[57] Subsequently, different groups described that DCs may also secrete exosomes expressing functional MHC class I–peptide complexes. DC exosomes have been described as being able to induce CD8+ T-lymphocyte-dependent immune responses, resulting in the regression of established tumors in immune-competent mice.[58]

Interestingly, exosomes released by immune cells at different differentiation stages showed diverse immunogenic properties. Segura et al.[59] demonstrated that both immature and mature murine DCs secrete morphologically similar exosomes. Proteomic analysis of the two populations showed equal overall protein composition but quantitative differences in protein levels. Moreover, functional T-cell stimulation assays demonstrated that exosomes secreted by mature DCs were more potent than those derived from immature DCs both *in vitro* and *in vivo*.[59] Exosomes derived from mature DCs have been also described to support anticancer therapies.[60] By contrast, exosomes secreted by immature DCs or by DCs overexpressing anti-inflammatory cytokines[61,62] promote the suppression of inflammation and autoimmune responses and may represent a novel therapy for the treatment of autoimmune diseases.[57]

Recent studies have established that miRNAs show a distinctive profile in cells of the immune system and play pivotal roles in the regulation of both cell development and function. In addition, the aberrant expression of miRNAs may contribute to pathological states involving the immune system, such as cancer and autoimmunity.[63] The role of miRNAs, shuttled by EVs, in immune system regulation has only been recently described.[57] The exchange of miRNAs, via exosomes, between T-cells and antigen presenting cells (APCs) has been demonstrated at the site of immune synapse. Inhibition of exosome production by targeting N-SMase reduces the transfer of miRNAs to APCs.[64] On the other hand, the cognate DC–T-cell interaction induced exosome release by

DCs,[65] supporting the idea of a dynamic exchange of information via exosomes between DCs and T-cells.[57]

3.4.2 Role in Tumor Cell Biology

Several studies indicate that tumor-derived EVs profoundly influence the cancer microenvironment by changing the phenotype of stroma cells and favoring tumor cell escape from immune surveillance.[66] Castellana et al.[67] demonstrated that tumor and normal stroma cells communicate through the mutual shedding of MVs, increasing tumorigenicity (Fig. 3.2). This reciprocal communication allows the exchange of receptors, active proteins, lipids, or genetic information via EVs.

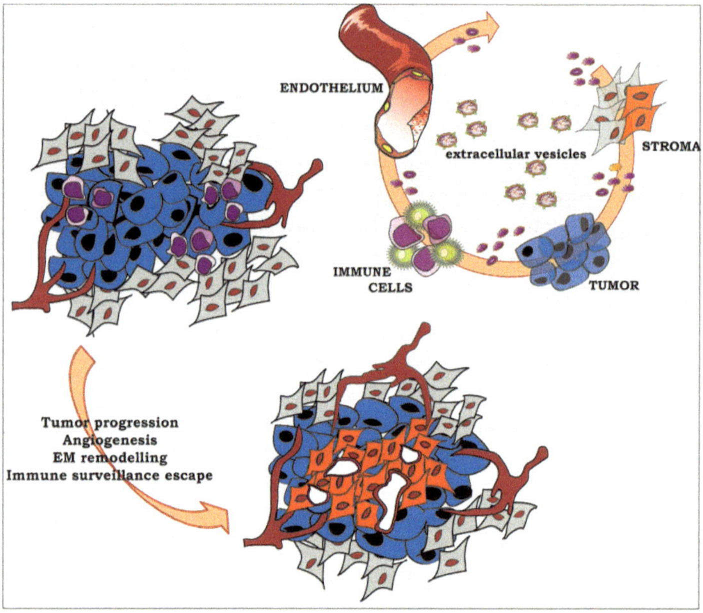

Figure 3.2 Extracellular vesicles mediate tumor–stromal cell interaction. EVs may reprogram the phenotype of stromal and immune cells within the tumor microenvironment. The reciprocal exchange of information between tumor and stromal/immune cells is mediated by the delivery of transcription factors, mRNA, and miRNA that may favor the formation of a favourable microenvironment for tumor development. Tumor-derived EVs stimulate angiogenesis and extracellular matrix remodeling and escape from immune surveillance.

An example of receptor-mediated transfer of information is the phenotypic change induced by the transfer of the membrane-bound CX3CL1/fractalkine ligand for the chemokine receptor CX3CR1 by MVs derived from fibroblasts of highly metastatic prostate carcinoma cells ensuing in increased migration and invasion.[67]

The delivery of biologically active proteins has been shown in several experimental conditions. For instance, tumor-derived MVs may deliver interleukin-8 and chemokines,[68] thus mimicking the effect of tumor cells on monocytes.[68] MVs from a prostate carcinoma may enhance tumor invasion by delivering matrix metalloproteinases (MMPs) and extracellular MMP inducers.[67] Another example of protein delivery by EVs is the transfer of oncogene products to neighboring cells. Al Nedawi et al.[10] demonstrated that MVs may transfer the oncogenic form of EGFRvIII to nonaggressive tumor cells; EGFRvIII is expressed in a subset of aggressive gliomas. The horizontal propagation of this oncogene product determines the activation of signaling pathways, such as mitogen-activated protein kinase (MAPK) and Akt, and changes the expression of genes modulated by EGFRvIII, such as *P27*, $Bcl\text{-}X_L$, and *VEGF*. The outcome is a morphological and functional transformation of the recipient cells that acquire anchorage-independent growth properties.[10] Antonyak et al. recently showed that breast carcinoma and glioma cells may release MVs capable of transforming fibroblast and epithelial cells by transference of tissue transglutaminase and fibronectin.[70] Delivery of proangiogenic factors such as CD147/extracellular MMP inducers, tetraspanins, and sphingomyelins by tumor-derived EVs may stimulate an altered phenotype of tumor endothelial cells, promoting angiogenesis, tumor growth, and invasion.[71] The MV-mediated transfer of the oncogenic EGFR from tumor to endothelial cells activates an angiogenic program by inducing an autocrine expression of VEGF.[30] Exosomes derived from tumor endothelial cells may transfer Dll4 to the normal endothelium, thus transmitting the angiogenic signal.[32] Changes induced by EVs may not only occur within the tumor microenvironment, but by entering the circulation they may also contribute to the formation of the so-called "premetastatic niche," to enhance the recruitment and engraftment of tumor cells ensuing in the development of metastasis (Fig. 3.3). Hood et al. demonstrated that melanoma exosomes can induce remodeling of tissue matrices and prepare sentinel lymph nodes for tumor metastasis.[72] In fact, they observed selective homing of melanoma cells to melanoma-derived exosome-rich premetastatic sites. Jung et al. reported that exosomes released from rat pancreatic adenocarcinoma cells, in cooperation with soluble matrix components, modify the lung microenvironment, thus favoring the development of lung metastases.[73] It has been also described that exosomes isolated from highly metastatic melanoma BL6-10 cells may modify poorly metastatic F1 melanoma cells, transferring molecules responsible for the BL6-10 metastatic phenotype.[74] These data further highlight the role of exosomes in reprogramming recipient cells to acquire a more aggressive phenotype.

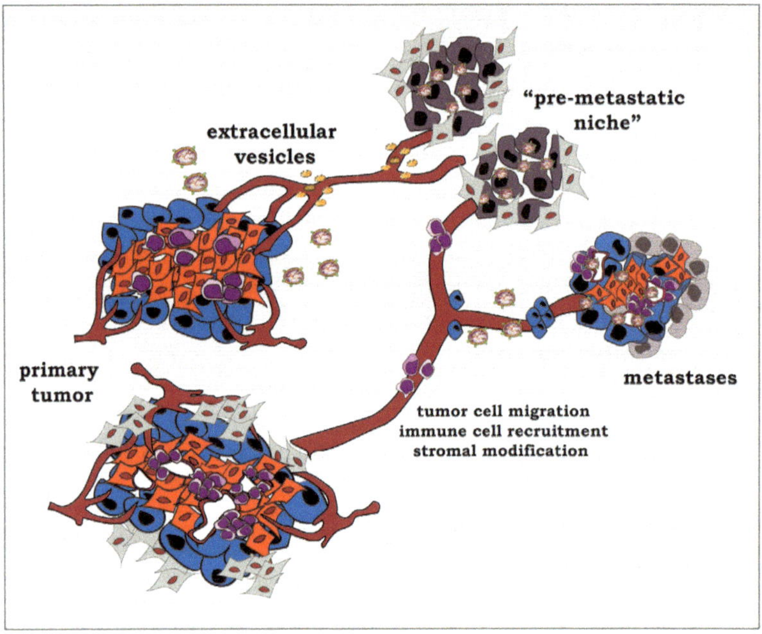

Figure 3.3 EVs favor premetastatic niche formation. EVs released from tumor cells may enter the blood or lymphatic circulation and reach distant sites, favoring the development of a microenvironment (premetastatic niche) that allows tumor cell diffusion and metastasis.

An increasing number of studies have reported that cancer cells have employed exosomes to evade the immune system by inducing apoptosis in immune effectors such as NK cells and cytotoxic T-lymphocytes (CTLs). Human prostate cancer cell line–derived exosomes have been shown to carry the Fas ligand, which may be shuttled to CTLs, mediating target cell apoptosis and promoting tumor immune evasion.[25] Interestingly exosomes released by cancer cells have also evolved mechanisms for transferring their drug resistance phenotype to target cells. Ciravolo et al.[75] found that carcinoma cells that overexpressed the human epidermal growth factor receptor 2 (HER2) release exosomes that affect the sensitivity of cells to HER2-targeted therapy: trastuzumab. HER2-positive exosomes have the unique ability to neutralize the cytotoxic efficacy of trastuzumab on SKBR3 cells.[75]

Only a few studies have described the active role of lipids in MV-mediated intercellular communication. Sphingomyelin expressed on the surface of tumor-derived MVs stimulates endothelial cell migration, invasion, and *in vitro* and *in vivo* angiogenesis.[76]

More recently, the transfer of genetic information by EVs has emerged as an important mechanism of epigenetic changes. By transferring mRNA and miRNAs to recipient cells, EVs may induce changes in the cancer microenvironment, leading to reprogramming of normal stromal and endothelial cells. Transfer of tumor mRNA to monocytes enhances their survival by inhibition of apoptotic pathways.[36]

Skog et al. showed that MVs released transport-specific and functional mRNA and miRNAs from a glioblastoma that are taken up by normal host cells, such as brain microvascular cells, with consequent activation of cell migration, angiogenesis, and proliferation.[77] Similarly, MVs produced by colorectal tumor cells contain several cell cycle–related mRNAs able to activate endothelial cell proliferation.[78] It has been recently shown by the group of Quesenberry that MVs derived from lung[79] and prostate cancer cells[80] may alter the genetic phenotype of normal cells by transferring a specific subset of mRNA and that in turn the tissue microenvironment may modify cancer cell gene expression.

In addition to mRNA, tumor-derived MVs shuttle retrotransposon elements and amplified oncogene sequences[81] that, once incorporated into stromal and endothelial cells, may support the development of a microenvironment that is favorable for cancer growth and progression.

Despite the fact that all cell types within the tumor mass have the potential to release EVs, we recently found that in renal cancer the MVs that contain proangiogenic properties, and that are able to induce a premetastatic niche in the lungs, were those derived from cancer stem cells.[82] MVs derived from renal cancer stem cells differ in their content of mRNA and miRNAs with respect to other renal cancer cells. In fact, they contained several proangiogenic mRNAs such as those for VEGF, FGF, angiopoietin 1, ephrin A3, MMP2, and MMP9. Besides the differences in mRNA, MVs from cancer stem cells were enriched for, and showed a differential pattern of, miRNAs with respect to the non-stem-cancer cell populations. These miRNAs are involved in several biological functions relevant for cell growth, regulation of transcription, cell matrix adhesion, and synthesis of macromolecules. Among the miRNAs we detected miR-200c, miR-92, and miR-141 that were described as being significantly up-regulated in patients with different types of cancers.[82] Moreover, MVs derived from cancer stem cells contained several miRNAs such as miR-29a, miR-650, and miR-151, known to be associated with tumor invasion and metastases or, as in the case of miR-19b, miR-29c, and miR-151, specifically up-regulated in patients with renal carcinomas.

Taken together, these data indicate that EVs derived from tumor cells may be instrumental in transferring genetic information to bystander cells, inducing epigenetic changes in their phenotype that may sustain an unfavorable outcome of the tumor.

3.4.3 Role in Stem Cell Biology

Ratajczak et al.[35] firstly hypothesized that the maintenance of pluripotency and undifferentiated propagation of stem cells *in vitro* may require effective intercellular signaling, mediated by MVs. MVs derived from ES cells were shown to express stem cell–specific molecules that may also support self-renewal and expansion of adult stem cells. ES-MVs express the Wnt-3 protein and are selectively highly enriched in mRNA for several pluripotent transcription factors when compared to parental ES cells. It has been shown that this mRNA can be delivered by ES-MVs to target cells and can be translated into the corresponding proteins. The authors demonstrated that mRNA transferred by murine ES-MVs may reprogram murine hematopoietic progenitors by enhancing survival and improving expansion, by inducing expression of early pluripotent (Oct-4, Nanog, and Rex-1) and early hematopoietic stem cell (Scl, HoxB4, and GATA 2) markers, and by inducing phosphorylation of mitogen-activated protein kinase (MAPK) p42/44 and serine-threonine kinase AKT.[35] Inactivation, by RNase, of MV-shuttled RNA inhibited these biological effects, indicating the involvement of horizontal transfer of mRNA in the observed phenomena. Murine ES-MVs also contain abundant amounts of miRNAs that can be transferred to mouse embryonic fibroblasts *in vitro*.[42] The ability of EVs to convey genetic information is also shared by adult stem/progenitor cells. MVs released from EPCs tend to shuttle a specific subset of cellular mRNA, such as mRNA associated with the PI3K/AKT signaling pathway.[38] Protein expression and functional studies have shown that PI3K and eNOS play a critical role in the angiogenic effect of MVs. These results suggest that EPCs may activate angiogenesis in endothelial cells by releasing MVs that are able to trigger an angiogenic program.[38]

Recently, Quesenberry and Aliotta suggested that MVs are an integral component in the exchange of information involved in the continuum model of stem cell biology.[83] In this context, exosome-/MV-dependent exchange of genetic information plays a critical role in modulating the plasticity of stem cells, as well as the response of differentiated cells to injury. The exchange of genetic information between stem and injured cells is bidirectional, with a reciprocal influence on cell phenotype.

3.4.3.1 EVs released from injured tissue inducing changes in the phenotype of bone marrow stem cells

Despite it being suggested that, following engraftment, bone marrow cells may give origin to nonmarrow cells in many injured tissues, the underlying mechanism is still debated. Transdifferentiation and cell fusion do not exhaustively explain stem cell plasticity. Aliotta et al.[41] demonstrated that murine bone marrow cells cocultured with injured lung cells expressed genes for lung-specific proteins such as Clara cell–specific protein, surfactant B, and surfactant C. Moreover, they found that the released MVs were responsible for such a phenomenon.[41] In this case, MVs were shown to contain high levels of lung-specific mRNA and to deliver this to bone marrow cells. This observation may explain bone marrow stem cell plasticity involved in physiologic tissue repair in the absence of transdifferentiation or fusion.[83]

3.4.3.2 MVs released from stem cells reprogramming injured cells to repair damaged tissues

A major contribution of tissue-resident cells rather than marrow-derived cells has been reported in the repair of many organs. Repopulation of the tubules during recovery from acute kidney injury has been shown to depend mainly on proliferation of tubular cells that have survived after an injury.[84] This process involves the de-differentiation of tubular cells to a mesenchymal phenotype, migration to the site of injury, re-entry into the cell cycle with proliferation, and finally redifferentiation into functional epithelial cells. Similarly liver regeneration mainly depends on the proliferation of resident cells. On the other hand, experiments based on the administration of exogenous MSCs demonstrate functional and morphologic recovery from renal, liver, or heart injury,[85] dependent on paracrine/endocrine mechanisms. In fact, MSCs only transiently accumulated within the injured tissues.[22] Bi et al. showed that an MSC-conditioned medium mimics the beneficial effects of the cells of origin, diminishing tubular cell apoptosis, increasing tubular cell survival, and limiting renal injury in cisplatin-induced acute kidney injury.[86] This was also observed in other organs such as the heart and liver.[85] In this context, MVs released from stem cells may play a critical role in reprogramming injured cells. MV accumulation at the site of tissue injury and their subsequent incorporation into resident tissue cells may induce their de-differentiation to a stem cell–like phenotype, with the resulting activation of tissue regenerative programs. We recently showed that MVs released from human MSCs stimulate *in vivo* proliferation and apoptosis resistance of renal tubular epithelial cells. When administered in SCID mice with acute kidney injury, MVs mimicked the functional and morphological recovery induced by MSCs.[39] RNA inactivation in MVs inhibited both *in vitro* and *in vivo* biological effects, suggesting a major role of RNA in microvesicle function. MVs released from MSCs also enhanced survival in SCID mice with cisplatin-induced lethal acute kidney injury.[87] Whereas the single administration of MVs improved survival and ameliorated renal function and morphology, but did not prevent chronic tubular injury, multiple injections of MVs led to complete restoration of normal morphology and function.[87] The mechanism of protection was mainly ascribed to changes in gene expression of the recipient cells that up-regulated antiapoptotic genes, such as *Bcl-xL*, *Bcl2*, and *BIRC8*, and down-regulated genes with a central role in the execution phase of cell apoptosis, such as *Casp1*, *Casp8*, and *LTA*. The change in the cell phenotype and the activation of regenerative programs have been also shown in other experimental models, including a model of renal ischemia/reperfusion injury[88] and a model of 70% hepatectomy in rats.[40] In the heart, Timmers et al. demonstrated a significant reduction of infarct size in pig and mouse models of ischemia/reperfusion injury by MSC-derived exosomes.[89] Evidence of a role of MV-mediated transfer of miRNAs in the renal-protective and proangiogenic effect of MVs derived from EPCs has been recently suggested by unspecific miRNA depletion of MVs derived from Dicer knockdown in EPCs and by MV depletion of the proangiogenic miRNAs miR-126 and miR-296 by EPC transfection with specific antagomiRs.[90] Because miRNAs are multifunctional molecules that regulate post-transcriptional gene expression, they are one of the major candidates for exosome-/MV-mediated epigenetic changes. Therefore, clarifying the role of miRNAs and other noncoding RNAs shuttled by EVs may be relevant for several physiological and pathological conditions.

3.5 Conclusions

EVs released from a given cell type may not only modify the phenotype of neighboring cells but also behave as hormones by acting at a distance from the site of secretion by entering the circulation and other biological fluids. These "messengers" deliver multiple and complex information that influences the behavior of recipient cells. In particular, the horizontal transfer of gene products emerges as a potential mechanism of epigenetic alterations with physiological and pathological relevance. Multiple actors are involved in the pleiotropic activity of EVs. Besides proteins and bioactive lipids, the nucleic acid transferred between cells may induce transient or stable phenotypic changes in recipient cells. In the context of cancer, tumor-derived EVs may influence the tumor microenvironment by reprogramming the phenotype and functions of stromal and immune cells, favoring tumor growth and diffusion. Thus, an understanding of signals that EVs deliver might be relevant not only for comprehending tumor biology but also for designing new therapeutic strategies.

In the context of stem cell biology, epigenetic changes induced by the signals of EVs may explain stem cell plasticity, especially in response to injury. Moreover, EVs released from stem cells may reprogram cells that have survived after an injury and thus favor tissue regeneration. Finally, since EVs retain several biological properties of the stem cell of origin, development of therapeutic strategies that avoid the administration of viable replicating cells can be envisaged.

Acknowledgments

Our research is supported by grants from Associazione Italiana per la Ricerca sul Cancro (AIRC), Regione Piemonte, Piattaforme Biotecnologiche, project PiSTEM and NanoIgT, Converging Technologies, and from Ministero dell'Istruzione, dell'Università e della Ricerca (MIUR) project PRIN08.

Conflict of Interest

All the authors declared no competing interests.

References

1. Peinado H, Lavotshkin S, Lyden D (2011) The secreted factors responsible for pre-metastatic niche formation: old sayings and new thoughts, *Semin. Cancer Biol.*, **21**(2), 139–146.
2. Sherer NM, Mothes W (2008) Cytonemes and tunnelling nanotubules in cell-cell communication and viral pathogenesis, *Trends Cell Biol.*, **18**(9), 414–420.
3. Heijnen HF, Schiel AE, Fijnheer R, Geuze HJ, Sixma JJ (1999) Activated platelets release two types of membrane vesicles: microvesicles by surface shedding and exosomes derived from exocytosis of multivesicular bodies and alpha-granules, *Blood*, **94**(11), 3791–3799.
4. Lee TH, D'Asti E, Magnus N, Al-Nedawi K, Meehan B, Rak J (2011) Microvesicles as mediators of intercellular communication in cancer-the emerging science of cellular "debris," *Semin. Immunopathol.*, **33**(5), 455–467.
5. Schara K, Jansa V, Sustar V, Dolinar D, Pavlic JI, Lokar M, Kralj-Iglic V, Veranic P, Iglic A (2009) Mechanisms for the formation of membranous

nanostructures in cell-to-cell communication, *Cell Mol. Biol. Lett.*, **14**(4), 636–656.

6. Del Conde I, Shrimpton CN, Thiagarajan P, Lopez JA (2005) Tissue-factor-bearing microvesicles arise from lipid rafts and fuse with activated platelets to initiate coagulation, *Blood*, **106**(5), 1604–1611.
7. Piccin A, Murphy WG, Smith OP (2007) Circulating microparticles: pathophysiology and clinical implications, *Blood Rev.*, **21**(3), 157–171.
8. Zwaal RF, Schroit AJ (1997) Pathophysiologic implications of membrane phospholipid asymmetry in blood cells, *Blood*, **89**(4), 1121–1132.
9. Dvorak HF, Weaver VM, Tlsty TD, Bergers G (2011) Tumor microenvironment and progression, *J. Surg. Oncol.*, **103**(6), 468–474.
10. Al-Nedawi K, Meehan B, Micallef J, Lhotak V, May L, Guha A, Rak J (2008) Intercellular transfer of the oncogenic receptor EGFRvIII by microvesicles derived from tumour cells, *Nat. Cell. Biol.*, **10**(5), 619–624.
11. Bianco F, Perrotta C, Novellino L, Francolini M, Riganti L, Menna E, Saglietti L, Schuchman EH, Furlan R, Clementi E, Matteoli M, Verderio C (2009) Acid sphingomyelinase activity triggers microparticle release from glial cells, *EMBO J.*, **28**(8), 1043–1054.
12. Buschow SI, Liefhebber JM, Wubbolts R, Stoorvogel W (2005) Exosomes contain ubiquitinated proteins, *Blood Cells Mol. Dis.*, **35**(3), 398–403.
13. Al-Nedawi K, Meehan B, Rak J (2009) Microvesicles: messengers and mediators of tumor progression, *Cell Cycle*, **8**(13), 2014–2018.
14. Trajkovic K, Hsu C, Chiantia S, Rajendran L, Wenzel D, Wieland F, Schwille P, Brügger B, Simons M (2008) Ceramide triggers budding of exosome vesicles into multivesicular endosomes, *Science*, **319**(5867), 1244–1247.
15. Nurminen TA, Holopainen JM, Zhao H, Kinnunen PK (2002) Observation of topical catalysis by sphingomyelinase coupled to microspheres, *J. Am. Chem. Soc.*, **124**(41), 12129–12134.
16. Mathivanan S, Ji H, Simpson RJ (2010) Exosomes: extracellular organelles important in intercellular communication, *J. Proteomics*, **73**(10), 1907–1920.
17. Cossetti C, Smith JA, Iraci N, Leonardi T, Alfaro-Cervello C, Pluchino S (2012) Extracellular membrane vesicles and immune regulation in the brain, *Front Physiol.*, **3**, 117.

18. Ratajczak J, Wysoczynski M, Hayek F, Janowska-Wieczorek A, Ratajczak MZ (2006) Membrane-derived microvesicles: important and underappreciated mediators of cell-to-cell communication, *Leukemia*, **20**(9), 1487–1495.
19. Wysoczynski M, Ratajczak MZ (2009) Lung cancer secreted microvesicles: underappreciated modulators of microenvironment in expanding tumors, *Int. J. Cancer*, **125**(7), 595–603.
20. Vickers KC, Remaley AT (2012) Lipid-based carriers of microRNAs and intercellular communication, *Curr. Opin. Lipidol.*, **23**(2), 91–97.
21. Morel O, Toti F, Hugel B, Freyssinet JM. (2004) Cellular microparticles: a disseminated storage pool of bioactive vascular effectors, *Curr. Opin. Hematol.*, **11**(3), 156–164.
22. Camussi G, Deregibus MC, Bruno S, Cantaluppi V, Biancone L (2010) Exosomes/microvesicles as a mechanism of cell-to-cell communication, *Kidney Int.*, **78**(9), 838–848.
23. Cocucci E, Racchetti G, Meldolesi J (2008) Shedding microvesicles: artefacts no more, *Trends Cell Biol.*, **19**(2), 43–51.
24. Quah BJ, Barlow VP, McPhun V, Matthaei KI, Hulett MD, Parish CR (2008) Bystander B cells rapidly acquire antigen receptors from activated B cells by membrane transfer, *Proc. Natl. Acad. Sci. U. S. A.*, **105**(11), 4259–4264.
25. Kim JW, Wieckowski E, Taylor DD, Reichert TE, Watkins S, Whiteside TL (2005) Fas ligand-positive membranous vesicles isolated from sera of patients with oral cancer induce apoptosis of activated T lymphocytes, *Clin. Cancer Res.*, **11**(3), 1010–1020.
26. Barry OP, Praticò D, Savani RC, FitzGerald GA (1998) Modulation of monocyte-endothelial cell interactions by platelet microparticles, *J. Clin. Invest.*, **102**(1), 136–144.
27. Rozmyslowicz T, Majka M, Kijowski J, Murphy SL, Conover DO, Poncz M, Ratajczak J, Gaulton GN, Ratajczak MZ (2003) Platelet- and megakaryocyte-derived microparticles transfer CXCR4 receptor to CXCR4-null cells and make them susceptible to infection by X4-HIV, *AIDS*, **17**(1), 33–42.
28. Mack M, Kleinschmidt A, Brühl H, Klier C, Nelson PJ, Cihak J, Plachý J, Stangassinger M, Erfle V, Schlöndorff D (2000) Transfer of the chemokine receptor CCR5 between cells by membrane-derived microparticles: a mechanism for cellular human immunodeficiency virus 1 infection, *Nat. Med.*, **6**(7), 769–775.
29. Yu J, May L, Milsom C, Anderson GM, Weitz JI, Luyendyk JP, Broze G, Mackman N, Rak J (2008) Contribution of host-derived tissue factor

to tumor neovascularization, *Arterioscler. Thromb. Vasc. Biol.*, **28**(11), 1975–1981.

30. Al-Nedawi K, Meehan B, Kerbel RS, Allison AC, Rak J (2009) Endothelial expression of autocrine VEGF upon the uptake of tumor-derived microvesicles containing oncogenic EGFR, *Proc. Natl. Acad. Sci. U. S. A.*, **106**(10), 3794–3799.
31. Sarkar A, Mitra S, Mehta S, Raices R, Wewers MD (2009) Monocyte derived microvesicles deliver a cell death message via encapsulated caspase-1, *PLOS ONE*, **4**(9), e7140.
32. Sheldon H, Heikamp E, Turley H, Dragovic R, Thomas P, Oon CE, Leek R, Edelmann M, Kessler B, Sainson RC, Sargent I, Li JL, Harris AL (2010) New mechanism for notch signaling to endothelium at a distance by Delta-like 4 incorporation into exosomes, *Blood*, **116**(13), 2385–2394.
33. Gould SJ, Booth AM, Hildreth JE (2003) The trojan exosome hypothesis, *Proc. Natl. Acad. Sci. U. S. A.*, **100**(19), 10592–10597.
34. Stroun M, Anker P, Beljanski M, Henri J, Lederrey C, Ojha M, Maurice PA (1978) Presence of RNA in the nucleoprotein complex spontaneously released by human lymphocytes and frog auricles in culture, *Cancer Res.*, **38**(10), 3546–3554.
35. Ratajczak J, Miekus K, Kucia M, Zhang J, Reca R, Dvorak P, Ratajczak MZ (2006) Embryonic stem cell-derived microvesicles reprogram hematopoietic progenitors: evidence for horizontal transfer of mRNA and protein delivery, *Leukemia*, **20**(5), 847–856.
36. Baj-Krzyworzeka M, Szatanek R, Weglarczyk K, Baran J, Urbanowicz B, Brański P, Ratajczak MZ, Zembala M (2006) Tumour-derived microvesicles carry several surface determinants and mRNA of tumour cells and transfer some of these determinants to monocytes, *Cancer Immunol. Immunother.*, **55**(7), 808–818.
37. Valadi H, Ekstrom K, Bossios A, Sjostrand M, Lee JJ, Lotvall JO (2007) Exosome-mediated transfer of mRNAs and microRNAs is a novel mechanism of genetic exchange between cells, *Nat. Cell. Biol.*, **9**(6), 654–659.
38. Deregibus MC, Cantaluppi V, Calogero R, Lo Iacono M, Tetta C, Biancone L, Bruno S, Bussolati B, Camussi G (2007) Endothelial progenitor cell derived microvesicles activate an angiogenic program in endothelial cells by a horizontal transfer of mRNA, *Blood*, **110**(7), 2440–2448.
39. Bruno S, Grange C, Deregibus MC, Calogero RA, Saviozzi S, Collino F, Morando L, Busca A, Falda M, Bussolati B, Tetta C, Camussi G (2009) Mesenchymal stem cell-derived microvesicles protect against acute tubular injury, *J. Am. Soc. Nephrol.*, **20**(5), 1053–1067.

40. Herrera MB, Fonsato V, Gatti S, Deregibus MC, Sordi A, Cantarella D, Calogero R, Bussolati B, Tetta C, Camussi G (2010) Human liver stem cell-derived microvesicles accelerate hepatic regeneration in hepatectomized rats, *J. Cell Mol. Med.*, **14**(6B), 1605–1618.

41. Aliotta JM, Pereira M, Johnson KW, de Paz N, Dooner MS, Puente N, Ayala C, Brilliant K, Berz D, Lee D, Ramratnam B, McMillan PN, Hixson DC, Josic D, Quesenberry PJ (2010) Microvesicle entry into marrow cells mediates tissue-specific changes in mRNA by direct delivery of mRNA and induction of transcription, *Exp. Hematol.*, **38**(3), 233–245.

42. Yuan A, Farber EL, Rapoport AL, Tejada D, Deniskin R, Akhmedov NB, Farber DB (2009) Transfer of microRNAs by embryonic stem cell microvesicles, *PLOS ONE*, **4**(3), e4722

43. Collino F, Deregibus MC, Bruno S, Sterpone L, Aghemo G, Viltono L, Tetta C, Camussi G (2010) Microvesicles derived from adult human bone marrow and tissue specific mesenchymal stem cells shuttle selected pattern of miRNAs, *PLOS ONE*, **5**(7), e11803.

44. Zhang Y, Liu D, Chen X, Li J, Li L, Bian Z, Sun F, Lu J, Yin Y, Cai X, Sun Q, Wang K, Ba Y, Wang Q, Wang D, Yang J, Liu P, Xu T, Yan Q, Zhang J, Zen K, Zhang CY (2010) Secreted monocytic miR-150 enhances targeted endothelial cell migration, *Mol. Cell*, **39**(1), 133–144.

45. Kosaka N, Iguchi H, Yoshioka Y, Hagiwara K, Takeshita F, Ochiya T (2012) Competitive interactions of cancer cells and normal cells via secretory microRNAs, *J. Biol. Chem.*, **287**(2), 1397–1405.

46. Pegtel DM, van de Garde MD, Middeldorp JM (2011) Viral miRNAs exploiting the endosomal-exosomal pathway for intercellular cross-talk and immune evasion, *Biochim. Biophys. Acta*, **1809**(11–12), 715–721.

47. Waldenström A, Gennebäck N, Hellman U, Ronquist G (2012) Cardiomyocyte microvesicles contain DNA/RNA and convey biological messages to target cells, *PLOS ONE*, **7**(4), e34653.

48. Ronquist KG, Ronquist G, Carlsson L, Larsson A (2009) Human prostasomes contain chromosomal DNA, *Prostate*, **69**(7), 737–743.

49. Guescini M, Genedani S, Stocchi V, Agnati LF (2010) Astrocytes and glioblastoma cells release exosomes carrying mtDNA, *J. Neural. Transm.*, **117**(1), 1–4.

50. Gibbings DJ, Ciaudo C, Erhardt M, Voinnet O (2009) Multivesicular bodies associate with components of miRNA effector complexes and modulate miRNA activity, *Nat. Cell Biol.*, **11**(9), 1143–1149.

51. Kosaka N, Iguchi H, Yoshioka Y, Takeshita F, Matsuki Y, Ochiya T (2010) Secretory mechanisms and intercellular transfer of microRNAs in living cells, *J. Biol. Chem.*, **285**(23), 17442–17452.

52. de Jong OG, Verhaar MC, Chen Y, Vader P, Gremmels H, Posthuma G, Schiffelers RM, Gucek M, van Balkom BWM (2012) Cellular stress conditions are reflected in the protein and RNA content of endothelial cell-derived exosomes, *JEV*, **1,** 18396.

53. Hedlund M, Nagaeva O, Kargl D, Baranov V, Mincheva-Nilsson L (2011) Thermal- and oxidative stress causes enhanced release of NKG2D ligand-bearing immunosuppressive exosomes in leukemia/lymphoma T and B cells, *PLOS ONE*, **6**(2), e16899.

54. Lv LH, Wan YL, Lin Y, Zhang W, Yang M, Li GL, Lin HM, Shang CZ, Chen YJ, Min J (2012) Anticancer drugs cause release of exosomes with heat shock proteins from human hepatocellular carcinoma cells that elicit effective natural killer cell antitumor responses *in vitro*, *J. Biol. Chem.*, **287**(19), 15874–15885.

55. Chen T, Guo J, Yang M, Zhu X, Cao X (2011) Chemokine-containing exosomes are released from heat-stressed tumor cells via lipid raft-dependent pathway and act as efficient tumor vaccine, *J. Immunol.*, **186**(4), 2219–2228.

56. Raposo G, Nijman HW, Stoorvogel W, Liejendekker R, Harding CV, Melief CJ, Geuze HJ (1996) B lymphocytes secrete antigen-presenting vesicles, *J. Exp. Med.*, **183**(3), 1161–1172.

57. Bobrie A, Colombo M, Raposo G, Théry C (2011) Exosome secretion: molecular mechanisms and roles in immune responses, *Traffic*, **12**(12), 1659–1668.

58. Zitvogel L, Regnault A, Lozier A, Wolfers J, Flament C, Tenza D, Ricciardi-Castagnoli P, Raposo G, Amigorena S (1998) Eradication of established murine tumors using a novel cell-free vaccine: dendritic cell-derived exosomes, *Nat. Med.*, **4**(5), 594–600.

59. Segura E, Nicco C, Lombard B, Veron P, Raposo G, Batteux F, Amigorena S, Thery C (2005) ICAM-1 on exosomes from mature dendritic cells is critical for efficient naive T-cell priming, *Blood*, **106**(1), 216–223.

60. Viaud S, Ploix S, Lapierre V, Théry C, Commere PH, Tramalloni D, Gorrichon K, Virault-Rocroy P, Tursz T, Lantz O, Zitvogel L, Chaput N (2011) Updated technology to produce highly immunogenic dendritic cell-derived exosomes of clinical grade: a critical role of interferon-γ, *J. Immunother.*, **34**(1), 65–75.

61. Kim SH, Lechman ER, Bianco N, Menon R, Keravala A, Nash J, Mi Z, Watkins SC, Gambotto A, Robbins PD (2005) Exosomes derived from IL-10-treated dendritic cells can suppress inflammation and collagen-induced arthritis, *J. Immunol.*, **174**(10), 6440–6448.

62. Peche H, Heslan M, Usal C, Amigorena S, Cuturi MC (2003) Presentation of donor major histocompatibility complex antigens by bone marrow dendritic cell-derived exosomes modulates allograft rejection, *Transplantation*, **76**(10), 1503–1510.

63. O'Connell RM, Rao DS, Chaudhuri AA, Baltimore D (2010) Physiological and pathological roles for microRNAs in the immune system, *Nat. Rev. Immunol.*, **10**(2), 111–122.

64. Mittelbrunn M, Gutierrez-Vazquez C, Villarroya-Beltri C, Gonzalez S, Sanchez-Cabo F, Gonzalez MA, Bernad A, Sanchez-Madrid F (2011) Unidirectional transfer of microRNA-loaded exosomes from T cells to antigen-presenting cells, *Nat. Commun.*, **2**, 282.

65. Buschow SI, Nolte-'t Hoen EN, van Niel G, Pols MS, Ten Broeke T, Lauwen M, Ossendorp F, Melief CJ, Raposo G, Wubbolts R, Wauben MH, Stoorvogel W (2009) MHC II in dendritic cells is targeted to lysosomes or T cell-induced exosomes via distinct multivesicular body pathways, *Traffic*, **10**(10), 1528–1542.

66. Castellana D, Kunzelmann C, Freyssinet JM (2009) Pathophysiologic significance of procoagulant microvesicles in cancer disease and progression, *Hamostaseologie*, **29**(1), 1–57.

67. Castellana D, Zobairi F, Martinez MC, Panaro MA, Mitolo V, Freyssinet JM, Kunzelmann C (2009) Membrane microvesicles as actors in the establishment of a favorable prostatic tumoral niche: a role for activated fibroblasts and CX3CL1-CX3CR1 axis, *Cancer Res.*, **69**(3), 69785–69793.

68. Baj-Krzyworzeka M, Weglarczyk K, Mytar B, Szatanek R, Baran J, Zembala M (2011) Tumour-derived microvesicles contain interleukin-8 and modulate production of chemokines by human monocytes, *Anticancer Res.*, **31**(4), 1329–1335.

69. Baj-Krzyworzeka M, Baran J, Weglarczyk K, Szatanek R, Szaflarska A, Siedlar M, Zembala M (2010) Tumour-derived microvesicles (TMV) mimic the effect of tumour cells on monocyte subpopulations, *Anticancer Res.*, **30**(9), 3515–3519.

70. Antonyak MA, Li B, Boroughs LK, Johnson JL, Druso JE, Bryant KL, Holowka DA, Cerione RA (2011) Cancer cell-derived microvesicles induce transformation by transferring tissue transglutaminase and

fibronectin to recipient cells, *Proc. Natl. Acad. Sci. U. S. A.*, **108**(12), 4852–4857.

71. Bussolati B, Grange C, Camussi G (2011) Tumor exploits alternative strategies to achieve vascularization, *FASEB J.*, **25**(9), 2874–2882.

72. Hood JL, San RS, Wickline SA (2011) Exosomes released by melanoma cells prepare sentinel lymph nodes for tumor metastasis, *Cancer Res.*, **71**(11), 3792–3801.

73. Jung T, Castellana D, Klingbeil P, Cuesta Hernández I, Vitacolonna M, Orlicky DJ, Roffler SR, Brodt P, Zöller M (2009) CD44v6 dependence of premetastatic niche preparation by exosomes, *Neoplasia*, **11**(10), 1093–1105.

74. Hao S, Ye Z, Li F, Meng Q, Qureshi M, Yang J, Xiang J (2006) Epigenetic transfer of metastatic activity by uptake of highly metastatic B16 melanoma cell-released exosomes, *Exp. Oncol.*, **28**(2), 126–131.

75. Ciravolo V, Huber V, Ghedini GC, Venturelli E, Bianchi F, Campiglio M, Morelli D, Villa A, Della Mina P, Menard S, Filipazzi P, Rivoltini L, Tagliabue E, Pupa SM (2012) Potential role of HER2-overexpressing exosomes in countering trastuzumab-based therapy, *J. Cell Physiol.*, **227**(2), 658–667.

76. Kim CW, Lee HM, Lee TH, Kang C, Kleinman HK, Gho YS (2002) Extracellular membrane vesicles from tumor cells promote angiogenesis via sphingomyelin, *Cancer Res.*, **62**(21), 6312–6317.

77. Skog J, Würdinger T, van Rijn S, Meijer DH, Gainche L, Sena-Esteves M, Jr., Curry WT, Carter BS, Krichevsky AM, Breakefield XO (2008) Glioblastoma microvesicles transport RNA and proteins that promote tumour growth and provide diagnostic biomarkers, *Nat. Cell. Biol.*, **10**(12), 1470–1476.

78. Hong BS, Cho JH, Kim H, Choi EJ, Rho S, Kim J, Kim JH, Choi DS, Kim YK, Hwang D, Gho YS (2009) Colorectal cancer cell-derived microvesicles are enriched in cell cycle-related mRNAs that promote proliferation of endothelial cells, *BMC Genomics*, **10**, 556–568.

79. Del Tatto M, Ng T, Aliotta JM, Colvin GA, Dooner MS, Berz D, Dooner GJ, Papa EF, Hixson DC, Ramratnam B, Aswad BI, Sears EH, Reagan J, Quesenberry PJ (2011) Marrow cell genetic phenotype change induced by human lung cancer cells, *Exp. Hematol.*, **39**(11), 1072–1080.

80. Renzulli JF 2nd, Del Tatto M, Dooner G, Aliotta J, Goldstein L, Dooner M, Colvin G, Chatterjee D, Quesenberry P (2010) Microvesicle induction of prostate specific gene expression in normal human bone marrow cells, *J. Urol.*, **184**(5), 2165–2171.

81. Balaj L, Lessard R, Dai L, Cho YJ, Pomeroy SL, Breakefield XO, Skog J (2011) Tumour microvesicles contain retrotransposon elements and amplified oncogene sequences, *Nat. Commun.*, **2,** 180.

82. Grange C, Tapparo M, Collino F, Vitillo L, Damasco C, Deregibus MC, Tetta C, Bussolati B, Camussi G (2011) Microvesicles released from human renal cancer stem cells stimulate angiogenesis and formation of lung premetastatic niche, *Cancer Res.*, **71**(15), 5346–5356.

83. Quesenberry PJ, Dooner MS, Aliotta JM (2010) Stem cell plasticity revisited: the continuum marrow model and phenotypic changes mediated by microvesicles, *Exp. Hematol.*, **38**(7), 581–592.

84. Humphreys BD, Valerius MT, Kobayashi A, Mugford JW, Soeung S, Duffield JS, McMahon AP, Bonventre JV (2008) Intrinsic epithelial cells repair the kidney after injury, *Cell Stem Cell*, **2**(3), 284–291.

85. Otto WR, Wright NA (2011) Mesenchymal stem cells: from experiment to clinic, *Fibrogenesis Tissue Repair*, **4,** 20.

86. Bi B, Schmitt R, Israilova M, Nishio H, Cantley LG (2007) Stromal cells protect against acute tubular injury via an endocrine effect, *J. Am. Soc. Nephrol.*, **18**(9), 2486–2496.

87. Bruno S, Grange C, Collino F, Deregibus MC, Cantaluppi V, Biancone L, Tetta C, Camussi G (2012) Microvesicles derived from mesenchymal stem cells enhance survival in a lethal model of acute kidney injury, *PLOS ONE*, **7**(3), e33115.

88. Gatti S, Bruno S, Deregibus MC, Sordi A, Cantaluppi V, Tetta C, Camussi G (2011) Microvesicles derived from human adult mesenchymal stem cells protect against ischaemia-reperfusion-induced acute and chronic kidney injury, *Nephrol. Dial. Transplant.*, **26**(5), 1474–1483.

89. Timmers L, Lim SK, Hoefer IE, Arslan F, Lai RC, van Oorschot AA, Goumans MJ, Strijder C, Sze SK, Choo A, Piek JJ, Doevendans PA, Pasterkamp G, de Kleijn DP (2011) Human mesenchymal stem cell-conditioned medium improves cardiac function following myocardial infarction, *Stem Cell Res.*, **6**(3), 206–214.

90. Cantaluppi V, Gatti S, Medica D, Figliolini F, Bruno S, Deregibus MC, Sordi A, Biancone L, Tetta C, Camussi G (2012) Microvesicles derived from endothelial progenitor cells protect the kidney from ischemia-reperfusion injury by microRNA-dependent reprogramming of resident renal cells, *Kidney Int.*, **82**(4), 412–427.

Chapter 4

An Overview of Novel and Conventional Methods to Detect Extracellular Vesicles

Edwin van der Pol,[a,b] Ton G. van Leeuwen,[b] and Rienk Nieuwland[a]
[a]*Laboratory of Experimental Clinical Chemistry, Academic Medical Center of the University of Amsterdam, Meibergdreef 9, 1105AZ Amsterdam, the Netherlands*
[b]*Biomedical Engineering and Physics, Academic Medical Center of the University of Amsterdam, Meibergdreef 9, 1105AZ Amsterdam, the Netherlands*
e.vanderpol@amc.uva.nl

Although the biological and clinical relevance of extracellular vesicles (EVs) is increasingly recognized, their isolation and detection have proven to be extremely difficult. On the one hand, due to the biological complexity of body fluids, physical separation of vesicles from similar-size particles and cells is complicated. Consequently, it is probable that in many studies not only vesicles were detected.[1] On the other hand, since the diameter of vesicles is typically ranging from 30 nm to 1 μm, vesicles are below the detection range of many currently used techniques.[2] As a result, after isolation the recovery and contamination cannot be reliably quantified, and isolation protocols have not been standardized (Chapter 5). The interrelated difficulties of the detection and isolation of EVs clearly expose one of the main issues to be solved by the research field. To fully understand

Extracellular Vesicles in Health and Disease
Edited by Paul Harrison, Christopher Gardiner, and Ian L. Sargent

ISBN 978-981-4411-98-1 (Hardcover), 978-981-4411-99-8 (eBook)
www.panstanford.com

and appreciate the content of this book, it is essential to point out the capabilities and limitations of methods to isolate and detect EVs. This chapter provides an overview of novel and conventional methods to detect EVs. For each technique, we will discuss its capability to assess relevant properties of vesicles, whether the information is obtained from individual or multiple vesicles and whether the technique can be applied directly in suspension. In addition, we will contemplate the applicability of each method in terms of measurement time, assuming the detection of 10,000 vesicles, a number that is common in flow cytometry. All findings are summarized in Table 4.1. The methods listed in Table 4.1 can often provide more detailed information than mentioned here if the instrument is operated by a specialist with foreknowledge of the sample.

4.1 Vesicle Isolation

Ideally, vesicles are detected directly in body fluids, but due to, for instance, the viscosity of body fluids, the presence of cells that dominate detection signals, the presence of enzymes causing vesicle degradation, the presence of particles within the size range of vesicles, and the presence of high-molecular-weight proteins, vesicles often need to be isolated before they can be reliably detected. The isolation of vesicles is complicated and depends not only on body fluids but also on preanalytical variables (Chapter 5) and the applied isolation method. In this section we will discuss three isolation methods because this knowledge is essential to understand the rest of this chapter.

4.1.1 Differential Centrifugation

The most common method to isolate vesicles is differential centrifugation. With centrifugation, a centrifugal force is induced by the spinning of the centrifuge rotor to sediment particulate matter, for instance, vesicles in suspension. Isolation of vesicles is based on differences in size and density, with smaller and less dense components migrating toward the axis of the centrifuge and larger and denser components migrating away from the axis. With differential centrifugation, multiple sequential centrifugation steps are applied, each step removing particulate matter such as cells and increasing the centrifugal force to separate less dense components than the previous step. Typically applied centrifugal accelerations range between 200 and 1,500 × g to remove cells, 1,500 and 3,000 × g to remove cellular debris, 10,000 and 20,000 × g to pellet vesicles with a diameter that is on average larger than 100 nm, and 100,000 and 200,000 × g to pellet vesicles with a diameter that is on average smaller than 100 nm. Because all physical properties of vesicles that are involved in differential centrifugation are heterogeneous, complete isolation of vesicles with a certain diameter or density remains unfeasible. In addition, centrifugation may also induce the release of vesicles or fragmentation from cells due to the applied mechanical forces.[3]

4.1.2 Filtration

With filtration, the isolation of vesicles is based on differences in size, shape, and deformability between types of vesicles and other particles. Although most filters have a well-defined pore size, the filtrate may contain larger vesicles than the pore size due to the deformation of vesicles. In addition, filters may bind subpopulations of vesicles, and increasing forces have to be applied with decreasing pore size. One recent advance is the availability of nanofabricated filtration sieves, which have pores with well-defined diameters as small as 100 nm.[4–6] Figure 4.1a shows a scanning electron micrograph of a filtration sieve containing pores with a diameter of 900 nm. Nanofabricated filters have the capability to separate plasma from whole blood by capillary forces only, as was recently demonstrated using a planar filter with a thickness of 500 nm.[7] However, the total obtained volume of plasma from 5 μL blood was merely 45 nL. Further investigation is necessary to employ nanofabricated filters for the isolation of EVs.

4.1.3 Flow-Field Flow Fractionation

Alternatively, vesicles can be isolated and fractionated by size using flow-field flow fractionation (FFFF). Figure 4.1b shows a schematic of the underlying principle of FFFF. In an open flow channel a laminar flow with a parabolic stream profile is formed. This main flow transports the sample through the channel to which a cross flow is applied perpendicularly. The cross flow is directed through a semipermeable membrane that is located at the bottom wall. The membrane allows the fluid to exit the channel but prevents the vesicles to pass through. Vesicles in the fluid continuously move in random directions due to collisions with solvent molecules, which is called Brownian motion. The mean squared displacement of the Brownian motion is given by the translational diffusion coefficient D

$$D = \frac{k_B T}{3\pi \eta d}$$

where k_B is Boltzmann's constant, T the absolute temperature, η the viscosity, and d the hydrodynamic vesicles' diameter. Thus, the smaller the particle, the larger the mean squared displacement. Consequently, under the influence of the Brownian motion of vesicles and the counteracting cross flow, different equilibrium layer heights are formed by different size fractions. Small vesicles with high diffusion coefficients are on average located more centrally in the flow channel in fast stream lines and elute first. Larger vesicles with lower diffusion coefficients are on average located in slow stream lines and elute later. This results in size-based fractionation with a resolution of up to 10 nm without exposing the vesicles to high shear stress.[8]

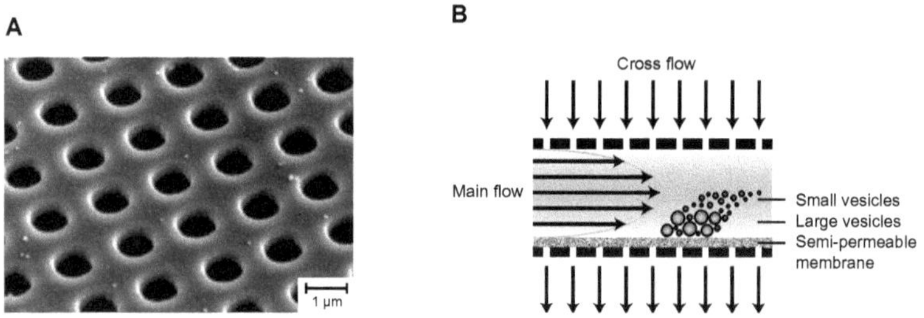

Figure 4.1 Advanced methods to isolate vesicles. (A) Scanning electron micrograph of a filtration sieve containing pores with a diameter of 900 nm. Image courtesy of C. J. M. van Rijn. (B) Schematic of the underlying principle of FFFF. Since relatively small vesicles (purple) have a large diffusion coefficient compared to relatively large vesicles (green), small vesicles are on average located more centrally in the flow channel than larger vesicles. Consequently, due to the parabolic flow profile of the channel flow, small vesicles elute faster than larger vesicles.

FFFF has been successfully applied to isolate exosomes from human neural stem cells,[9] but FFFF is not widely applied, because it requires extensive optimization of the instrument settings and it is relatively expensive. The total elution time is typically ranging from 30 min to 1 h.

In sum, vesicles require isolation before detection due to the high complexity of body fluids, in particular blood. Although many techniques are available to isolate vesicles, it is unknown what the most effective method will be.

4.2 Properties of Vesicles

After vesicles have been isolated, relevant properties are preferably assessed from single vesicles. Relevant properties of vesicles are *size*, *density*, *morphology*, *biochemical composition*, and *zeta potential*.[2,10] From the *size* information a differential size distribution can be obtained, providing insight into the number of vesicles of one particular size relative to another size. We define *concentration* as the number of vesicles per unit volume. If both the differential size distribution and the concentration are known, an absolute size distribution can be obtained, which gives the number of vesicles of one particular size per unit volume. *Morphology* refers to shape and ultrastructure, which are best illustrated in Fig. 4.2b, where vesicles differ in shape, contrast, and surface pattern. *Biochemical composition* points to the building blocks whereof vesicles consist, for example, lipids, proteins, deoxyribonucleic acid (DNA), and ribonucleic acid (RNA). Since the outer surface of the phospholipid membrane of a vesicle can be negatively charged, it is important to realize that this negative charge is counterbalanced by ions of opposite charge that are present in the surrounding solution. These counterions are spread in a diffuse layer of thicknesses of several nanometers surrounding the phospholipid membrane of the vesicle. The *zeta potential* is the electric potential difference between the stationary layer of ions that is bound to the vesicles and the medium.

4.2.1 Vesicle Properties Revealing the Origin of a Vesicle

When vesicles of one particular type have unique (combinations of) properties, their *cellular* and *subcellular origin* may be deduced.

The *cellular origin* refers to the cell type from which vesicles originate. Since vesicles originate from their parent cell, we expect that vesicles will have at least some properties resembling their parent cell, for instance, the presence of characteristic membrane proteins. Furthermore, these features may reflect the status of the parent cell, for instance, whether the parent cell was undergoing programmed cell death or was activated when releasing the vesicles. The *subcellular origin* indicates whether a vesicle originates from an intracellular compartment or the plasma membrane.

4.2.2 Information from Multiple Vesicles

Useful information on vesicles can also be obtained by studying total populations of vesicles. This includes functional as well as antigenic assays (Chapters 6 and 7). An example of a functional assay is an assay used to study the coagulant properties of vesicles in human body fluids. In such assays, the coagulant properties of the total population of vesicles are measured (Chapter 6). Well-known examples of assays to detect a particular or multiple proteins in total populations of vesicles are Western blot or mass spectroscopy, respectively. A limitation of these methods is, however, that contaminants may be present that affect the information obtained from such assays.

4.3 Optical Detection Methods

Optical methods have the potential to obtain properties of single vesicles in solution with high throughput. Two important parameters in optics are the *wavelength* λ of light and the *refractive index* n of particles relative to the suspending medium. The refractive index is defined as the ratio of the speed of light in vacuum to that in the material. Optical phenomena depend on the refractive index of the material, which in turn depends on the wavelength of light. In practice, the higher the difference is between the refractive index of a vesicle and the surrounding medium, the more light will be scattered by the vesicle. In this chapter we have selected the optical methods that are most applicable for detection of vesicles. We have made a distinction between optical detection methods based on light scattering, fluorescence, or both light scattering and fluorescence.

4.3.1 Light Scattering

Light that illuminates a vesicle is partly scattered and partly absorbed. It is useful to know how much light is scattered by a single vesicle, because many optical setups are based on the detection of scattered light. The quantity of light scattered by a single vesicle is proportional to the scattering cross section σ and can be calculated by Mie theory, which involves the diameter and refractive index of the particle, the refractive index of the surrounding medium, and the wavelength and polarization of light.[11] In contrast to Fraunhofer diffraction and Rayleigh scattering, Mie theory applies to any ratio of the diameter of the particle to the wavelength. Figure 4.2a, in a semi-log representation, shows the scattering cross section versus the diameter of a polystyrene bead, a silica bead, and a vesicle at a wavelength of 488 nm, which is commonly used in flow cytometers. The scattering cross section decreases tremendously with decreasing diameter for all particles. For example, the scattering cross section and thus the scattered amount of light of a single polystyrene bead with a diameter of 60 nm are 64,000-fold lower than the scattering cross section of a 600 nm polystyrene bead. From Fig. 4.2a it becomes also clear that the scattering cross section strongly depends on the refractive index, n, of the material. For instance, a 300 nm polystyrene bead (n = 1.605) in water (n = 1.33) scatters seven times more light than a silica bead (n = 1.445) of similar size, which in turn scatters four times more light than a vesicle, again due to differences in the refractive index. Since the refractive index of vesicles is not exactly known, and may also differ between different types of vesicles or between vesicles of different cellular origin, the light green area in Fig. 4.2a represents an estimated confidence interval, which is based on the assumption that vesicles are spherical and have a refractive index of 1.38 ± 0.02 inside and 1.48 at the 10 nm thick phospholipid membrane.[12–15] This estimation of the inner refractive index of vesicles is based on the measured inner refractive index of cells and bacteria.[12] Since vesicles originate from their parent cell, we expect that vesicles have an inner refractive index that should be equivalent to their parent cell.

To study light scattered from all vesicles of a population, we isolated a population of vesicles from human urine. Urine can be used to prepare a relatively high concentration of vesicles without substantial contamination.[2] Vesicles from fresh cell-free urine of five overnight-fasting healthy male subjects were isolated by high-speed centrifugation (Fig. 4.2b; 30 min at 18,890 g), followed by ultracentrifugation of the supernatant (Fig. 4.2c; 1 h at 154,000 g). We imaged vesicles by transmission electron microscopy (TEM) and measured the diameter of 1,000 vesicles in each fraction. The combined size distributions are shown in Fig. 4.2d. According to recent estimates with a novel detection method, the concentration of vesicles and other particles in plasma ranges between 7.0×10^8 mL^{-1} and 1.2×10^{10} mL^{-1}.[16,17] As our next calculation demands an absolute size distribution as input, we arbitrarily multiplied our relative vesicle size distribution by 5×10^6 mL^{-1} to obtain a realistic vesicle concentration of 10^{10} mL^{-1}.

Although one has to bear in mind that the obtained absolute size distribution can be affected by isolation procedures, it is very useful to provide insight into the complex mechanisms underlying light scattering from multiple vesicles. If we multiply the concentration of the vesicles obtained from the size distribution (Fig. 4.2d) by the scattering cross section (Fig. 4.2a), the scattering coefficient per vesicle diameter is obtained (Fig. 4.2e). The scattering coefficient, depicting the average number of scattering events of light per unit length, is a measure of the quantity of light scattered by all vesicles per diameter. Please note that the contribution of light scattered by vesicles smaller than 100 nm is surprisingly small (Fig. 4.2e), given their high concentration (Fig. 4.2d). Consequently, smaller vesicles require far more sensitive optical detection than larger vesicles, and light scattering of small vesicles can easily be overwhelmed by scattering of large vesicles.

4.3.1.1 Dynamic light scattering

Dynamic light scattering (DLS), also known as quasi elastic light scattering or photon correlation spectroscopy (Chapter 10), determines the differential size distribution of multiple particles in suspension, ranging in diameter from 1 nm to 6 μm, by measuring the intensity fluctuations of scattered light caused by vesicles undergoing Brownian motion.[18,19] The differential size distribution is obtained from the intensity fluctuations by applying a mathematical algorithm following from diffusion and light-scattering theory. DLS performs well on the size determination of monodisperse samples, that is, samples containing particles of one particular size.[20–22] However, as scattering is detected from

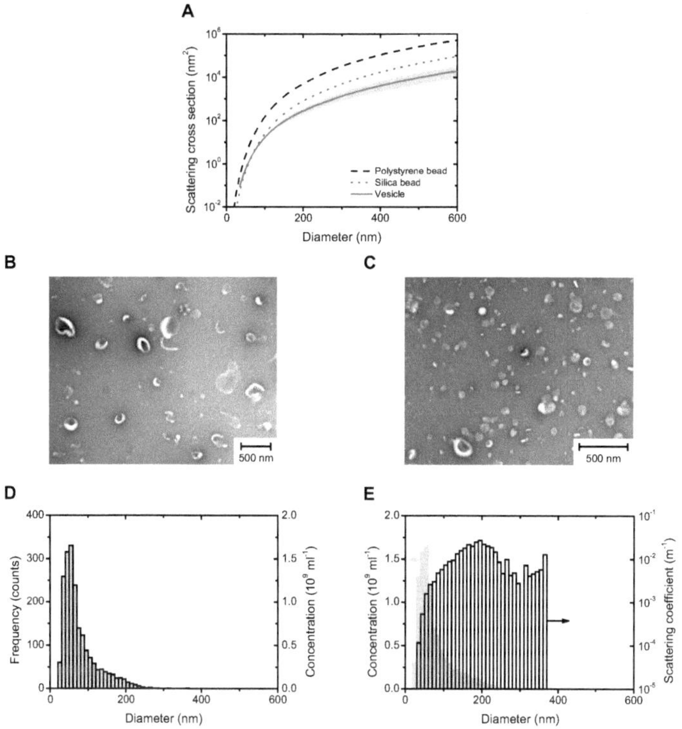

Figure 4.2 Light-scattering properties of vesicles. (A) Scattering cross section vs. diameter (logarithmic scale) at a wavelength of 488 nm for a polystyrene bead (n = 1.605), a silica bead (n = 1.448), and a vesicle calculated using Mie theory. The refractive index for vesicles is assumed to be 1.38 ± 0.02 inside and 1.48 at the 10 nm thick phospholipid membrane. The scattering cross section, and thus the quantity of light scattered by a particle, strongly decreases with decreasing diameter and decreasing refractive index. (B) Vesicles isolated from cell-free human urine by centrifugation (30 min at 18,900 × g). (C) Vesicles isolated from EV-depleted urine by ultracentrifugation (1 h at 154,000 × g). (D) Size distribution of vesicles measured by TEM. The size distribution is arbitrarily multiplied by 5×10^6 mL^{-1} to obtain a realistic vesicle concentration of 10^{10} mL^{-1}. The plot shows that most vesicles are smaller than 100 nm. (E) Scattering coefficient vs. diameter (logarithmic scale). The scattering coefficient, which depicts the mean number of scattering events that light encounters per unit length, is given by the product of the concentration of the standard population (D) and the scattering cross-section (A). The scattering coefficient strongly increases with increasing diameter, indicating that the contribution of light scattered by vesicles smaller than 100 nm is relatively small.

multiple particles simultaneously, size distributions of polydisperse samples, that is, samples containing different-size particles, are less accurate and require foreknowledge of the sample to apply the most suitable mathematical algorithm.[20,21] For example, size distributions of polydisperse samples are likely to be biased toward small numbers of larger particles such as platelets,[21,22] which scatter light more efficiently than small vesicles, as shown in Fig. 4.2e. Two populations can be resolved only if the particle diameter differs at least twofold.[20–23] The typical measurement time ranges from 30 s to 1 min.

Many commercial DLS instruments can also determine the zeta potential, which is measured by applying an electric field across the suspension. Consequently, charged vesicles will migrate toward the electrode of opposite charge with a velocity proportional to the magnitude of their zeta potential. The mean velocity of vesicles is measured using laser Doppler velocimetry, whereupon the zeta potential of vesicles can be determined by taking the Smoluchowski approximation of the Henry equation.

4.3.1.2 Raman spectroscopy

Raman spectroscopy is based on the detection of inelastic light scattering and is used to study the chemical composition and structure of macromolecules inside single living cells.[24] The sample is illuminated by monochromatic laser light. When the light is scattered by the sample, the wavelength may change due to an energy loss or gain, which is caused by vibrations of molecules in the sample. This wavelength shift is molecule specific and can be detected by sensitive spectrometers. Since vesicles contain many different biomolecules, which all have unique Raman spectra, the chemical composition can be investigated without labeling.

Figure 4.3 shows the Raman spectrum of a single vesicle isolated from platelet concentrates by differential centrifugation. This spectrum was obtained using a confocal Raman microspectrometer, in which a krypton ion laser operating at a wavelength of 647 nm was focused to a probe volume of 0.3 μm^3.[25] Due to the high illumination intensity, the vesicle was trapped in the laser beam. The peaks in the spectrum are specific to the chemical bonds and symmetry of the molecules. For instance, the pronounced peak at 2,947 cm^{-1} is characteristic of hydrocarbon stretching. Since the signal amplitude is linearly proportional to the number of molecules,

Raman microspectroscopy is a quantitative technique. For a vesicle that fits within the probe volume, the magnitude of the Raman signal strength is proportional to the volume of a single vesicle and therefore the relative *size* can be estimated, which is a method that warrants further investigation. The concentration of vesicles can be determined if the probe volume is known. Since the intensity of Raman scattering is weak in comparison to Rayleigh scattering, most detectors require acquisition times in the order of seconds. Thus, with the current state of the art, obtaining Raman spectra from 10,000 vesicles would take at least several hours.

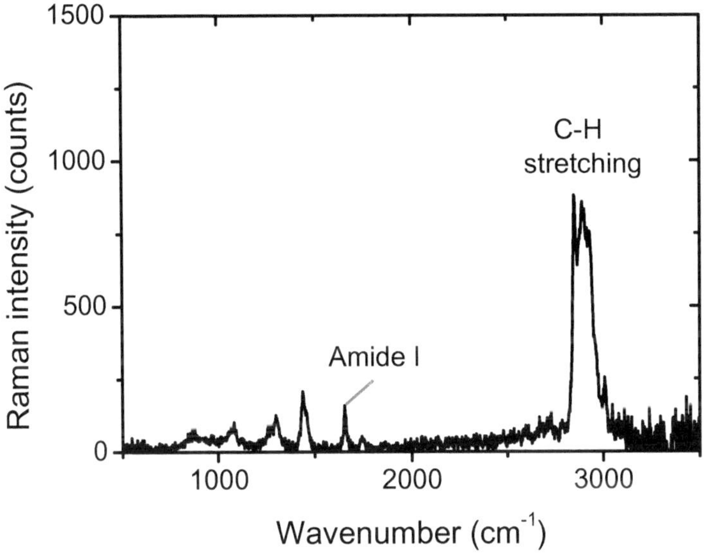

Figure 4.3 Raman spectrum of a single vesicle from platelet concentrates after subtraction of the background spectrum of the medium. The peaks reveal specific chemical bonds that are present in this vesicle. For instance, the peak at 2,947 cm^{-1} is characteristic of hydrocarbon (C–H) stretching, and the peak at 1,660 cm^{-1} is characteristic of amide I.

4.3.2 Fluorescence

Fluorescence is the emission of light by a material that has absorbed light at a usually shorter wavelength. Since most vesicles exhibit no intrinsic fluorescence, they are usually labeled using conjugates of antibodies or proteins with fluorophores.[26] Common fluorophores are quantum dots (QDs) and organic dye molecules. QDs can be seen as artificial atoms that are generally stabler and brighter than organic dye molecules or fluorescent proteins. The emission wavelength of QDs can be controlled precisely, because this depends on their size, which can be regulated during fabrication. QDs have a typical diameter of 2–20 nm, and they have been successfully used to label vesicles.[17,27] Since vesicles usually expose antigens from the parental cells, all methods based on fluorescence detection potentially provide information on the biochemical composition and origin of vesicles. Fluorescence also offers opportunities to acquire additional chemical information, since the fluorescence wavelength, intensity, and fluorescence lifetime depend on the molecular environment.[12,28] Nevertheless, fluorescent multilabeling is not easy to perform and involves several practical problems, including nonspecific binding and aggregate formation of labeled antibodies.[2,29,30]

4.3.2.1 Fluorescence correlation spectroscopy

Fluorescence correlation spectroscopy (F-CS) was originally developed to study molecular diffusion.[31] The absolute size distribution can be obtained from fluorescence intensity fluctuations caused by one or more particles moving by Brownian motion through a well-characterized illuminated volume,[32] which may take several minutes to 1 h, depending on the sample. As the probe volume is typically only 1 μm^3, F-CS is a very sensitive technique, capable of detecting single fluorescent molecules. In terms of vesicle detection, potentially even the smallest vesicle can be detected as long as it is labeled. Although a small numbers of larger vesicles may influence the size distribution substantially, it is expected that this effect is smaller in comparison to DLS because the probe volume is much smaller.

4.3.2.2 Stimulated emission depletion microscopy

In practice, stimulated emission depletion microscopy (STED) is fluorescence microscopy with a spatial resolution of up to 6 nm, which is sufficiently small to size vesicles.[33–35] In addition, the high resolution may be used to gain information on morphology and to determine the distribution of labeled receptors at the surface of larger vesicles. The concentration can be determined if the detection area and binding efficiency of vesicles to the surface are known. The measurement time is in the order of hours.

4.3.3 Light Scattering and Fluorescence

4.3.3.1 Flow cytometry

Flow cytometry is probably the most commonly applied optical method to detect vesicles in clinical samples.[36] A flow cytometer guides cells and vesicles through one or more laser beams in a hydrodynamically focused fluid stream (Chapter 8). One detector is placed in line with the laser beam and measures the forward-scattered light (FSC). Other detectors measure the side-scattered light (SSC) and fluorescence intensity perpendicular to the beam. The ability to simultaneously detect light scattering and fluorescence at multiple channels from single particles makes flow cytometry a powerful detection method, especially for the characterization of single cells in suspension.

The applicability of flow cytometry to detect single vesicles, however, is still limited.[37] In an effort to standardize vesicle detection, the Scientific Standardization Committee of the International Society on Thrombosis and Haemostasis proposed a protocol that used 500 nm and 900 nm polystyrene beads from Megamix to define a vesicle size gate.[36] However, a difficulty with flow cytometry is that the relation between the measured light scattering and the diameter of vesicles is unknown, resulting in a still ongoing discussion about the standardization of vesicle detection.[38–40] From Fig. 4.2a it is clear that the use of 500 nm and 900 nm polystyrene beads results in the gating of vesicles that are significantly larger due to differences in the refractive index of beads and vesicles. Furthermore, the exact relation between the applied gating and the selected vesicle diameter is instrument specific and depends on the illumination wavelength and collection angle. Neverthcless, FSC and SSC detectors can be calibrated by combining Mie calculations with measurements on polystyrene beads and silica beads of known diameter, concentration, and refractive index, so the measured light-scattering power can be related to the diameter of single vesicles.[41]

The same study showed that vesicle detection by flow cytometry is attributed not only to large *single* vesicles but also to swarm detection of smaller vesicles, that is, *multiple* vesicles that are illuminated simultaneously by the laser beam and that are being counted as a single event signal. Like any detector, the FSC and SSC detectors of a flow cytometer have a lower detection limit, which is the minimum amount of light that is scattered by a single particle to generate an event signal. We define a particle scattering *more* or *less* light than this detection limit as a *large* or a *small* particle, respectively. Figure 4.4a shows the SSC histogram of *large* 610 nm silica beads at a concentration of 10^7 mL^{-1} as measured with a FACSCalibur (Becton Dickinson, Franklin Lakes, NJ, USA) flow cytometer at a high flow rate (60 μL min^{-1}). At this concentration, on average a single bead is illuminated by the laser beam at a time, and since 610 nm silica beads scatter more light than the detection limit of the flow cytometer, every single bead passing the laser beam is detected. For this flow cytometer, the smallest detectable silica beads are 204 nm in diameter,[41] so it is expected that smaller beads cannot be detected as a single event. However, Fig. 4.4b shows that silica beads with a diameter of 89 nm can be successfully detected when the vesicles are present at a sufficiently high concentration, in this particular experiment 10^{11} mL^{-1}. At such a high concentration, multiple 89 nm silica beads are simultaneously present within the laser beam, and because the total amount of light that is scattered simultaneously by these vesicles exceeds the detection limit, the vesicles will be counted by the flow cytometer as a single event signal.

To create a sample with physical properties resembling vesicles in plasma,[16,17] a mixture of 89 nm and 610 nm silica beads was prepared. Figure 4.4c shows the SSC histogram of a mixture of *large* 610 nm silica beads at a concentration of 10^7 mL^{-1} and *small* 89 nm beads at a concentration of 10^{11} mL^{-1}. The two peaks confirm that the presence of a single *large* or multiple *small* particles in the laser beam both contribute to the count rate. In this particular case, the count rate was close to the maximum acquisition rate such that the counts corresponding to 89 nm or 610 nm beads are lower than expected depending on their individual histograms.

Taking into account the geometry of the FACSCalibur flow cytometer and a realistic concentration of vesicles, it is obvious that swarm detection also plays a role in vesicle detection. Since the sample core diameter is 56 μm at a high flow rates and the laser beam is elliptically focused to a cross-sectional area of 22 × 66 μm^2,[42] the effective beam volume is $22 \times \pi \times 28^2 = 5.4 \times 10^4$ μm^3, or 54 pL.

Assuming a realistic vesicle concentration of 10^{10} mL^{-1} [16,17], the mean amount of vesicles that are simultaneously present in the laser beam will be approximately 540.

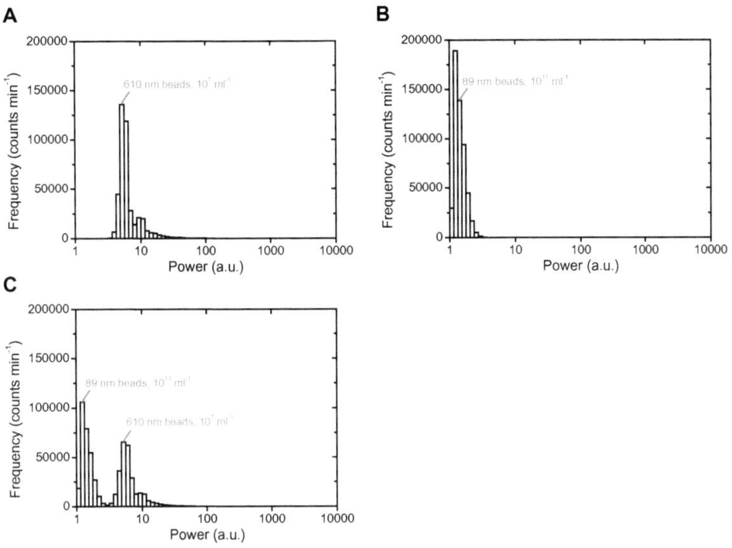

Figure 4.4 Relative contributions of mechanisms underlying vesicle detection by flow cytometry. SSC histogram (logarithmic horizontal scale) for (A) 610 nm silica beads at a concentration of 1×10^7 mL^{-1}, (B) 89 nm silica beads at a concentration of 1×10^{11} mL^{-1}, and (C) a mixture of 89 nm and 610 nm silica beads at concentrations of 1×10^{10} mL^{-1} and 1×10^7 mL^{-1}, respectively.

Although swarm detection allows the detection of smaller vesicles than previously thought possible and explains why flow cytometry underestimates the concentration of vesicles, the simultaneous presence of multiple vesicles of various sizes in the laser beam is a very complex process. For example, to what extent swarm detection plays a role in the detection of fluorescence signals from vesicles is unknown. Preferably, vesicles are detected by an instrument with a sample core diameter of ~2 μm and an effective beam volume of ~10 fL, which is over 5,000-fold smaller than FACSCalibur and technically feasible. Novel flow cytometers have higher sensitivity than the FACSCalibur, and the smallest detectable *single* polystyrene bead by a commercial instrument is currently 100 nm,[40,43,44] corresponding to a single vesicle with a diameter of at least 150 nm, depending on the optical configuration and the exact refractive index of the vesicle. Nevertheless, the illumination geometry of these "new generation" flow cytometers does not show much improvement. For example, the new Beckman Coulter Gallios flow cytometer has a rectangular flow channel of 150 × 460 μm, and the violet laser beam is elliptically focused to a cross-sectional area of 9 × 70 μm^2. These large dimensions may still warrant swarm detection and will potentially complicate the detection of single vesicles with diameters below 100 nm.

4.3.3.2 Nanoparticle tracking analysis

Nanoparticle tracking analysis (NTA), also known as single-particle tracking, measures the absolute size distribution of vesicles with a diameter ranging from 50 nm to 1 μm (Chapter 11). Vesicles in suspension are illuminated by a laser beam and therefore scatter light or exhibit fluorescence in case the vesicles are fluorescently labeled. The light is collected by a dark-field microscope, thereby revealing the position of the vesicles, which are continuously moving due to Brownian motion (Eq. 4.1). For each vesicle in the field of view of the microscope, the movements are tracked during one to several minutes and the mean squared displacement is calculated by digital image processing. Since the mean squared displacement of the Brownian motion depends on the viscosity, temperature, and (hydrodynamic) particle diameter, an absolute size distribution of vesicles in suspension can be obtained after the system has been calibrated with beads of known size and concentration.[17]

An advantage of NTA is that samples are visualized, providing real-time feedback on sample aggregation and monitoring the possible presence of cells after vesicle isolation. Furthermore, since the fluorescence from individual QDs can be detected, NTA is an extremely sensitive method capable of detecting single vesicles with a diameter of 50 nm.[17] A limitation of NTA is that the moderate sensitivity to size differences, due to the uncertainty in the measured diffusion coefficient, leads to broadening of the obtained size distribution. Consequently, two populations can only be resolved if the particle diameter differs at least 1.5-fold. In addition, especially for polydisperse samples the obtained concentration of vesicles is influenced by the size and refractive index of the vesicles.[45] For example, at the less intense edge of the laser beam large vesicles may remain visible, whereas the intensity of light scattered by the smaller vesicles may be below the detection threshold. However, this issue may be solved by selecting a proper particle concentration and optimizing the camera gain.[46]

Like DLS, NTA can also determine the zeta potential by applying an electric field across the suspension. The velocity of vesicles due to electrophoresis can be measured by tracking the individual vesicles, and after correction for electro-osmosis, the zeta potential can be determined.

4.4 Nonoptical Methods

4.4.1 Atomic Force Microscopy

Atomic force microscopy (AFM) was developed in 1986 and provides subnanometer-resolution topography imaging.[47]An atomic force microscope consists of a cantilever with a sharp tip at its end that scans a sample surface without physical contact (Chapter 9). Movements of the tip are measured, and a three-dimensional image is created by software. Owing to a lateral resolution of 3 nm and a vertical resolution of less than 0.1 nm,[47] AFM can measure the relative size distribution of vesicles in their physiological state[16,48] and performs better than DLS on polydisperse samples.[49] Due to the high resolution of AFM, vesicles must be bound to an extremely flat surface, such as mica. Antibodies can be used to bind vesicles to the surface, so biochemical information can also be obtained.[16] Because the efficiency of vesicle binding to a surface using antibodies is unknown, the concentration of vesicles cannot be determined accurately. Furthermore, the surface binding may affect the morphology of vesicles, which may hamper the determination of the real diameter. Due to extensive sample preparation and scanning speed, the measurement time is in the order of hours.

4.4.2 Electron Microscopy

With electron microscopy, electrons are used to create an image. Since the de Broglie wavelength of electrons is more than three orders of magnitude shorter than the wavelength of visible light, the imaging resolution can be subnamometer, depending on the spatial stability of the electron beam and the chemical stability of the sample. Due to this high resolution, it is possible to determine the size and morphology of single vesicles. The measurement time is in the order of hours.

With TEM, the electron beam is interacting with a thin (typically <500 nm) specimen as it passes through. The resulting image is magnified and focused onto an imaging device. Since TEM is performed in vacuum, vesicles require fixation and dehydration, followed by adhesion to a surface and negative staining to enhance the imaging contrast. After visualization by TEM, vesicles typically appear as cup shaped.[50–52] Using immunogold labeling, it is also possible to provide biochemical information.[53] However, the concentration of vesicles cannot be reliably determined due to the unknown binding efficiency of vesicles to the surface.

To exploit TEM to study vesicles in their native environment, vesicles are rapidly frozen in liquid nitrogen or ethane, followed by fracturing of the specimen.[54] Then, a replica of the frozen fractured surface is made by vacuum deposition of platinum and carbon. After cleaning, the replica is imaged by TEM. This procedure of sample preparation and imaging is called freeze-fracture electron microscopy (FEM). Alternatively, ultrathin (<200 nm) sections of specimens can be directly imaged at –80°C using cryo–electron microscopy (cryo-EM).[55] Both FEM and cryo-EM can reveal the size, morphology, and distribution and organization of integral membrane proteins. However, the concentration of vesicles in most biological samples needs to be increased before freezing, and both techniques require considerable know-how and expertise.

Unlike aforementioned techniques, scanning electron microscopy (SEM) can be exploited to study relatively thick samples, such as cells undergoing apoptosis.[56] With SEM, the electron beam is scanned in a raster pattern to image the sample. Cells and vesicles require fixation before analysis by SEM, but in addition they are coated with gold to enhance contrast and prevent the buildup of charge. SEM was successfully combined with Raman microspectroscopy to correlate morphology with the composition of dehydrated biological materials[57] and polystyrene beads at cryogenic temperatures.[58]

4.4.3 Resistive Pulse Sensing

Resistive pulse sensing (RPS), also known as scanning ion occlusion sensing, is based on the Coulter principle and determines the absolute size distribution of vesicles in suspension ranging in diameter from 70 nm to 10 μm.[59] Figure 4.5a shows a schematic representation of the instrument, which consists of two fluid cells divided by an insulating membrane containing a single pore. In each fluid cell, an electrode is immersed to drive an ionic current through the pore, which is typically less than 1 μm in diameter. Figure 4.5b shows the current versus time for cell-free urine diluted 1:10 with phosphate buffered saline (PBS) as measured by RPS (Izon Ltd, qNano,, Christchurch, New Zealand). When a single vesicle passes through the pore, it causes a reduction in current, which appears as a downward spike in the graph. Since the relative change in current, $\Delta I/I$, is approximately proportional to the volume of the vesicle, RPS is very sensitive to changes in the diameter of vesicles. Beads of known size are used to relate $\Delta I/I$ to the real diameter of vesicles, a procedure that has been verified using liposomes of known size (personal communication with Izon).

Four mechanisms cause the vesicles to pass through the pore: pressure-driven flow, electrophoresis, electro-osmosis, and diffusion. Because these mechanisms are interdependent, a pressure of 7 cm H_2O was applied on fluid cell 2 (Fig. 4.5a) to dominate the transport by pressure-driven flow, resulting in a linear relationship between the concentration of vesicles and the blockade frequency. Since the flow is mainly pressure driven, determination of the concentration is independent of the size and zeta potential of the vesicles. Figure 4.5c shows the cumulative counts versus time for vesicles from cell-free urine diluted 1:10 with PBS, as well as polystyrene beads with a diameter of 203 nm and a concentration of 10^9 mL^{-1}. After 242 s, 1,000 polystyrene beads were counted, whereas it took 658 s to count 1,000 vesicles. Consequently, the concentration of vesicles was $10^9/658 \times 242 \times 10 = 3.7 \times 10^9$ mL^{-1}. Now that the differential size distribution and concentration are derived, an absolute size distribution can be composed, as shown in Fig. 4.5d.

Besides size and concentration, RPS is also capable of assessing the zeta potential. Figure 4.5e shows the cumulative counts versus time for polystyrene beads with a diameter of 206 nm and a concentration of 2.2×10^9 mL^{-1}. During the measurement, a pressure was applied on fluid cell 2 (Fig. 4.5a) ranging from −6 cm H_2O to 5.7 cm H_2O, as indicated by the top axis. At −0.5 cm H_2O, the count rate reaches a minimum, indicating that the forces acting upon the vesicles are in equilibrium. The pressure applied at this minimum count rate can be used to calculate the zeta potential of vesicles if the dimensions of the pore are well known.[60]

Although the Coulter principle has been used for detection of cells since the 1950s, the application of RPS to detect vesicles is new and has some drawbacks. For example, the sensitivity to changes in the diameter of vesicles is limited by the dynamic size range, as clearly visible in Fig. 4.5d, revealing a cutoff at 100 nm due to the lower detection limit. Obviously, the upper detection limit is defined by the pore size, which is estimated at ~500 nm for this pore (personal communication with Izon). To increase the dynamic range of a pore, flexible pores or pores with different diameters can be used. A second disadvantage of measuring biological samples with RPS is pore clogging. Due to the presence of high-molecular-weight proteins such as von Willebrand factor or fibrinogen, or particles larger than the pore, for example, vesicles, small cells, or aggregates thereof, the pore may get clogged. Consequently, the measurement has to be paused and the pore needs to be unclogged. Sample preparation is therefore critical. Particles larger than the pore can be removed by filtration or centrifugation, whereas the concentration of proteins can be reduced by washing or dilution. However, all these preanalytical steps will affect the presence of vesicles within the sample (Chapter 5) and will increase the measurement time.

4.5 Isolation Methods Combined with Optical Detection Methods

The size distribution of polydisperse samples as determined by optical detection methods is often broadened (e.g., NTA) or dominated by the presence of a low concentration of relatively large particles (e.g., DLS). Since vesicle populations are polydisperse, it may be beneficial to fractionate the vesicles on the basis of their size before detection. Two methods that perform size-based fractionation before analysis are analytical ultracentrifugation (A-UC) and FFFF.

4.5.1 Analytical Ultracentrifugation

With A-UC, the size distribution of particles in suspension ranging from 1 nm to 10 μm in diameter can be determined.[61,62] Similar to centrifugation, dispersed particles are exposed to centrifugal forces with the advantage that the particles are followed in real time. There are several variants of A-UC, and many different experiment types exist. Therefore, we selected the disk centrifuge, also known as centrifugal liquid sedimentation or differential centrifugal sedimentation, as the most intuitive method to explain the operation

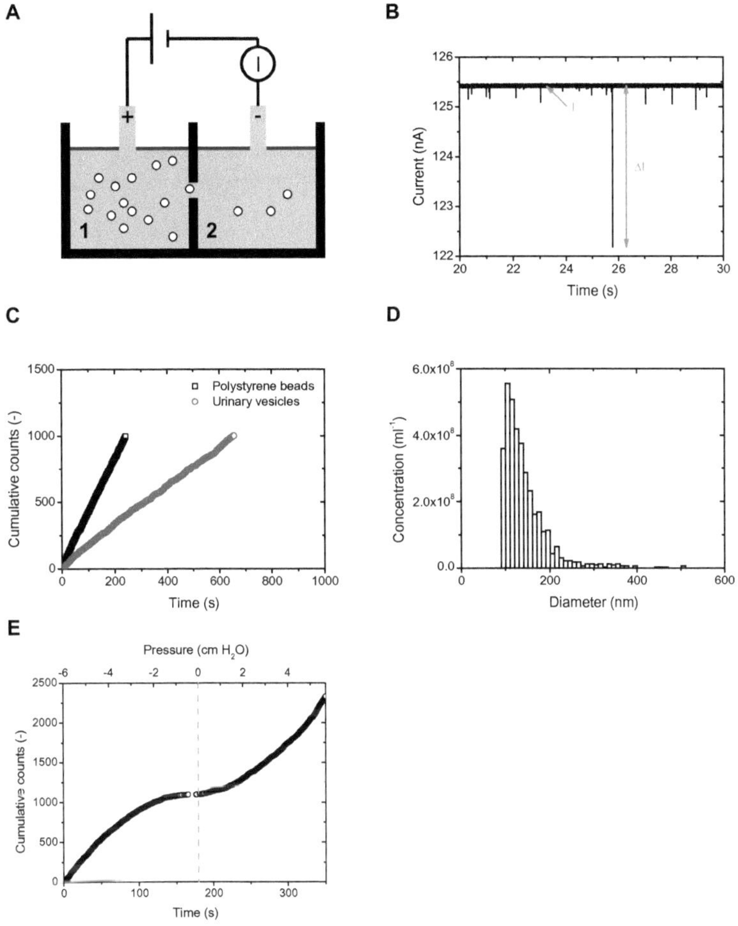

Figure 4.5 Schematic representation of RPS. The instrument consists of two fluid cells divided by an insulating membrane containing a single pore. In each fluid cell, an electrode is immersed to drive an ionic current through the pore. (B) Current I vs. time for vesicles from cell-free urine diluted 1:10 with PBS as measured by RPS. The downward spikes ΔI are due to single vesicles passing through the pore. (C) Cumulative counts vs. time for vesicles from cell-free urine diluted 1:10 with PBS (black squares), as well as polystyrene beads with a diameter of 203 nm and a concentration of 10^9 mL^{-1} (red circles). (D) Absolute size distribution of vesicles from cell-free urine by RPS. (E) Cumulative counts vs. time and pressure for polystyrene beads with a diameter of 206 nm and a concentration of 2.2×10^9 mL^{-1} to determine the zeta potential.

principle of A-UC. With disk centrifugation, vesicles are injected into the hollow center of a spinning disk containing a fluid through which the vesicles sediment to the outer edge. According to Stokes's law, larger vesicles have a higher sedimentation velocity than smaller vesicles. The diameter d of vesicles located at a distance R_d from the center of the disk is therefore time dependent and can be described as follows:

$$d = \sqrt{\frac{18 \cdot \eta \cdot \ln(R_d/R_i)}{(\rho_v - \rho_f) \cdot \omega^2 \cdot t}}$$

where η is the viscosity of the fluid, R_i the distance between the injected sample and the center of rotation, ρ_v the density of vesicles, ρ_f the density of the fluid, ω the angular velocity, and t the sedimentation time.[62] Beads of known size and density are used to determine the numerator of Eq. 4.2. Then, under the assumption that the density of the vesicles is known, it is known at what time vesicles with a certain diameter pass an optical detector, which is placed near the outer edge of the disk. The optical detector monitors the concentration of vesicles, N, by measuring the attenuation of light, which can be quantitatively described by the Lambert–Beer law:

$$\frac{I}{I_0} = e^{-N\sigma(d)h}$$

where I_0 is the intensity of incident light, I the intensity of light after passing through the sample with thickness h, and σ the scattering cross section.[11,63] For spherical particles, such as vesicles, the scattering cross section can be calculated by Mie theory and involves the diameter and refractive index of vesicles. Since the refractive index of vesicles is not exactly known, the determination of the number concentration of vesicles warrants further investigation.

Due to the size-based fractionation, A-UC can accurately determine the size and concentration of a mixture of nine different polystyrene beads ranging from 94 μm to 4.9 μm in diameter using interference optics.[64] The technique is also successfully applied to study the size of vesicles from breast cancer cells.[65] An advantage of A-UC is that multiple samples can be measured simultaneously if the instrument is computer controlled and equipped with multiple tubes. In addition, the technique can be expanded with other optical techniques, for example, fluorescence and Raman detection, to obtain biochemical information.[61] A drawback of A-UC is that the technique cannot assess information from single vesicles and that relatively high expertise is needed to run the apparatus and analyze the data.[61] The measurement time depends on the sedimentation time and is comparable to the centrifugation time required to isolate the vesicles of interest.[62]

4.5.2 FFFF, DLS, and Multiangle Light Scattering

FFFF can separate vesicles on the basis of their size (section 4.1.3). Since size-based fractions are monodisperse, the size of vesicles can be determined with DLS or multiangle light scattering (MALS).[9] MALS is based on angle-resolved light scattering and is used for mean particle size and molar mass determination. Combined with an optical detector that measures the attenuation of light, the concentration of vesicles can also be estimated using the Beer–Lambert law (Eq. 4.3) under the assumption that the refractive index of vesicles is known. An advantage of FFFF above A-UC is that the size-based fractions can be used for further analysis. A drawback of FFFF combined with DLS or MALS is that no information from single vesicles can be obtained and that relatively high expertise is needed to optimize the flow conditions and analyze the data. The measurement time is typically ranging from 30 min to 1 h.

4.6 Discussion and Conclusion

This chapter gives an overview of novel and conventional methods to detect and characterize extracellular vesicles. Table 4.1 lists the assessed possibilities and limitations of each method, depending on the underlying physical parameters of each method.

Since physical separation of vesicles from similar-size particles and cells (section 4.1) is complicated and since vesicles are below the detection range of many currently used techniques, the criteria to identify different types of vesicles have only been partially elucidated so far.[66] Therefore, a detection method is required capable of measuring the size, concentration, morphology, composition, and zeta potential of single vesicles in their native environment within a reasonable time (section 4.2). Despite rapid technological

Table 4.1 Novel and conventional methods to detect EVs

Method	Single-vesicle detection [a]	Free in suspension	Detection limit[c]	Size	Concentration	Morphology	Composition	Zeta potential	Measurement time[b]
Optical methods									
Light scattering									
DLS	–	+	1 nm–6 μm	+/–	–	–	–	+	S
Raman spectroscopy	+	+	?	?	+/–	–	+	–	H
Fluorescence									
F-CS	+	+	Single molecule/QD	+/–	+/–	–	+	–	M
STED	+	–	Single molecule/QD	+	+/–	?	+	–	H
Light scattering and fluorescence									
Flow cytometry	+/–	+	>~150 nm/ single QD	–	–	–	+/–	–	S
NTA	+	+	50 nm–1 μm	+/–	+/–	–	+	+	M

Method	Single-vesicle detection [a]	Free in suspension	Detection limit[c]	Size	Concentration	Morphology	Composition	Zeta potential	Measurement time[b]
Nonoptical methods									
AFM	+	–	<1 nm	+	+/–	+	+/–	–	H
TEM	+	–	~1 nm	+	–	+	+/–	–	H
FEM	+	+/–	~1 nm	+	+/–	+	+	–	H
Cryo-EM	+	+/–	~1 nm	+	–	+	+	–	H
SEM	+	–	~1 nm	+	–	+	+/–	–	H
RPS	+	+	70 nm–10 μm	+/–	+/–	–	–	+	M
Isolation methods combined with detection methods									
A-UC	–	+	1 nm–10 μm	+	+/–	–	+/–	–	M
FFFF, MALS, DLS	–	+	1 nm–50 μm	+	+/–	–	+/–	–	M

[a] A method that is capable of, capable of but with limitations, or incapable of providing information on single vesicles, vesicles free in suspension, size distribution, particle concentration, morphology, biochemical composition, or zeta potential is indicated by +, +/–, or –, respectively.

[b] The measurement time is indicated by H, M, or S, which means longer than 1 h, between 1 min and 1 h, or shorter than 1 min, respectively.

[c] A question mark means "to be investigated." The detection limit of flow cytometry is >~150 nm based on light scattering and a single QD for luorescence detection. The detection limit of all microscopy methods is determined by the spatial resolution of the technique.

developments, there is currently no such technique available, according to Table 4.1. However, by combining multiple detection methods and improving techniques, all relevant properties of single vesicles may become assessable in the future. An improvement may be to isolate and analyze vesicles on disposable microfluidic chips, on which small volumes (~pL) of fluids can be controlled to allow faster reaction times, increased sensitivity, and reduced costs.[67]

At present, diameter criteria are often used to distinguish between vesicles originating from multivesicular endosomes and vesicles originating directly from the plasma membrane.[66] However, the size of a vesicle can also be expressed in terms of surface area or volume. We speculate that the number of *receptors* exposed on a vesicle is proportional to the surface area, whereas the amount of *cargo* a vesicle can transport is related to the volume of the vesicle. Figure 4.6 shows that vesicles with a diameter smaller than 100 nm have a relatively large contribution to the transport of surface receptors (Fig. 4.6a) but hardly contribute to cargo transport (Fig. 4.6b). This finding may have major implications. For example, many properties currently attributed to transport of cargo by vesicles smaller than 100 nm, such as the presence of micro-RNA (miRNA), cannot be justified as long as samples may also contain larger vesicles. However, to provide definite answers the assessment of relevant properties from single vesicles is essential.

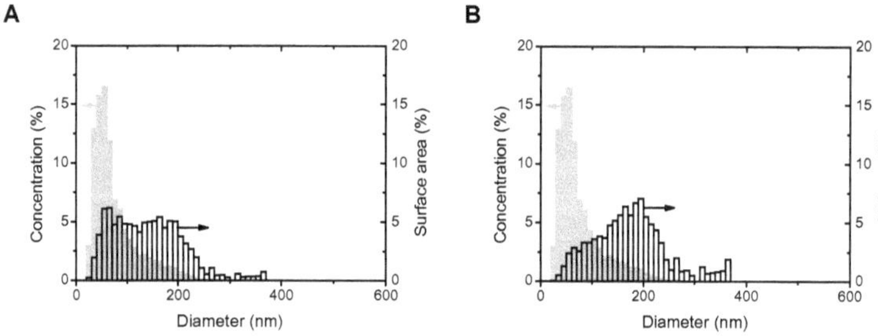

Figure 4.6 (a) Surface area and (b) volume distribution (black bars) of the size distribution (blue bars) shown in Fig. 4.2d. The concentration of the full population is 10^{10} vesicles mL^{-1}, the total surface area (100%) is 257 mm^2 mL^{-1}, and the total volume (100%) is 6 nL mL^{-1}. The surface area distribution is shifted to the right in comparison to the diameter distribution, and the volume distribution is shifted to the right in comparison to the surface area distribution.

Acknowledgments

We would like to acknowledge C. J. M. van Rijn for providing an image of a filtration sieve and A. N. Böing and M. C. L. Schaap (Laboratory of Experimental Clinical Chemistry, University of Amsterdam), A. T. M. Lenferink and C. Otto (Medical Cell BioPhysics, University of Twente), and H. A. van Veen (Department of Cell Biology and Histology, University of Amsterdam) for experimental support. We thank M. J. C. van Gemert (Department of Biochemical Engineering and Physics, University of Amsterdam), J. Hardij (Department of Pharmacy, University of Namur), A. G. Hoekstra (Computational Science Research Group, University of Amsterdam), and A. Sturk (Laboratory of Experimental Clinical Chemistry, University of Amsterdam) for valuable discussions.

References

1. Gyorgy B, et al. (2011) Detection and isolation of cell-derived microparticles are compromised by protein complexes resulting from shared biophysical parameters, *Blood*, **117**, E39–E48.
2. van der Pol E, Hoekstra AG, Sturk A, Otto C, van Leeuwen TG, Nieuwland R (2010) Optical and non-optical methods for detection and characterization of microparticles and exosomes, *J. Thromb. Haemost.*, **8**, 2596–2607.
3. Sustar V, et al. (2011) Nanoparticles isolated from blood: a reflection of vesiculability of blood cells during the isolation process, *Int. J. Nanomed.*, **6**, 2737–2748.
4. Tsutsui H, Ho CM (2009) Cell separation by non-inertial force fields in microfluidic systems, *Mech. Res. Commun.*, **36**, 92–103.
5. van Dalum G, Coumans FAW, Beck M, Terstappen LWMM (2012) Key factors for filtration of tumor cells from whole blood, *Cancer Res.*, **72**(8 Suppl), Abstract No. 2396.
6. Van Rijn CJM (2004) *Nano and Micro Engineered Membrane Technology*, 1st ed, Amsterdam: Elsevier Science.
7. Crowley TA, Pizziconi V (2005) Isolation of plasma from whole blood using planar microfilters for lab-on-a-chip applications, *Lab Chip*, **5**, 922–929.
8. Korgel BA, van Zanten JH, Monbouquette HG (1998) Vesicle size distributions measured by flow field-flow fractionation coupled with multiangle light scattering, *Biophys. J.*, **74**, 3264–3272.

9. Kang DJ, Oh S, Ahn SM, Lee BH, Moon MH (2008) Proteomic analysis of exosomes from human neural stem cells by flow field-flow fractionation and nanoflow liquid chromatography-tandem mass spectrometry, *J. Proteome Res.*, **7**, 3475–3480.
10. Thery C, Ostrowski M, Segura E (2009) Membrane vesicles as conveyors of immune responses, *Nat. Rev. Immunol.*, **9**, 581–593.
11. Bohren CF, Huffman DR (1983) *Absorption and Scattering of Light by Small Particles,* New York, NY: Wiley.
12. van Manen HJ, Verkuijlen P, Wittendorp P, Subramaniam V, van den Berg TK, Roos D, Otto C (2008) Refractive index sensing of green fluorescent proteins in living cells using fluorescence lifetime imaging microscopy, *Biophys. J.*, **94**, L67–L69.
13. Beuthan J, Minet O, Helfmann J, Herrig M, Muller G (1996) The spatial variation of the refractive index in biological cells, *Phys. Med. Biol.*, **41**, 369–382.
14. Foladori P, Quaranta A, Ziglio G (2008) Use of silica microspheres having refractive index similar to bacteria for conversion of flow cytometric forward light scatter into biovolume, *Water Res.*, **42**, 3757–3766.
15. Briggs C, Harrison P, Samuel JM (2006) Platelet counting, in *Platelets,* 2nd ed., San Diego, CA: Academic Press, 475–483.
16. Yuana Y, Oosterkamp TH, Bahatyrova S, Ashcroft B, Garcia RP, Bertina RM, Osanto S (2010) Atomic force microscopy: a novel approach to the detection of nanosized blood microparticles, *J. Thromb. Haemost.*, **8**, 315–323.
17. Dragovic RA, Gardiner C, Brooks AS, Tannetta DS, Ferguson DJP, Hole P, Carr B, Redman CWG, Harris AL, Dobson PJ, Harrison P, Sargent IL (2011) Sizing and phenotyping of cellulars vesicles using nanoparticle tracking analysis, *Nanomedicine*, **7**, 780–788.
18. Clark NA, Lunacek JH, Benedek GB (1970) A study of Brownian motion using light scattering, *Am. J. Phys.*, **38**, 575–585.
19. Dieckmann Y, Colfen H, Hofmann H, Petri-Fink A (2009) Particle size distribution measurements of manganese-doped ZnS nanoparticles, *Anal. Chem.*, **81**, 3889–3895.
20. Bryant G, Abeynayake C, Thomas JC (1996) Improved particle size distribution measurements using multiangle dynamic light scattering. 2. Refinements and applications, *Langmuir*, **12**, 6224–6228.
21. Filella M, Zhang JW, Newman ME, Buffle J (1997) Analytical applications of photon correlation spectroscopy for size distribution measurements

of natural colloidal suspensions: capabilities and limitations, *Colloids Surf. A*, **120**, 27–46.

22. Hoo CM, Starostin N, West P, Mecartney ML (2008) A comparison of atomic force microscopy (AFM) and dynamic light scattering (DLS) methods to characterize nanoparticle size distributions, *J. Nanopart. Res.*, **10**, 89–96.
23. Korgel BA, van Zanten JH, Monbouquette HG (1998) Vesicle size distributions measured by flow field-flow fractionation coupled with multiangle light scattering, *Biophys. J.*, **74**, 3264–3272.
24. Puppels GJ, Demul FFM, Otto C, Greve J, Robertnicoud M, Arndtjovin DJ, Jovin TM (1990) Studying single living cells and chromosomes by confocal raman microspectroscopy, *Nature*, **347**, 301–303.
25. Pully VV, Lenferink ATM, Otto C (2011) Time-lapse Raman imaging of single live lymphocytes, *J. Raman Spectrosc.*, **42**, 167–173.
26. McCarthy DA (2007) Fluorochromes and fluorescence, in *Flow Cytometry: Principles and Applications*, 1st ed., New York, NY: Humana Press, 59–112.
27. Zhang Q, Li YL, Tsien RW (2009) The dynamic control of kiss-and-run and vesicular reuse probed with single nanoparticles, *Science*, **323**, 1448–1453.
28. Cesa Y, Blum C, van den Broek JM, Mosk AP, Vos WL, Subramaniam V (2009) Manipulation of the local density of photonic states to elucidate fluorescent protein emission rates, *Phys. Chem. Chem. Phys.*, **11**, 2525–2531.
29. Radbruch A (2000) *Flow Cytometry and Cell Sorting,* New York, NY: Springer.
30. Baumgarth N, Roederer M (2000) A practical approach to multicolor flow cytometry for immunophenotyping, *J. Immunol. Methods*, **243**, 77–97.
31. Webb WW (1976) Applications of fluorescence correlation spectroscopy, *Q. Rev. Biophys.*, **9**, 49–68.
32. Starchev K, Buffle J, Perez E (1999) Applications of fluorescence correlation spectroscopy: polydispersity measurements, *J. Colloid Interface Sci.*, **213**, 479–487.
33. Westphal V, Hell SW (2005) Nanoscale resolution in the focal plane of an optical microscope, *Phys. Rev. Lett.*, **94**, 143903.
34. Willig KI, Rizzoli SO, Westphal V, Jahn R, Hell SW (2006) STED microscopy reveals that synaptotagmin remains clustered after synaptic vesicle exocytosis, *Nature*, **440**, 935–939.

35. Rittweger E, Han KY, Irvine SE, Eggeling C, Hell SW (2009) STED microscopy reveals crystal colour centres with nanometric resolution, *Nat. Photonics*, **3**, 144–147.

36. Lacroix R, Robert S, Poncelet P, Kasthuri RS, Key NS, Dignat-George F (2010) Standardization of platelet-derived microparticle enumeration by flow cytometry with calibrated beads: results of the International Society on Thrombosis and Haemostasis SSC collaborative workshop, *J. Thromb. Haemost.*, **8**, 2571–2574.

37. Robert S, Poncelet P, Lacroix R, Arnaud L, Giraudo L, Hauchard A, Sampol J, Dignat-George F (2009) Standardization of platelet-derived microparticle counting using calibrated beads and a Cytomics FC500 routine flow cytometer: a first step towards multicenter studies? *J. Thromb. Haemost.*, **7**, 190–197.

38. Chandler WL, Yeung W, Tait JF (2011) A new microparticle size calibration standard for use in measuring smaller microparticles using a new flow cytometer, *J. Thromb. Haemost.*, **9**, 1216–1224.

39. Mullier, F, Bailly, N, Chatelain, C, Dogne, JM, Chatelain, B. (2011). More on: calibration for the measurement of microparticles: needs, interests, and limitations of calibrated polystyrene beads for flow cytometry-based quantification of biological microparticles. *J. Thromb. Haemost.* **9**, 1679–1681.

40. Robert, S, Poncelet, P, Lacroix, R, Raoult, D, Dignat-George, F. (2011). More on: calibration for the measurement of microparticles: value of calibrated polystyrene beads for flow cytometry-based sizing of biological microparticles. *J. Thromb. Haemost.*, **9**, 1676–1678.

41. van der Pol, E, Van Gemert, MJC, Sturk, A, Nieuwland, R, Van Leeuwen, TG (2012). Single vs. swarm detection of microparticles and exosomes by flow cytometry. *J. Thromb. Haemost.*, **10**, 919–930.

42. Collier JL (2000) Flow cytometry and the single cell in phycology, *J. Phycol.*, **36**, 628–644.

43. Vorauer-Uhl K, Wagner A, Borth N, Katinger H (2000) Determination of liposome size distribution by flow cytometry, *Cytometry*, **39**, 166–171.

44. Lacroix R, Robert S, Poncelet P, Dignat-George F (2010) Overcoming limitations of microparticle measurement by flow cytometry, *Semin. Thromb. Haemost.*, **36**, 807–818.

45. Malloy A, Carr B (2006) Nanoparticle tracking analysis—the Halo (TM) system, *Part. Part. Syst. Char.*, **23**, 197–204.

46. Du SF, Kendall K, Morris S, Sweet C (2010) Measuring number-concentrations of nanoparticles and viruses in liquids on-line, *J. Chem. Technol. Biotechnol.*, **85**, 1223–1228.

47. Binnig G, Quate CF, Gerber C (1986) Atomic force microscope, *Phys. Rev. Lett.*, **56**, 930–933.

48. Siedlecki CA, Wang IW, Higashi JM, Kottke-Marchant K, Marchant RE (1999) Platelet-derived microparticles on synthetic surfaces observed by atomic force microscopy and fluorescence microscopy, *Biomaterials*, **20**, 1521–1529.

49. Hoo CM, Starostin N, West P, Mecartney ML (2008) A comparison of atomic force microscopy (AFM) and dynamic light scattering (DLS) methods to characterize nanoparticle size distributions, *J. Nanopart. Res.*, **10**, 89–96.

50. Raposo G, Nijman HW, Stoorvogel W, Leijendekker R, Harding CV, Melief CJM, Geuze HJ (1996) B lymphocytes secrete antigen-presenting vesicles, *J. Exp. Med.*, **183**, 1161–1172.

51. Marzesco AM, Janich P, Wilsch-Brauninger M, Dubreuil V, Langenfeld K, Corbeil D, Huttner WB (2005) Release of extracellular membrane particles carrying the stem cell marker prominin-1 (CD133) from neural progenitors and other epithelial cells, *J. Cell Sci.*, **118**, 2849–2858.

52. Heijnen HFG, Schiel AE, Fijnheer R, Geuze HJ, Sixma JJ (1999) Activated platelets release two types of membrane vesicles: microvesicles by surface shedding and exosomes derived from exocytosis of multivesicular bodies and alpha-granules, *Blood*, **94**, 3791–3799.

53. Pisitkun T, Shen RF, Knepper MA (2004) Identification and proteomic profiling of exosomes in human urine, *Proc. Natl. Acad. Sci. U. S. A.*, **101**, 13368–13373.

54. Severs NJ (2007) Freeze-fracture electron microscopy, *Nat. Prot.*, **2**, 547–576.

55. Conde-Vancells J, Rodriguez-Suarez E, Embade N, Gil D, Matthiesen R, Valle M, Elortza F, Lu SC, Mato JM, Falcon-Perez JM (2008) Characterization and comprehensive proteome profiling of exosomes secreted by hepatocytes, *J. Proteome Res.*, **7**, 5157–5166.

56. Coleman ML, Sahai EA, Yeo M, Bosch M, Dewar A, Olson MF (2001) Membrane blebbing during apoptosis results from caspase-mediated activation of ROCK I, *Nat. Cell Biol.*, **3**, 339–345.

57. Van Apeldoorn AA, Aksenov Y, Stigter M, Hofland I, De Bruijn JD, Koerten HK, Otto C, Greve J, Van Blitterswijk CA (2005) Parallel high-

resolution confocal Raman SEM analysis of inorganic and organic bone matrix constituents, *J. R. Soc. Interface*, **2**, 39–45.

58. Hazekamp J, Reed MG, Howard CV, Van Apeldoorn AA, Otto C (2011) The feasibility of Cryo In-SEM Raman microspectroscopy, *J. Microsc.*, **244**, 122–128.
59. Vogel R, Willmott G, Kozak DM, Roberts GS, Anderson W, Groenewegen L, Glossop B, Barnett A, Turner A, Trau M (2011) Quantitative sizing of nano/microparticles with a tunable elastomeric pore sensor, *Anal. Chem.*, **83**, 3499–3506.
60. Vogel R, Anderson W, Eldridge J, Glossop B, Willmott G (2012) A variable pressure method for characterizing nanoparticle surface charge using pore sensors, *Anal. Chem.*, **84**, 3125–3131.
61. Mächtle W, Börger L (2006) *Analytical Ultracentrifugation of Polymers and Nanoparticles,* Berlin: Springer-Verlag.
62. Merkus HG (2009) Sedimentation techniques, in *Particle Size Measurements*, Delft: Springer Science; Business Media BV, 319–348.
63. Coll H, Searles CG (1987) Particle-size analysis with the Joyce-Loebl disk centrifuge—a comparison of the line-start with the homogeneous-start method, *J. Colloid Interface Sci.*, **115**, 121–129.
64. Müller HG (2004) Determination of very broad particle size distributions via interference optics in the analytical ultracentrifuge, *Prog. Colloid Polym. Sci.*, **127**, 9–13.
65. Hardij J, et al. (2011) The CPS disc centrifuge: a new original method to detect microparticles (MPS), *Int. Soc. Thromb. Haemost.*, **July**, Abstract No. 01827.
66. van der Pol E, Böing AN, Harrison P, Sturk A, Nieuwland R (2012) Classification, functions and clinical relevance of extracellular vesicles, *Pharmacol. Rev.*, **64**(3), 676–705.
67. Chen C, Skog J, Hsu CH, Lessard RT, Balaj L, Wurdinger T, Carter BS, Breakefield XO, Toner M, Irimia D (2010) Microfluidic isolation and transcriptome analysis of serum microvesicles, *Lab Chip*, **10**, 505–511.

Chapter 5

Preanalytical Variables

Romaric Lacroix,[a,b] Coralie Judicone,[b] Stéphane Robert,[a] Laurent Arnaud,[b] Florence Sabatier,[a,b] and Françoise Dignat-George[a,b]

[a]*VRCM, UMR-S1076, UFR de Pharmacie, INSERM/Aix-Marseille Université, Marseille, France*

[b]*Laboratoire d'hématologie, et de biologie vasculaire, CHU Conception, AP-HM, Marseille, France*

Romaric.Lacroix@univ-amu.fr

5.1 Introduction

Microvesicles (MVs) are submicron-size vesicles released from cell membranes in response to activation or apoptosis.[1] They are generally defined as 0.1–1 μm membrane fragments resulting from the remodeling of membrane phospholipids in response to cell activation or apoptosis. As a result, blood MVs generally express the anionic phospholipid phosphatidylserine (PS) and bear membrane antigens representative of their parental cells so that their origin can be determined by using the specific antibodies against these antigens.[2,3] The majority of MVs detected in the blood originate from platelets and erythrocytes, but other blood cells such as leucocytes, endothelial cells, and even malignant cells may also shed MVs.[4] MVs differ

Extracellular Vesicles in Health and Disease
Edited by Paul Harrison, Christopher Gardiner, and Ian L. Sargent

ISBN 978-981-4411-98-1 (Hardcover), 978-981-4411-99-8 (eBook)
www.panstanford.com

from other types of extracellular vesicles (EVs), including apoptotic bodies and exosomes. These different vesicles are distinguished one from another on the basis of their subcellular origin, their size, their content, the mechanisms leading to their formation, and, from a practical point of view, their mode of obtention. Exosomes (less than 100 nm in diameter) are produced in multivesicular bodies during endocytosis, and they play a role in antigen presentation. Unlike MVs they do not externalize PS, and they express specific exosomal markers such as lysosome-associated membrane protein 1 (LAMP1), CD63, and TSG101; they also contain messenger ribonucleic acid (mRNA) and micro-RNAs.[5,6] Apoptotic bodies are larger than MVs or exosomes and are characterized by externalized PS, and unlike MVs, they may also contain a permeable membrane facilitating propidium iodide staining of the nuclear material. Several reports indicate that apoptotic bodies are passive cargos delivering their nuclear content (oncogenes, deoxyribonucleic acid [DNA], micro-RNA) to phagocytes by horizontal transfer.[7] However, there is no consensus on the phenotypic features and precise size distribution of these different types of EVs, and in reality there is probably significant overlap between the three different populations in size, content, and function. This is not only due to limitations of the methods used for their characterization but also due to preanalytical variables.

It is now well recognized that EVs behave as conveyors of bioactive molecules, playing a key role in the exchange of cellular information. They are considered as messengers and mediators in coagulation, inflammation, angiogenesis, immune response, and cancer spreading[8–12] but also as emergent biomarkers that may be useful in clinical practice. Indeed, an overview of the literature from the last 10 years evidences that circulating MVs originating from blood and vascular cells are elevated in a variety of prothrombotic and inflammatory disorders, cardiovascular diseases, autoimmune, infectious diseases, and cancer.[13–15] In these clinical settings, MV counts may be useful for identifying patients at risk for vascular disorders and for monitoring response to treatment.[16,17] The accurate measurement of any circulating EVs is therefore of potential major importance. However, among the issues limiting the full definition of the clinical relevance of EVs, preanalytical conditions represent an important source of variability and may contribute to significant potential artifacts in their analysis. Accordingly, the first methodological studies and reviews published on EV determination have stressed the major impact on preanalytical parameters for the standardization of EV determination[3,18–21] and the need to establish recommendations.

The objective of this chapter is to review the different preanalytical parameters influencing EV measurement in plasma samples, including phlebotomy, collection tubes, transportation, plasma centrifugation, and storage. Issues of EV purification won't be addressed. In 2003, the Vascular Biology Group of the Scientific and Standardization Committee (SSC-VB) of the International Society of Thrombosis and Haemostasis (ISTH) decided to address issues related to the measurement of MVs. Accordingly, we will make some general recommendations on preanalytical conditions in MV assays and will discuss some aspects related to standardization.

Provided that cells can be easily activated during sample processing and that storage is generally needed, the aim of the optimal preanalytical protocol for measurement of circulating MVs would be to avoid *ex vivo* MV generation during all steps and to obtain complete removal of cells, including platelets, without loss of MVs.

5.2 Blood Collection

In hemostasis laboratories, needles ranging from 19 gauge (G) to 22 G are commonly used for venepuncture.[22] Theoretically, when a small-diameter needle (high G number) is connected to an evacuated tube, shear stress and excessive pressure on the vein during blood withdrawal might cause *in vitro* hemolysis[23] with possible formation of erythrocyte microparticles (EryMPs).[24] It is recommended that we avoid prolonged placement of a tourniquet during venepuncture.[25] However, a system using a small-diameter butterfly needle is commonly used on small children and patients with difficult venous access.[26] This system reduces the blood flow and consequently increases the risk of platelet activation.[22] To sum up, the different studies analyzing the impact of routine hemostasis testing have shown that blood collection performed with a large-size, 21-gauge needle avoids *in vitro* hemolysis, EryMP generation, and platelet activation.[3,14,22,27] The first 2–3 mL of blood is discarded as already recommended to limit the effects of the vascular damage caused by venepuncture.[26,28,29] To compare the influence of tourniquet and needle type during venepuncture, a recent study measured MVs in healthy volunteers when blood was collected successively from both arms of the same donor.[28] Interestingly, application of a light (20 mmHg) compared to a strong (80 mmHg) tourniquet did not significantly differ in terms of MV counts and procoagulant activity. Moreover, evaluating preanalytical conditions for routine hemostasis analysis did not show a significant difference between a straight needle or a butterfly device for blood collection.[28,30] In conclusion, provided blood collection is performed by an experienced phlebotomist with a standardized protocol and conditions, this step should not impact extensively MV analysis.

5.3 Collection Tubes

Several anticoagulants are used to collect blood for MV analysis (Table 5.1, adapted from Ref. 20). These anticoagulants should limit platelet activation during blood collection and plasma preparation, and artifactual release of platelet-derived MVs (PMVs).[31-33] Sodium citrate (concentration of 0.105 M or 0.129 M) is the most commonly used anticoagulant in hemostasis laboratories and for MV analysis.[2] It acts by chelation of free calcium ions, thus making the *ex vivo* calcium unavailable to the coagulation system,[23,34] and prevents leukocytes and platelets from significant activation and degranulation.[35] Because calcium is a key player in membrane phospholipid remodeling, citrate is believed to prevent, at least partially, the vesiculation process.[36] In the literature, other anticoagulants have been used for MV determination, including acid-citrate-dextrose (ACD),[37-39] ethylenediaminetetraacetic acid (EDTA),[40-42] a strong chelator of calcium ions, and heparin,[43,44] which will preserve extracellular calcium. Even if platelet reactivity may vary among different tube brands,[45] citrated anticoagulant generated less artifactual MVs as compared to heparin or EDTA. A comparison between blood collection in heparin by reference to sodium citrate demonstrated that the levels of annexin V–positive MVs were significantly higher.[46] One study simultaneously compared the effects of citrate, heparin, and EDTA on plas-

Table 5.1 Representative panel of the heterogeneity in preanalytical parameters used in the recent literature for microparticles analysis

Blood collection		Plasma preparation				Storage	Thawing	Ref.
Needle (G)	Anti-coagulant	Type	Speed (g)	Time (min)	T (°C)	T (°C)	T (°C)	
19	EDTA	PFP	1st: 1,200 2nd: 12,000	15 12	20 20	−70	n.s	39
21	Citrate	PPP	2,600	15	4	–	n.s	38
n.s	Citrate	PPP	1,550	20	20	−80	4	72
21	Citrate	PFP	1st: 1,200 2nd: 13,000	15 2	20 20	−80	37	73
19 (butterfly needle)	Citrate	PPP	3,000	15	n.s	−40	n.s	74
n.s	Citrate	PPP	2,000	15	8	−80	n.s	75
21	Citrate	PFP	1st: 200 2nd: 20,000	15 15	20 20	−80	n.s	76
n.s	Citrate	PFP	1st: 1,500 2nd: 13,400	15 2	n.s	−80	n.s	77
21	EDTA-ACD	PPP	8,000	5	20	-40	n.s	36
21	ACD	[a]PPP [b]PFP	[a]2 × 1,550 [b]1st: 2 × 1,550 2nd: 10,000	20 20 10	n.s	−80	n.s	26
n.s	Citrate	PPP	2 × 2,100	20	n.s	−80	n.s	17
21	Citrate	PPP	2 × 2,000	10	20	–	n.s	78
21	Citrate	PPP	a: 2 × 5,000 b: 5,000 c: 1,500	5 15 15	20 – –	−80	n.s	79
21	Citrate	PFP	2 × 2,500	15	20	80	37	28
21	Citrate	PFP	1st: 1,500 2nd: 13,000	15 2	20 4	−80	n.s	80
19 (butterfly needle)	EDTA, CTAD, heparin, hirudin	PFP	2 × 3,000	15	–	–	n.s	81
n.s	Citrate	PPP	2 × 3,000	2 × 15	20	−80	37	82
19	Citrate	PPP	1,550	20	20	−80	4	23

Source: Adapted from Ref. 20

ma MV determination measured on the same healthy donors.[28] Total MV counts measured by flow cytometry (FCM) were dramatically increased in EDTA and heparin as compared to the citrated tube. Consistently, EDTA is known to induce a P-selectin-dependent platelet activation process[18] that may also result in pseudo-thrombocytopenia and platelet aggregates on blood smears. Moreover, EDTA tubes contain extremely high concentrations of potassium,[47] with an unknown impact on vesiculation. However, contradictory results have been published in the literature on the relative impact of these three anticoagulants.[48] The use of citrate-theophylline-adenosine-dipyridamole (CTAD) has also been recommended because CTAD inhibits platelet activation *in vitro* by increasing the cytosolic cyclic adenosine monophosphate (cAMP) concentration without affecting platelet function.[19,39,49] Therefore, the platelet activation inhibitors contained in CTAD tubes may be more effective in inhibiting *ex vivo* PMV formation.[50–52] Accordingly, PMV counts were reported to increase twofold in plasma from citrated blood but to remain relatively stable in plasma isolated from CTAD blood,[44] during a period of 3 h after blood collection. However, experimental data on a series of healthy donors did not demonstrate any significant difference in MV counts or MV-dependent procoagulant activity between citrate and CTAD tubes.[28] As a whole, although additional experiments need to be performed on the other 109 mM citrate tubes and also on pathological samples, comparison of the most commonly used anticoagulants identifies citrate or CTAD as most appropriate for MV analysis.

5.4 Transportation

Conditions of routine blood tube transportation may significantly vary in terms of generated agitation. One recent study investigated the impact of gentle and strong agitation by comparison with complete absence of any agitation.[28] There was no evidence of a significant impact of gentle agitation on MV counts and procoagulant activity. In contrast, strong agitation induces a critical, although highly variable, increase in MV counts and procoagulant potential. To better mimic the real situation in hospitals, blood tubes transported either unsupported as usual in double plastic bags or purposely kept immobile using transportation boxes were compared to reference tubes that remained motionless in the lab (in both horizontal or vertical positions). In contrast to what is observed when red blood cells are vortexed,[53] results of this study based on high-sensitivity FCM did not demonstrate a significant impact on the generation of EryMVs by gentle agitation. Importantly, the impact of agitation induced by a common hospital transportation system was similar to the effect of strong agitation. Interestingly, transporting blood tubes in a vertical rather than a horizontal position, via the use of special transportation boxes, also limited the extent of *in vitro* MV generation.[28] The data identify agitation during transportion as a crucial step and suggest that such transportation boxes should be used for MV studies when patients' samples are not directly collected in the local environment.

In this study, as blood samples were collected from healthy volunteers directly in the laboratory, it was also possible to study the impact of delay between blood sampling and the first centrifugation step with a short delay of 5 min serving as the reference control sample. In such conditions, results showed that a 1 h time delay before the first centrifugation influences total and PMV counts as well as procoagulant activity, consistent with a PMV increase already described in other studies.[50,52,54] However, the increase during the two first hours remains moderate in comparison to other preanalytical parameters such as centrigution speed or agitation during transportation, suggesting that a short delay, which remains compatible with current laboratory practice, may be acceptable, provided control and test samples are handled in an identical manner. Noteworthy, we did not evidence a significant change on EryMVs with such a short time delay, contrary to what may happen during storage of blood units for several days.[53]

5.5 Plasma Processsing and Microvesicle Preparation

An appropriate centrifugal speed should be applied for plasma preparation in order to eliminate cells, including platelets, to freeze the samples regardless of the potential of cellular fragmentation during thawing and consequent artifactual formation of MVs. According to the Clinical and Laboratory Standards Institute (CLSI), plasma with a platelet count less than 10×10^9/L is obtained by centrifuging blood at 1,500 g for 15 min at room temperature (RT).[55] To obtain platelet-poor plasma (PPP), the CLSI also recommends recentrifuging the plasma for another 10 min at 1,500 g. However, platelet-free plasma (PFP) obtained by performing two-step centrifugation, each at 2,000 g for 10 min, can also significantly reduce the level of any residual platelets.[20] Some studies indicate that MV counts measured by conventional FCM are lower in PFP than in PPP, suggesting that a percentage of MVs could also be lost during this step.[29,56] However, careful removal of the majority of platelets is also required prior to freezing the samples. The ideal equilibrium remains to be found. The literature indicates that centrifugation conditions, speed, and time all vary widely among different studies and laboratories (see Table 5.1, adapted from Ref. 20). The initial centrifugation intended to generate cell-free plasma is 1,500–2,500 × g for 15–20 min in most studies. However, a significant number of platelets also persist after a single centrifugation step.[18] An additional centrifugation step of 13,000 × g for 2 min ensures the generation of PFP.[2,57,58] Following this second centrifugation, the plasma should be carefully aspirated but leaving the bottom 1 cm undisturbed.[59]

Because this high-speed centrifugation is not always easily available in some laboratories, a recent study compared the currently recommended protocol of 1,500 g 15 min + 13,000 g 2 min[57,60] to a more routine laboratory-adapted protocol, that is, 2 × 2,500 g 15 min.[61] The effect of these two different centrifugation protocols, both intended to entirely remove platelets, was therefore compared on healthy donors. When fresh samples were analyzed, no significant difference was observed between both protocols. In contrast, after a freeze–thaw cycle, data showed a significant increase in MV counts and procoagulant activity with the protocol of 1,500 g 15 min + 13,000 g 2 min compared to the 2 × 2,500 g 15 min protocol.[28] This was also confirmed by another team.[62] Although both protocols create PFP with very low levels of contaminating platelets, the routine laboratory-adapted protocol generates less artifactual PMV after freezing, due to a lower level of residual platelets. The more complex tube handling involved in the high-speed centrifugation may partly explain such a difference. Altogether, these studies demonstrated the best efficiency of the routine laboratory-adapted protocol in terms of platelet removal efficiency and thus stability after deep-freeze storage.

To limit platelet activation, the temperature is generally kept at RT (20–25°C) for centrifuging citrated blood.[3] However, the effect of centrifugation temperature on MVs has not yet been thoroughly investigated. Furthermore, during centrifugation the use of brakes should be avoided. A pertinent question about this plama preparation step using centrifugation is: does centrifugation itself generate artifactual MVs MV-free plasma was prepared by filtration at 0.1 µm of PFP from the same donor. Purified platelets and MV-free plasma were mixed and analyzed by impedance-based FCM for PMV and platelet content after annexin V–fluorescein isothiocyanate (FITC) and CD41-PE labeling according to Zwicker et al.[17] The residual PMVs were very low. After centrifugation for 2 min at 13,000 g, each successive 100 µL layer of plasma (1 to 10) was analyzed for platelet and PMV content using the same protocol. As a result, after centrifugation, all platelets were pelleted, but no significant PMV increase was observed in the plasma layers. In contrast, as expected, a large number of PMVs was found in the pellet indicating that platelet activation occurs at the bottom of the tube. Altogether these results show that the artifactual PMV generation induced by centrifugation is located in the pellet area only. Thus, platelet activation during centrifugation should not artificially affect PMV counts, provided that (i) the pellet remains undisturbed and (ii) the PFP is sampled shortly after centrifugation.

5.6 Storage

The use of freshly isolated plasma is recommended for MV measurements to avoid changes in MV number and characteristics.[59] In particular, remaining platelets present in depleted plasma are fragmented during freezing and thawing, leading to an increase in the number of PMVs and annexin V–positive MVs. However, the logistics of blood sampling, transport time to the in-house laboratory, or even transport to laboratories elsewhere are all variables that are difficult to control and keep uniform across different centers. Moreover, in multicenter studies and large prospective trials it is often inevitable to freeze and store the plasma samples before performing the assay.

Because samples must be stored prior to analysis, several teams have studied the impact of a freeze–thaw cycle on MV measurements. However, freezing and thawing protocols reported in the literature are variable, as illustrated in Table 5.1. Studies where platelets were insufficiently depleted of course resulted in obvious differences in the MV counts after thawing.[63] In proper conditions, however, although repeated freeze–thaw cycles are known to significantly alter the number of PMVs[54,64] no major impact of deep-freezing was observed, as shown in studies on PMVs[28,64] and EryMVs.[53] One study showed a decrease in MVs after long-term storage.[65] Results about the other MV subpopulations were often inconclusive because of their smaller quantity within healthy subjects. The literature remains contradictory on the influence of freezing on endothelial-derived MVs (EMVs), ranging from no variation in EMV counts after freezing[66] to important changes.[29] Analysis of the stability of MV determination up to 12 months after storage at –80°C also demonstrated that MV-dependent procoagulant activity remained constant. Some teams advocate snap-freezing of PFP in liquid nitrogen before storage at –80°C to reduce ice crystal formation.[59] On the basis of MV counts by FCM, thrombin generation, and phospholipid-dependent coagulation time, no significant difference comparing the previous snap-freezing step with direct storage at –80°C has been reported.[28] However, more subtle structural changes or modifications in the intracytoplasmic MV content cannot be excluded.

Different methods regarding thawing have been described for thawing MV samples, including thawing slowly on ice.[67] Thawing for several minutes at 37°C may prevent formation of ice crystals and cryoprecipitation.[67–69] One study reported that thawing on ice resulted in significant artefactual generation of procoagulant activity.[28] Therefore, when frozen-thawed plasma is used for MV measurements, especially PMVs and annexin V–positive MVs, and measured, MV-dependent procoagulant activity should be interpreted with caution.

To conclude, analysis of the relative impact of preanalytical parameters on circulating MVs has been performed in a study that demonstrates that the most critical steps are (i) the delay before the first centrifugation, (ii) agitation of the tubes during transportation, and (iii) the protocol of centrifugation.

5.7 Recommendations

Ideally, MV measurement should be performed in fresh plasma, prepared directly after blood venepuncture. Nevertheless, the logistics of blood sampling are difficult to control, explaining why frozen samples are most popular. Therefore, platelet contamination remains the major preanalytical issue in processing plasma for MV determination. During blood collection, it is necessary to prevent *ex vivo* activation of blood, leading to inadvertent production of PMVs. Therefore, the choice of anticoagulant, the duration and condition of blood transportation, and the centrifugation speeds used for plasma preparation are critical. Accordingly, to progress toward the standardization of MV determination, it is important to set up guidelines for preparing plasma samples for MV determination.

Collectively, results from the literature and our own experience led us to propose the following protocol for MV analysis (Table 5.2), including (a) the use of plastic citrated tubes, (b) a blood collection device with a large needle size (21 G), (c) discarding of the first milliliters of blood, (d) a time delay of maximum 2 h before the first centrifugation, (e) transportation in appropriate boxes that maintain tubes vertically, (e) a centrifugation protocol consisting of two successive centrifugations of 2,500 g 15 min at RT. And (f) storage of PFP at –80°C with rapid thawing in a water bath at 37°C.

This protocol is currently being evaluated in a multicenter study between several teams belonging to the SSC-VB working group within the ISTH.

5.8 Standardization of Microvesicle Preanalytical Variables

Preliminary results of this multicenter study aimed to standardize MV preanalytical steps have been presented in the minutes of the SSC-VB working group that took place in Kyoto in July 2011. The objective of the study was to compare the interlaboratory variability of MV measurement using a nonstandardized protocol versus a standardized routine laboratory-adapted preanalytical protocol consisting of two successive centrifugations of 2,500 g 15 min at RT, as summarized in Table 5.2. Fourteen laboratories participated in this multicenter study. Determinations were made on 15 healthy male volunteers aged 20 to 40 years. In this first workshop report, only PMVs and related functional properties (most sensitive to preanalytical variables) were analyzed. Preliminary results

Table 5.2 Proposed preanalytical protocol for the analysis of circulating MPs

Preanalytical parameters	Recommendations
Nature of tube	Citrated tubes
Blood collection	Large needle size (21G) and discarding of the first milliliters of blood
Time delay before first centrifugation	< or = 2 hr
Tubes transportation	Appropriate boxes that maintain tubes vertically
Centrifugation Protocol	Two successive centrifugations: 2,500 g 15 min at RT
Storage/transportation of PFP	−80°C or in dry ice
Thawing	Fast in water bath at 37°C

Abbreviation: MP, microparticle.

showed that (1) a standardized protocol results in a reduction of the interlaboratory variability for PMVs (more markedly for FCM analysis) and (2) centrifugation conditions are the major preanalytical parameter that impact MV analysis. Recommendations to achieve this step in a standardized way are illustrated in Fig. 5.1. The impact of this standardized preanalytical protocol on EryMV, leukocyte-derived MV, and EMV subpopulations is also being currently analyzed. The next steps will be to extend this multicenter study to larger series of patients with specific disease states.

In conclusion, there is a great need for internationally validated guidelines for the preparation and storage of plasma for MV measurement. This step is a prerequisite to delineate the usefulness of MVs as a new clinically relevant biomarker.

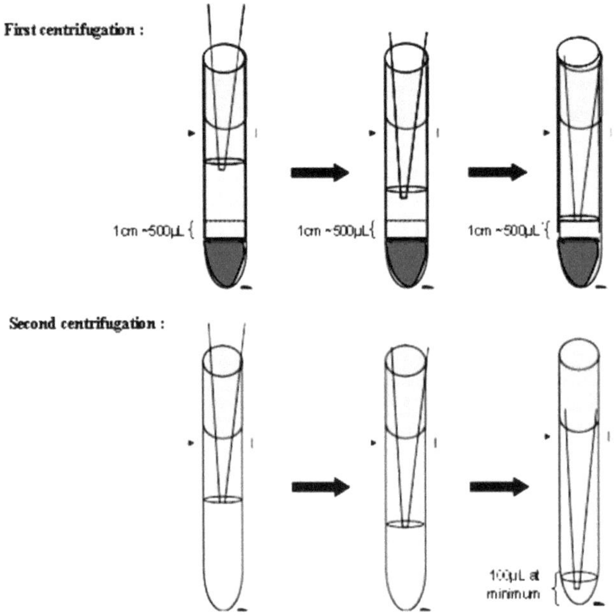

Figure 5.1 Instructions for plasma sampling after centrifugation steps for MP analysis. After the first centrifugation at 2,500 g 15 min at RT, the PPP is gently collected using a 1,000 µL pipette with tips until 1 cm above the buffy coat layer. Plasma is transferred in a plastic tube. After the second centrifugation at 2,500 g 15 min at RT, the PFP is gently collected using a 200 µL pipette. PFP is homogenized before being aliquoted and frozen at –80°C (snap-freezing in nitrogen may be beneficial).

References

1. Freyssinet JM, Dignat-George F (2005) More on: measuring circulating cell-derived microparticles, *J. Thromb. Haemost.*, **3**(3), 613–614.
2. Burnier L, Fontana P, Kwak BR, Angelillo-Scherrer A (2009) Cell-derived microparticles in haemostasis and vascular medicine, *Thromb. Haemost.*, **101**(3), 439–451.
3. Jy W, Horstman LL, Jimenez JJ, Ahn YS (2004) Measuring circulating cell-derived microparticles, *J. Thromb. Haemost.*, **2**(10), 1842–1851.
4. Navasiolava NM, Dignat-George F, Sabatier F, Larina IM, Demiot C, Fortrat JO, Gauquelin-Koch G, Kozlovskaya IB, Custaud MA (2010)

Enforced physical inactivity increases endothelial microparticle levels in healthy volunteers, *Am. J. Physiol. Heart Circ. Physiol.*, **299**(2), H248–H256.

5. Thery C, Ostrowski M, Segura E (2009) Membrane vesicles as conveyors of immune responses, *Nat. Rev. Immunol.*, **9**(8), 581–593.
6. Toth B, Lok CA, Boing A, Diamant M, Van Der Post JA, Friese K, Nieuwland R (2007) Microparticles and exosomes: impact on normal and complicated pregnancy, *Am. J. Reprod. Immunol.*, **58**(5), 389–402.
7. Bergsmedh A, Szeles A, Henriksson M, Bratt A, Folkman MJ, Spetz AL, Holmgren L (2001) Horizontal transfer of oncogenes by uptake of apoptotic bodies, *Proc. Natl. Acad. Sci. U. S. A.*, **98**(11), 6407–6411.
8. Al-Nedawi K, Meehan B, Rak J (2009) Microvesicles: messengers and mediators of tumor progression, *Cell Cycle*, **8**(13), 2014–2018.
9. Ardoin SP, Shanahan JC, Pisetsky DS (2007) The role of microparticles in inflammation and thrombosis, *Scand. J. Immunol.*, **66**(2–3), 159–165.
10. Davizon P, Lopez J. A (2009) Microparticles and thrombotic disease, *Curr. Opin. Hematol.*, **16**(5), 334–341.
11. Morel O, Toti F, Hugel B, Bakouboula B, Camoin-Jau L, Dignat-George F, Freyssinet JM (2006) Procoagulant microparticles: disrupting the vascular homeostasis equation? *Arterioscler. Thromb. Vasc. Biol.*, **26**(12), 2594–2604.
12. Mause SF, Weber C (2010) Microparticles: protagonists of a novel communication network for intercellular information exchange, *Vasc. Biol.*, **107**(9), 1047–1057.
13. Diamant M, Tushuizen ME, Sturk A, Nieuwland R (2004) Cellular microparticles: new players in the field of vascular disease? *Eur. J. Clin. Invest.*, **34**(6), 392–401.
14. Piccin A, Murphy WG, Smith OP (2007) Circulating microparticles: pathophysiology and clinical implications, *Blood Rev.*, **21**(3), 157–171.
15. Sabatier F, Camoin-Jau L, Anfosso F, Sampol J, Dignat-George F (2009) Circulating endothelial cells, microparticles and progenitors: key players towards the definition of vascular competence, *J. Cell Mol. Med.*, **13**(3), 454–471.
16. Boulanger CM, Amabile N, Tedgui A (2006) Circulating microparticles: a potential prognostic marker for atherosclerotic vascular disease, *Hypertension*, **48**(2), 180–186.
17. Zwicker JI, Liebman HA, Neuberg D, Lacroix R, Bauer KA, Furie BC, Furie B (2009) Tumor-derived tissue factor-bearing microparticles

are associated with venous thromboembolic events in malignancy, *Clin. Cancer Res.*, **15**(22), 6830–6840.

18. Enjeti AK, Lincz LF, Seldon M (2007) Detection and measurement of microparticles: an evolving research tool for vascular biology, *Semin. Thromb. Haemost.*, **33**(8), 771–779.
19. Shet A. S (2008) Characterizing blood microparticles: technical aspects and challenges, *Vasc. Health Risk Manag.*, **4**(4), 769–774.
20. Yuana Y, Bertina RM, Osanto S (2010) Pre-analytical and analytical issues in the analysis of blood microparticles, *Thromb. Haemost.*, **105**(3), 396–408.
21. Zwicker JI, Lacroix R, Dignat-George F, Furie BC, Furie B (2012) Measurement of platelet microparticles, *Methods Mol. Biol.*, **788**, 127–139.
22. Polack B, Schved JF, Boneu B (2001) Preanalytical recommendations of the "Groupe d'Etude sur l'Hemostase et la Thrombose" (GEHT) for venous blood testing in hemostasis laboratories, *Haemostasis*, **31**(1), 61–68.
23. Lippi G, Franchini M, Montagnana M, Salvagno GL, Poli G, Guidi GC (2006) Quality and reliability of routine coagulation testing: can we trust that sample? *Blood Coagul. Fibrinolysis*, **17**(7), 513–519.
24. Van Beers EJ, Schaap MC, Berckmans RJ, Nieuwland R, Sturk A, Van Doormaal FF, Meijers JC, Biemond BJ (2009) Circulating erythrocyte-derived microparticles are associated with coagulation activation in sickle cell disease, *Haematologica*, **94**(11), 1513–1519.
25. Lippi G, Salvagno GL, Montagnana M, Franchini M, Guidi GC (2006) Venous stasis and routine hematologic testing, *Clin. Lab. Haematol.*, **28**(5), 332–337.
26. Hefler L, Grimm C, Leodolter S, Tempfer C (2004) To butterfly or to needle: the pilot phase, *Ann. Intern. Med.*, **140**(11), 935–936.
27. Lippi G, Salvagno GL, Montagnana M, Poli G, Guidi GC (2006) Influence of the needle bore size on platelet count and routine coagulation testing, *Blood Coagul. Fibrinolysis*, **17**(7), 557–561.
28. Lacroix R, Judicone C, Poncelet P, Robert S, Arnaud L, Sampol J, Dignat-George F (2011) Impact of pre-analytical parameters on the measurement of circulating microparticles: towards standardization of protocol, *J. Thromb. Haemost*, **10**(3), 437–46.
29. Van Ierssel SH, Van Craenenbroeck EM, Conraads VM, Van Tendeloo VF, Vrints CJ, Jorens PG, Hoymans VY (2010) Flow cytometric detection of endothelial microparticles (EMP): effects of centrifugation and storage alter with the phenotype studied, *Thromb. Res.*, **125**(4), 332–339.

30. Lippi G, Salvagno GL, Brocco G, Guidi GC (2005) Preanalytical variability in laboratory testing: influence of the blood drawing technique, *Clin. Chem. Lab. Med.*, **43**(3), 319–325.

31. Ahnadi CE, Sabrinah Chapman E, Lepine M, Okrongly D, Pujol-Moix N, Hernandez A, Boughrassa F, Grant AM (2003) Assessment of platelet activation in several different anticoagulants by the Advia 120 Hematology System, fluorescence flow cytometry, and electron microscopy, *Thromb. Haemost.*, **90**(5), 940–948.

32. Macey M, Azam U, Mccarthy D, Webb L, Chapman ES, Okrongly D, Zelmanovic D, Newland A (2002) Evaluation of the anticoagulants EDTA and citrate, theophylline, adenosine, and dipyridamole (CTAD) for assessing platelet activation on the ADVIA 120 hematology system, *Clin. Chem.*, **48**(6 Pt 1), 891–899.

33. Neufeld M, Nowak-Gottl U, Junker R (1999) Citrate-theophylline-adenine-dipyridamol buffer is preferable to citrate buffer as an anticoagulant for flow cytometric measurement of platelet activation, *Clin. Chem.*, **45**(11), 2030–2033.

34. Lippi G, Guidi GC, Mattiuzzi C, Plebani M (2006) Preanalytical variability: the dark side of the moon in laboratory testing, *Clin. Chem. Lab. Med.*, **44**(4), 358–365.

35. Gritters M, Grooteman MP, Schoorl M, Bartels PC, Scheffer PG, Teerlink T, Schalkwijk CG, Spreeuwenberg M, Nube MJ (2006) Citrate anticoagulation abolishes degranulation of polymorphonuclear cells and platelets and reduces oxidative stress during haemodialysis, *Nephrol. Dial. Transplant.*, **21**(1), 153–159.

36. Morel O, Toti F, Jesel L, Freyssinet JM (2010) Mechanisms of microparticle generation: on the trail of the mitochondrion! *Semin. Thromb. Haemost.*, **36**(8), 833–844.

37. Piersma SR, Broxterman HJ, Kapci M, De Haas RR, Hoekman K, Verheul HM, Jimenez CR (2009) Proteomics of the TRAP-induced platelet releasate, *J. Proteomics*, **72**(1), 91–109.

38. Smalley DM, Root KE, Cho H, Ross MM, Ley K (2007) Proteomic discovery of 21 proteins expressed in human plasma-derived but not platelet-derived microparticles, *Thromb. Haemost.*, **97**(1), 67–80.

39. Garcia BA, Smalley DM, Cho H, Shabanowitz J, Ley K, Hunt DF (2005) The platelet microparticle proteome, *J. Proteome Res.*, **4**(5), 1516–1521.

40. Ueba T, Haze T, Sugiyama M, Higuchi M, Asayama H, Karitani Y, Nishikawa T, Yamashita K, Nagami S, Nakayama T, Kanatani K, Nomura S (2008) Level, distribution and correlates of platelet-derived

microparticles in healthy individuals with special reference to the metabolic syndrome, *Thromb. Haemost.*, **100**(2), 280–285.

41. Nomura S, Shouzu A, Taomoto K, Togane Y, Goto S, Ozaki Y, Uchiyama S, Ikeda Y (2009) Assessment of an ELISA kit for platelet-derived microparticles by joint research at many institutes in Japan, *J. Atheroscler. Thromb.*, **16**(6), 878–887.
42. Hron G, Kollars M, Weber H, Sagaster V, Quehenberger P, Eichinger S, Kyrle PA, Weltermann A (2007) Tissue factor-positive microparticles: cellular origin and association with coagulation activation in patients with colorectal cancer, *Thromb. Haemost.*, **97**(1), 119–123.
43. Aras O, Shet A, Bach RR, Hysjulien JL, Slungaard A, Hebbel RP, Escolar G, Jilma B, Key NS (2004) Induction of microparticle- and cell-associated intravascular tissue factor in human endotoxemia, *Blood*, **103**(12), 4545–4553.
44. Breimo ES, Osterud B (2005) Generation of tissue factor-rich microparticles in an *ex vivo* whole blood model, *Blood Coagul. Fibrinolysis*, **16**(6), 399–405.
45. Philippe J, De Logi E, Baele G (2004) Comparison of five different citrated tubes and their *in vitro* effects on platelet activation, *Clin. Chem.*, **50**(3), 656–658.
46. Shah MD, Bergeron AL, Dong JF, Lopez JA (2008) Flow cytometric measurement of microparticles: pitfalls and protocol modifications, *Platelets*, **19**(5), 365–372.
47. Lippi G, Salvagno GL, Adcock DM, Gelati M, Guidi GC, Favaloro EJ (2010) Right or wrong sample received for coagulation testing? Tentative algorithms for detection of an incorrect type of sample, *Int. J. Lab. Hematol.*, **32**(1 Pt 2), 132–138.
48. Jayachandran M, Miller VM, Heit JA, Owen WG (2012) Methodology for isolation, identification and characterization of microvesicles in peripheral blood, *J. Immunol. Methods*, **375**(1–2), 207–214.
49. Contant G, Gouault-Heilmann M, Martinoli JL (1983) Heparin inactivation during blood storage: its prevention by blood collection in citric acid, theophylline, adenosine, dipyridamole-C.T.A.D. mixture, *Thromb. Res.*, **31**(2), 365–374.
50. Mody M, Lazarus AH, Semple JW, Freedman J (1999) Preanalytical requirements for flow cytometric evaluation of platelet activation: choice of anticoagulant, *Transfus. Med.*, **9**(2), 147–154.
51. Kim HK, Song KS, Lee ES, Lee YJ, Park YS, Lee KR, Lee SN (2002) Optimized flow cytometric assay for the measurement of platelet

microparticles in plasma: pre-analytic and analytic considerations, *Blood Coagul. Fibrinolysis*, **13**(5), 393–397.

52. Pearson L, Thom J, Adams M, Oostryck R, Krueger R, Yong G, Baker R (2009) A rapid flow cytometric technique for the detection of platelet-monocyte complexes, activated platelets and platelet-derived microparticles, *Int. J. Lab. Hematol.*, **31**(4), 430–439.
53. Rubin O, Crettaz D, Canellini G, Tissot JD, Lion N (2008) Microparticles in stored red blood cells: an approach using flow cytometry and proteomic tools, *Vox Sang.*, **95**(4), 288–297.
54. Keuren JF, Magdeleyns EJ, Govers-Riemslag JW, Lindhout T, Curvers J (2006) Effects of storage-induced platelet microparticles on the initiation and propagation phase of blood coagulation, *BJH*, **134**(3), 307–313.
55. Wayne PA (2008) *Collection, Transport, and Processing of Blood Specimens for Testing Plasma-Based Coagulation Assays and Molecular Haemostasis Assays; Approved Guideline*, 5th ed., CLSI document H211-A5, Clinical and Laboratory Standards Institute.
56. Horstman LL, Jy W, Jimenez JJ, Bidot C, Ahn YS (2004) New horizons in the analysis of circulating cell-derived microparticles, *Keio J. Med.*, **53**(4), 210–230.
57. Dignat-George F, Sabatier F, Camoin-Jau L, Sampol J (2004) Measuring circulating cell-derived microparticles, *J. Thromb. Haemost.*, **2**(10), 1842–1851.
58. Robert S, Poncelet P, Lacroix R, Arnaud L, Giraudo L, Hauchard A, Sampol J, Dignat-George F (2009) Standardization of platelet-derived microparticle counting using calibrated beads and a Cytomics FC500 routine flow cytometer: a first step towards multicenter studies? *J. Thromb. Haemost.*, **7**(1), 190–197.
59. Biro E, Nieuwland R, Sturk A (2004) Measuring circulating cell-derived microparticles, *J. Thromb. Haemost.*, **2**(10), 1842–1851.
60. Hugel B, Zobairi F, Freyssinet JM (2004) Measuring circulating cell-derived microparticles, *J. Thromb. Haemost.*, **2**(10), 1842–1851.
61. Nantakomol D, Imwong M, Soontarawirat I, Kotjanya D, Khakhai C, Ohashi J, Nuchnoi P (2009) The absolute counting of red cell-derived microparticles with red cell bead by flow rate based assay, *Cytometry B*, **76**(3), 191–198.
62. Stagnara J, Garnache Ottou F, Angelot F, Mourey G, Seilles E, Biichle S, Saas P, Racadot E (2012) Correlation between platelet-derived microparticle enumeration by flow cytometry and phospholipid-

dependent procoagulant activity in microparticles: the centrifugation step matters! *Thromb. Haemost.*, **107**(6), 1185–1187.

63. Ayers L, Kohler M, Harrison P, Sargent I, Dragovic R, Schaap M, Nieuwland R, Brooks SA, Ferry B (2011) Measurement of circulating cell-derived microparticles by flow cytometry: sources of variability within the assay, *Thromb. Res.*, **127**(4), 370–377.

64. Shet AS, Aras O, Gupta K, Hass MJ, Rausch DJ, Saba N, Koopmeiners L, Key NS, Hebbel RP (2003) Sickle blood contains tissue factor-positive microparticles derived from endothelial cells and monocytes, *Blood*, **102**(7), 2678–2683.

65. Dey-Hazra E, Hertel B, Kirsch T, Woywodt A, Lovric S, Haller H, Haubitz M, Erdbruegger U (2010) Detection of circulating microparticles by flow cytometry: influence of centrifugation, filtration of buffer, and freezing, *Vasc. Health Risk Manag.*, **6**, 1125–1133.

66. Simak J, Holada K, Risitano AM, Zivny JH, Young NS, Vostal JG (2004) Elevated circulating endothelial membrane microparticles in paroxysmal nocturnal haemoglobinuria, *BJH*, **125**(6), 804–813.

67. Biro E, Sturk-Maquelin KN, Vogel GM, Meuleman DG, Smit MJ, Hack CE, Sturk A, Nieuwland R (2003) Human cell-derived microparticles promote thrombus formation *in vivo* in a tissue factor-dependent manner, *J. Thromb. Haemost.*, **1**(12), 2561–2568.

68. Simak J, Gelderman MP (2006) Cell membrane microparticles in blood and blood products: potentially pathogenic agents and diagnostic markers, *Transfus. Med. Rev.*, **20**(1), 1–26.

69. Trummer A, De Rop C, Tiede A, Ganser A, Eisert R (2009) Recovery and composition of microparticles after snap-freezing depends on thawing temperature, *Blood Coagul. Fibrinolysis*, **20**(1), 52–56.

Chapter 6

Procoagulant Assays

Barry J. Woodhams
HaemaCon Ltd., Bromley, Kent.
barry@haemacon.co.uk

6.1 Introduction

Extracellular vesicles (EVs) comprising both nanovesicles (or exosomes) and microvesicles (MVs) (or microparticles) are found in plasma and other body fluids. The much smaller nanovesicles (exosomes) are formed internally inside the cell and are secreted by all cells and were first described in hematopoietic cells.[1] The larger MVs[2] are usually derived from blebbing of the outer membrane of the cell.[3] Although they are both derived from a number of different cellular sources, the key difference between these EVs are their source and the mechanism of formation. These vesicles also have an important number of physiological roles that are still not fully defined, and this has now become an area of active research. A procoagulant function has been associated with these EVs, particularly the larger MVs. The objective of this chapter is to describe the different procoagulant assays that are available to measure this activity.

Extracellular Vesicles in Health and Disease
Edited by Paul Harrison, Christopher Gardiner, and Ian L. Sargent

ISBN 978-981-4411-98-1 (Hardcover), 978-981-4411-99-8 (eBook)
www.panstanford.com

6.1.1 Nanovesicles (Exosomes)

Nanovesicles (50–90 nm)[1] are created intracellularly and after fusion with the cell membrane are released extracellularly. Their functions and properties are not well defined, although they do occur naturally in most body fluids, including plasma, ascites, urine, breast milk, and saliva.[3] They are released by a wide variety of cell types, including B-cells, macrophages, mast cells, T-cells, epithelial cells, platelets, reticulocytes, tumor cells, and placenta trophoblasts.

Nanovesicles are thought to have a biological function in cell–cell communication by transferring cellular material between different cells/cell types. They also act as a means of removing waste materials from the cell, and they have been shown to play an important role in stimulation of the immune system.[4]

Their molecular composition is dependent on their cellular origin. They differ from MVs in that they are rich in cholesterol and sphingomyelin.[5] They can be harvested by ultracentrifugation (at 110,000 × g). They occur naturally in plasma, ascites, urine, breast milk, and saliva.[3] They cannot be easily monitored without some prepurification. They are too small to be measured directly by present-day flow cytometric techniques.

6.1.2 Microvesicles

MVs are small membrane-coated EVs derived from the outer membrane of their parent cells. They are heterogeneous in size (100–1,000 nm) and formed after cell activation or cell apoptosis.[2] They are of diverse cellular origin and lack a nucleus or any synthetic capability but have a membrane skeleton and may have varying amounts of surface-exposed phosphatidylserine (PS). They are released from their parent cells by a budding process in response to cell activation or apoptosis. For example, activation of platelets with collagen, thrombin, or a mix of both results in the formation of platelet-derived MVs. Similarly, stimulation of leukocytes or endothelial cells with lipopolysaccharide (LPS) or cytokines such as tumor necrosis factor-α (TNFα) or interleukin (IL)-1 also causes the release of MVs from these cells. In blood, MVs are most commonly derived from platelets, erythrocytes, neutrophils, endothelial cells, or cancer cells. In normal plasma the majority (70–90%) are platelet derived.[2] The integrin αIIbβ3 plays a central role in the formation of platelet-derived MVs,[6] hence in Glanzmann thrombasthenia platelets are unable to vesiculate in response to thrombin, adenosine diphosphate (ADP), or collagen.[6]

Their composition and function depend upon their cellular origin. EVs are generated by all cell types, even bacteria, and released into most body fluids, including blood, tears, saliva, and urine. EVs carry markers on their external membrane surface derived from their parent cell type, and these can be used to capture, measure, and identify them—PS, tissue factor (TF), or antigens specific to the parent cell type, for example, endothelial cells (CD31 or CD146), platelets (αIIbβ3, GpIb, and P-selectin), and leukocytes (CD4, CD3, or CD8). In a number of different disease states their levels are elevated and the proportion derived from the different cell types altered.

EVs isolated from venous blood of healthy individuals (originating predominantly from platelets but also from erythrocytes, leukocytes, and endothelial cells) cause a low but significant *in vitro* activation of coagulation that is independent of TF/factor VII (FVII).[7] EVs from the pericardial blood of patients undergoing cardiac surgery (originating mainly from erythrocytes but also from other cell types such as platelets and leukocytes) and those isolated from synovial fluid of arthritic patients (derived mainly from leukocytes and only to a small extent from erythrocytes and platelets), for example, initiate coagulation in a TF-/FVII-dependent manner.[8]

Flow cytometric techniques are the most commonly used methods for measurement of EVs. Most of the published literature is based on this technique. Flow cytometry has one major flaw in that it is not sensitive enough to detect smaller particles. It also gives no information on EV function or activity. The limit of detection of most flow cytometers is around 500 nm, so smaller EVs are often missed, although more modern instruments are beginning to push down this limit of detection. Modern techniques in development are becoming available that theoretically allow measurement of some of the smaller particles, for example, atomic force microscopy (see Chapter 9), the Izon nanopore system (see Chapter 4), dynamic light scattering (see Chapter 10), and nanoparticle tracking analysis (see Chapter 11). Functional EV assays, measuring specific functions of MVs and exosomes, can measure the contribution of all the different particles (both nano and micro) present in the sample.

6.1.3 History

Although not identified as such at the time, cell-derived EVs were observed experimentally as early as 1946. Chargaff and West[9] found that the clotting time of plasma, obtained by centrifugation at 1,900 × g, is prolonged after further centrifugation at 31,000 × g and that the "clotting factor" removed by the high-speed centrifugation is found in the sedimented pellet. In 1967, Wolf[10] published a study providing evidence for the occurrence, in normal plasma, of a coagulant material that was sedimented by high-speed centrifugation that had originated from platelets but was distinguishable from intact platelets. He first described these MVs as "platelet dust" as they were not only rich in phospholipid but demonstrated platelet factor 3 (PF3)-like procoagulant activity.[11]

During the last 20–30 years, extensive research has been performed on cell-derived EVs within the blood. We now know that they can be released not only from platelets but also from erythrocytes, leukocytes, endothelial cells, and other diverse cell types.

Historically, most assays have been based on flow cytometry (see Chapter 8), and these are still regarded as the gold standard methods. More recently immunological capture assays or procoagulant assays have been developed. Present-day assays for procoagulant activity are based either on a clotting/chromogenic assay principle[12] or on a thrombin generation principle.[13] Measurement makes use of specific properties of the different types of vesicles. Their monitoring is rapidly gaining importance in disease detection, disease progression monitoring, and treatment. All these assay types have advantages and disadvantages (Table 6.1). The user must choose the best assay to suit his or her requirements.

6.2 Procoagulant Assays

All of the procoagulant activity assays are nonspecific and give no information on the cellular origin, type, or physical properties of EVs. If the type and source of a procoagulant activity is required alternative techniques such as flow cytometry or electron microscopy must be used. As procoagulant assays are unable to discriminate the type or source of a procoagulant activity in the sample being tested, all EVs, including nanovesicles and MVs, will be measured and cannot

Table 6.1 Advantages and disadvantages of different measurement techniques for EVs

	Flow cytometry	Capture assay	Global assay
Pros	Can determine: – Cell origin and type – Count	- Captures on ELISA plates - Functions with specific antibodies or ligands - (Cell origin?)	- Is a functional assay - Uses a procoagulant - Is based on thrombin generation - Can be automated - Is close to physiology - Involves no sample dilution or washing - Assays whole blood
Cons	- Needs instrument - Needs skilled staff - Is not automated - Is not standardized - Can miss small vesicles	- Is manual - Can only measure captured vesicles - Is far from physiology - Uses sample dilution, washing, etc.	Gives no information on cell origin or type

be discriminated from each other in these assays. As the majority are thought to be MVs this term will be used in the rest of this chapter as the generic name for all vesicle types in a sample.

MVs can expose either PS alone, thereby providing a negatively charged phospholipid surface for coagulation, or in combination with TF, which can then activate the coagulation cascade via FVIIa. For any significant physiological thrombin generation to occur a procoagulant phospholipid surface is obligatory either alone or in combination with TF. This provides the basis for their procoagulant properties. It is the procoagulant activity of either or both of these factors that is measured in procoagulant assays.[2]

All cell types contain anionic phospholipids on the inner surface of their membranes, PS being the most important for supporting coagulation. These translocate to the outer membrane surface when cells are activated or undergo apoptosis. During the formation of MVs there is loss of asymmetry in the plasma membrane, and with the anionic phospholipids being transferred to the outer membrane, the presence of PS significantly increases this procoagulant activity of the surface.[2] In addition to these anionic phospholipids, TF may also be associated with MVs. Geisen et al.[14] showed that blood contained very low levels of active TF (blood-borne TF) mainly associated with neutrophils and monocytes, but they also found that some of this activity was found in the plasma fraction. Circulating TF is very thrombogenic, and its presence in circulating blood remains controversial. Some workers consider that there are no significant levels of active circulating TF in the blood of healthy individuals.[15,16] Other workers have found low levels circulating in blood, but these levels are close to the detection limit of the assays used.[17]

The procoagulant activity of MVs can be monitored either by their procoagulant phospholipid activity or by measuring the TF activity associated with the procoagulant phospholipids on the MV surface (blood-borne TF). TF is generally expressed by cell types not exposed to blood and is strongly expressed in cells around tissues and blood vessels,[16] such as smooth muscle cells and fibroblasts. Their TF is generally only exposed to the blood when injury occurs. Monocytes and endothelial cells can express TF[16] but only when they are exposed to inflammatory molecules such as TNF-α or LPS. The other main source of TF is cancer cells.[18] Platelets have also been shown to potentially express TF, especially after activation, but this remains an area of controversy.[19,20] All of these cell types can release MVs into the circulation, giving rise to blood-borne TF.[21]

Functional TF assays must be differentiated from immunological assays for TF. These types of assays (unless they have an MV capture step) measure all types of TF, including the inactive (truncated forms) and alternative spliced forms of TF (asTF), which can also occur naturally in plasma.

6.2.1 Procoagulant Phospholipid Activity

The measurement of procoagulant activity on MVs is certainly not new. Over 60 years ago[9] it was shown that the clotting time of recalcified plasma increased after ultracentrifugation. In 1955 O'Brien found that there was a platelet-like activity in serum. It was this material that Wolf[10] described as platelet dust in 1967.

The first true assays that were in effect measuring this procoagulant activity were the early assays for PF3. The name "PF3" was first coined by van Creveld and Paulssen in 1953[11] and was defined as the phospholipid containing the coagulation factor derived from platelets that acts with plasma factors to convert prothrombin to thrombin. A variety of tests were developed in the laboratory to measure this PF3 activity—the prothrombin consumptive index, the PF3 availability assay, thrombin-based chromogenic assays[22] based on the kaolin clotting time test,[23] and Russell viper–based assay methods.[24]

Marie-Claire Boffa[25] described the first specific assay to measure this procoagulant activity in plasma in 1980. In this assay she measured the phospholipid-related procoagulant activity in a simple clotting assay and could show that this type of assay may have some clinical relevance. This assay used phospholipase A2 from the venom of *Vipera berus* to degrade phospholipids. In the presence of the venom the recalcification time increased and was inversely related to the amount of procoagulant phospholipids in the sample. A calibration curve was obtained using brain phospholipids and the concentration of phospholipids in the sample read from this curve. She found high levels of procoagulant activity in patients with cardiovascular disease and almost exclusively only in those patients with a recent or recurrent thromboembolism. In 1994 Wagenvoord et al.[26] developed a chromogenic assay to measure phospholipid-related procoagulant activity and showed that patients who had a thrombotic disorder had significantly higher levels than controls. This assay used whole blood or platelet-rich plasma (PRP) and measured both MV and platelet procoagulant activity. Neither of these assays was exploited. It was only after 2000 when interest really started focusing on the potential role of MVs that more functional assays, based on either capture or global assay principles, gained wider use.

6.2.1.1 Procoagulant phospholipid activity assay

The simple clotting assay originally described by Exner in 2003[12] has been developed into a commercial product. This is a simple clotting assay to measure procoagulant phospholipids in plasma or whole blood. The assay is extremely simple and easily automated (see Fig. 6.1): 0.025 mL of patient plasma is mixed with 0.025 mL of phospholipid-depleted normal plasma, incubated at 37°C for

2 min, and then coagulation initiated by the addition of 0.10 mL of a mixture containing 0.001 IU bovine factor Xa in 0.025 M calcium chloride and then timed to a clot endpoint. The shorter the clotting time obtained, the higher the procoagulant phospholipid content of the sample.

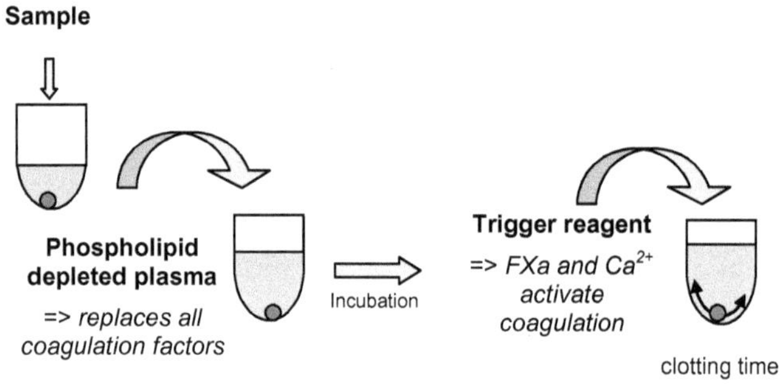

Figure 6.1 The experimental procedure for the STA® Procoag PPL assay.

The assay is now commercially available from two different manufacturers, STA® Procoag PPL from Diagnostica Stago (Asnieres, France) and XaCT from Haematex (Hornsby, Australia). These two assays mainly differ in their source of deficient plasma—the PPL assay uses human-source plasma, while XaCT uses porcine-source plasma.

The assay is extremely reproducible and readily automatable on coagulation analysers.[27] With normal plasma samples the Procoag PPL assay has a clotting time of 61–83 s with an intra-assay variation of <1.0% and an interassay variation of <2.5%.

The source of the phospholipid-depleted plasma can be either animal or human. The use of animal-source-deficient plasma slightly reduces the interference of antiphospholipid antibodies in the assay. Because of this interference this assay has to be used with caution in patients with antiphospholipid syndrome.[12,28] The assay can also be performed on whole blood.[12,29]

Like all the other MV assays, the Achilles heel is the preanalytical variables, especially the collection and processing of patient blood.[30] The Vascular Biology Scientific Subcommittee (SSC) of the International Society on Thrombosis and Haemostasis (ISTH) has a working party actively working on the standardization of the preanalytical steps.[31]

It is essential that blood be carefully collected and processed before analysis if meaningful and reproducible results are to be obtained. It is recommended that samples be centrifuged as soon as possible after collection, preferably within 1 h. Double centrifugation (2 × 15 min at 2,500 g) is recommended. Double centrifugation is essential if samples are to be stored frozen (see Fig. 6.2 below). Without the double centrifugation procedure any remaining cellular material will lyse on freeze/thawing, releasing MVs/phospholipids into the plasma, which by definition interfere with the assay. Figure 6.2 clearly shows the effect of freeze/thawing on the assay results if care is not taken to ensure platelet removal by a second centrifuge cycle. The effect of residual platelets on the PPL assay can also be clearly seen in the figure if only one centrifuge cycle is used to prepare the sample. A single centrifugation and freeze/thawing causes a shortening of clotting time in the PPL assay of at least 10 s.

Plasma samples prepared by the double centrifugation procedure are stable for at least 1 month at –20°C.

The recommended anticoagulant used for blood collection is 109 mmol/L (3.2%) sodium citrate anticoagulant (9 volumes of blood in 1 volume of citrate anticoagulant), although 129 mmol/L sodium citrate or disodium ethylenediaminetetraacetatic acid (EDTA) can also be used.

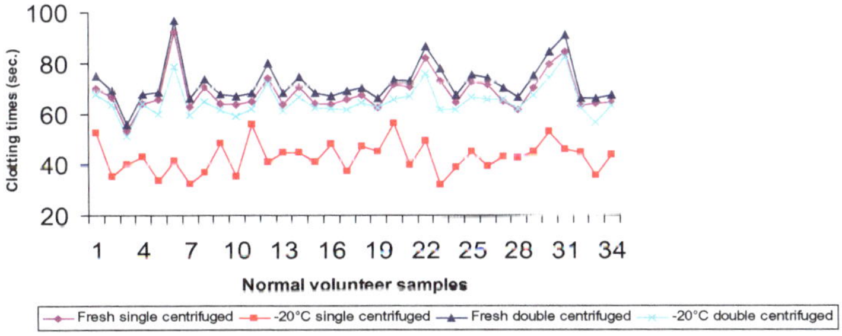

Figure 6.2 Effect on plasma, obtained from normal volunteers, of single or double centrifugation and freeze/thawing on procoagulant phospholipid activity. All assays performed using the STA® Procoag PPL assay on the STA-R instrument. Data generated internally in R&D, Diagnostica Stago, Gennevilliers, France.

Any hemolysis of the sample will affect the results as red blood cells (RBCs) produce very active MVs. This can be clearly seen in patients in sickle cell crisis where much of the increase in procoagulant activity is due to RBC-derived MVs.[32] The presence of RBC-derived MVs also strongly correlates with plasma markers for hemolysis.[33]

This assay has shown some interesting clinical results and has the potential to be used as a routine tool in the clinical laboratory. A significant inverse relation was shown by Connor et al.[34] between the annexin V–binding count and this procoagulant phospholipid assay. As with the annexin V flow cytometric–based assay, it is responsive to the same charged surfaces, so it has the potential to be used as a first-line screening test to detect abnormalities that can then be further investigated by other techniques, if required.

Another commercial assay based on the same basic principle is the Zymuphen MP activity assay (Hyphen Biomed, Neuville-sur Oise, France) (see Chapter 7). This assay differs from the above method in that it uses a capture step first described by Aupeix et al. in 1997.[35] The PS-bearing microparticles are captured by annexin V bound to the walls of microplates. The captured MV activity is then measured by incubating with factor V (FV), factor X (FX), and factor II (FII). The prothrombin complex is formed resulting in the formation of thrombin that is then measured chromogenically.[36] The system allows the capture of most MVs, even when the proportion of PS is very low.

A thrombin generation test (TGT) can also be used to measure MVs using reagents with no added or a reduced phospholipid concentration. The calibrated automated thrombogram (CAT) method (Thrombinoscope BV, Maastricht, Holland) uses a reagent that contains TF and a minimal amount of phospholipids (PRP reagent from Thrombinoscope). This can be used in PRP to measure platelet procoagulant activity but can also be used in platelet-poor plasma to detect MV procoagulant activity. Similarly the thrombin generation assay (TGA) using Ceveron Alpha (Technoclone Gmbh, Vienna, Austria) can be used with the Ceveron® TGA RC Low reagent. Technoclone also has a filtration method (Cevron MFU 500) that can be used to remove all MVs above 200 nm from a plasma sample. These methods are more suited to measuring TF activity and are described below in section 6.2.2.1.

6.2.1.2 Clinical studies with the procoagulant phospholipid assay

A number of results from clinical studies with this type of reagent are now appearing in the press. In one of the first publications using this assay, Exner et al.[37] showed that patients with acute coronary syndrome had shortened clotting times with the procoagulant phospholipid activity assay. It must be remembered that in this assay, when expressing results as clotting times, the shorter the time, the greater the procoagulant phospholipid activity in the sample.

In a study on patients who had suffered a myocardial infarction (MI) it was shown[38] that the procoagulant phospholipid activity fell from 59.8 s in controls (n = 34) to 29.8 s in patients who had suffered an MI (n = 46). Samples were collected within 6 h of the patients' arrival in the casualty department. The MI patients therefore had much greater procoagulant activity than the control patients. Interestingly if the patient groups were broken down into survivors and nonsurvivors the procoagulant phospholipid clotting time fell from 32.1 s and 21.7 s, respectively, showing much greater activity in the group that died (n = 10).

In a small study of patients with thyroid cancer[27] much shorter clotting times were seen in patients with cancer than normal controls. The clotting time fell from 57.6 s (range 50.9–74.9 s) in control subjects (n = 30) to 47.7 s (range 34–69 s) in patients with cancer and no metastasis (n = 8) to 28.1 s (range 19–56 s) in patients with metastasis (n = 12). These differences between patients and controls and the two patient groups were significant ($p < 0.01$).

The assay has also shown some utility in sickle cell disease (SCD). In a preliminary study[27] it was shown that the clotting time shortened slightly from 57.6 s (controls) to 52.3 s in steady-state SCD patients but shortened even further to 45.6 s in patients in SCD crisis. All of these differences were significantly different ($p <$ 0.05). In a subsequent study it could be shown that treatment with hydroxyurea, in addition to the regulation of hemolysis, also lowers procoagulant MVs and attenuates thrombin generation.[39]

The procoagulant phospholipid assay has also been used in a study of children undergoing bone marrow transplant.[40] In this study the researchers concluded that decreased levels of procoagulant phospholipids correlated with poor prognosis. In a study of patients with essential thrombocythemia it was shown that both MV-associated thrombin generation and PPL activity were significantly higher in patients than in controls.[41]

Although larger studies are needed in the disease states discussed it is clear that the procoagulant phospholipid assay is a very useful research tool and in the future will have a role in the routine laboratory, perhaps as a screening test for abnormal levels of MVs.

6.2.2 Tissue Factor Assays

In the normal state only very low levels of TF activity are found in the plasma. The levels detected are very low and close or below the detection level of most assays. Butenas et al.[15] calculated that in blood from normal individuals the level of TF is probably lower than 20 fM. The assay of TF is further complicated in that TF is considered by some groups to exist in two forms, an encrypted (very little procoagulant activity) and a de-encrypted (procoagulant active) form. The mechanism for this conversion remains controversial. It is considered that TF exists primarily in the encrypted form and is converted to the de-encrypted form when required. Some groups consider that the enzyme protein disulphide isomerase (PDI) is involved;[42,43] others consider that PDI plays no role in TF activation[44] and that activation may be due to membrane phospholipids.[45] No matter what the mechanism of action is, this issue potentially complicates TF assays. If an immunological-based assay is used, unless it uses an antibody with well-defined specificity, both forms of TF will be measured. To complicate the story even further the specificity of many antibodies used to measure blood-borne TF has been challenged and could explain some of the anomalies in the literature.[21]

For activity assays the major issue is the sensitivity of the assays. The levels of TF found in biological fluids of normal individuals are very low,[15] making functional assays very difficult unless the sample is first ultracentrifuged to concentrate the TF-bearing MVs.[2]

As with procoagulant activity assays preanalytical variables have an effect on all these assay types, and all samples must be carefully prepared if accurate and reproducible results are to be obtained.[30,46]

A number of assays are reported in the literature and fall into two main classes, thrombin generation–based assays and direct assays to measure TF activity.

6.2.2.1 Thrombin generation–based assays

The first TGTs were described in 1953.[47,48] These assays were complicated and difficult to perform, and it was not until 1986 that Hemker et al.[49] introduced a simpler two-stage chromogenic assay. The test procedure was further simplified into a single-step chromogenic assay by Hemker et al. in 1993[50] and then into the fluorogenic-type assay still in use today.[51] Three commercial formats of the TGT assay exist today: one based on use of a chromogenic substrate (Endogenous Thrombin Potential (ETP) Assay, Siemens Healthcare, Munich, Germany) and the other two based on a fluorogenic substrate (CAT method, Thrombinoscope BV, Maastricht, Holland, and the TGA, Technoclone Gmbh, Vienna, Austria). All three of these tests can be used to measure MVs. Thrombinoscope and Technoclone have commercial applications for monitoring both MV and TF activity.

None of these assays are quantitative and thus cannot measure the actual concentration of active TF in a sample. They all give some measure of the TF activity by its effect on the thrombin generation curve in the sample being tested. All produce thrombograms similar to that shown below (Fig. 6.3.). The presence of increased TF will cause a dose-dependent shift of the curve, shortening the lag time (LT) and time to peak (TTP) and increasing the ETP and peak of thrombin generation (PTG).[52]

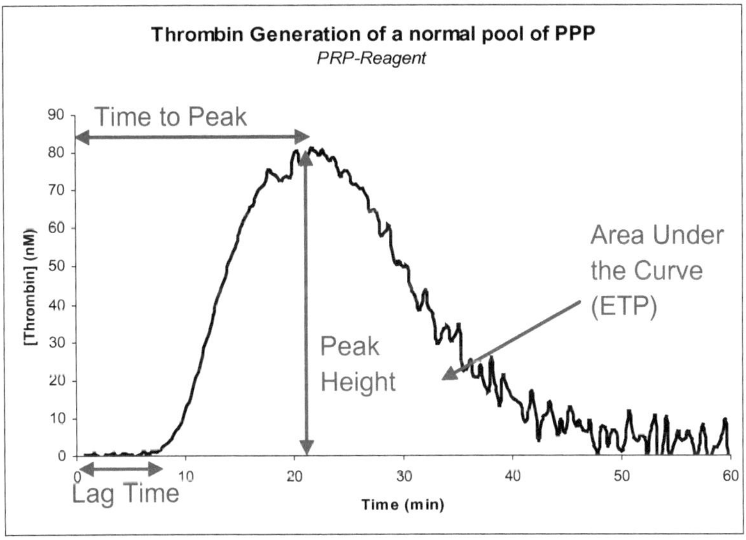

Figure 6.3 Example of a normal thrombogram obtained using the CAT method. The four commonly measured parameters are shown: (1) Lag time: the time in minutes from the start of the assay to the initial generation of thrombin; (2) time to peak: the time in minutes required to reach maximum thrombin generation; (3) peak of thrombin generation: the maximum thrombin concentration expressed in nmol/L; and (4) ETP: area under the curve expressed in nmol/L of thrombin.

One of the major issues with TGT is standardization.[53,54] The preanalytical variables have to be carefully controlled if reproducible results are to be obtained. It is recommended that samples be prepared as was described in section 6.2.1.1 above for procoagulant phospholipid activity assays. Positive controls can also be used to induce TF activity upon cells and MVs.

6.2.2.2 Tissue factor activity assays

As described in section 6.2.2, the level of TF found in normal plasma is very low, which complicates assay development. Two direct commercial assays are on the market, Actichrome TF (American Diagnostica Inc., Stamford, USA) and AssaySense Human Tissue Factor Chromogenic Activity Assay (AssayPro, St. Charles, USA). Both assays are chromogenic and measure the activation of FX, via FVIIa, by the TF in the sample. Both assays claim that they can be used both with plasma and with tissue culture samples. The lower end of sensitivity claimed by these assays is 2 pM (85 pg/ mL) for American Diagnostica and <1.5 pM for the AssayPro products. However, this low-end sensitivity is much higher than the levels that Butenas et al.[15] have calculated to be present in normal plasma, that is, <20 fM.

A third assay,[55] using a capture step with an anti-TF antibody coated into microtitre plate wells and a subsequent chromogenic measurement step to measure FXa generation by the sample TF and FVIIa, is commercially available (Zymuphen MP-TF from Hyphen Biomed, Neuville-sur-Oise, France). This assay claims to have a lower level of sensitivity ≤ 1 pM of MP-TF.

One problem with all TF-based assays is standardization, as no internationally accepted TF standard exists at present. Because of this, the levels claimed by each procedure are dependent on the TF used by each manufacturer, and the values obtained by the different assays may not be directly comparable. There is very little data on the performance of these three commercial assays in the literature yet.

The specificity of the Actichrom TF assay has been challenged.[56] The generation of FXa in the assay may not be totally dependent on the exogenous FVIIa and the sample TF.

Tesselaar et al.[57] developed an assay based on the same principle as the Actichrom TF, but to make the assay sensitive to TF they performed the assay in the presence and absence of anti–human TF antibodies. The TF activity was calculated from the difference in FXa activity in the presence and absence of the anti-FXa antibodies.

A number of other noncommercial assays have been described in the literature. The first clot-based assay for TF was published in 1980.[58] Assays can either have a clot-based endpoint[58,59] or use chromogenic[57,60–62] or fluorogenic substrates.[63] Most of these assays are based on the same principle—activation of FX and its measurement by either measuring clot formation or by using a chromogenic or fluorogenic substrate. Van Dreden et al.[61] describe an assay that they have used to measure blood-borne TF in plasma samples. This assay is also based on the activation of FX by TF, but it differs from many of the other tests in that it uses FVII and not FVIIa in the reagents and it includes a specific antibody to block the inhibitory effect of tissue factor pathway inhibitor (TFPI) on the assay.[61]

Many of these assays are designed for use only with MVs collected from a sample by ultracentrifugation or for use on cell culture supernatants.[57,58,61] The most commonly used method to harvest MVs is by ultracentrifugation at 20,000 g for 150 min at 4°C and then washing with a 4(2-hydroxyethyl)piperazine-1-ethanesulfonic acid (HEPES) buffer mix.[62] Using enrichment and a washing procedure allows basal normal levels of activity to be measured and also eliminates the specificity issues raised with some of these assays.

6.2.2.3 Clinical studies with tissue factor assays

The number of publications on clinical studies using either a TGT or a specific TF assay is growing. Raised levels have been reported in many different disease states[64] such as cancer, diabetes, cardiovascular complications, and miscarriage.

Geisen et al.[12] presented data showing that blood-borne TF is inherently thrombogenic and may be involved in thrombus formation at the injury site.

In cancer TF plays a pivotal role[18,57] and is involved in tumor progression and cancer coagulopathies. Patients with disseminated breast or pancreatic cancer who had a raised level of MV-associated TF activity had a lower survival rate than those with lower levels.[56] In pancreatic cancer raised levels of blood-borne TF have been associated with an increased rate of venous thromboembolism.[61,62] Monitoring TF activity may prove to be a very useful tool in monitoring cancer progression and treatment. TF has been identified as a possible therapeutic target in cancer treatment.

Raised levels of plasma TF[59] and rates of thrombin generation[39] have been reported in patients with SCD. This may play an important role in triggering the activation of coagulation seen in patients with SCD.

In cardiac disease, Van Dreden et al.[38] reported raised levels of TF in patients with acute myocardial infarction (AMI). They also found that those patients with a poor prognosis at a 2 month follow-up had much higher levels of TF (6.80 pmol.) than those classified as having a good prognosis (3.24 pmol). The TF activity/TFPI ratio was significantly modified in patients with AMI (no disease, 0.014; good prognosis, 0.119; and poor prognosis, 0.242) and could be of prognostic significance.

In pregnancy, as is seen with many other coagulation factors, blood-borne TF levels increase with advancing pregnancy.[65] The TF levels were significantly higher in patients who had either early or late pregnancy loss.[64,65] As was seen in AMI,[38] the TF activity/TFPI ratio seems to be significantly modified in patients with early or late pregnancy loss and could be of prognostic significance.[65] Nonpregnancy TF activity levels were 0.24 pmol, rising to 1.68 pmol at >22 weeks in normal pregnancy. Patients who experienced either early or late pregnancy loss had TF activity levels of 3.65 pmol and 3.92 pmol, respectively. It has also recently been shown that EVs derived from pre-eclampsia placentas trigger more thrombin generation in a TF-dependent manner than those from normal placentas, which may contribute to the activation of the coagulation system seen in this disorder of pregnancy.[66]

What is still lacking are good simple routine assays for measuring blood-borne TF activity for use in both the research and the routine laboratory. Much larger studies are required to prove the clinical utility of measuring blood-borne TF, but the preliminary data would indicate that there is potential clinical relevance in measuring it.

6.3 Conclusion

This chapter has described the procoagulant phospholipid assays and TF activity assays that are available for use in the laboratory today. Only a few of these assays are commercially available at present. No doubt this number will increase in the next few years.

Although much of the clinical data are preliminary and most need confirmation in larger clinical studies, it is clear that these assays do have some clinical relevance and appear to have a utility both as an aid in diagnosis and for following disease progression and treatment.

In the future, assays in this area will become routine automated laboratory tests. The data available to date would indicate that they may give clinically relevant and useful results on the hemostatic state of the patient and as an aid in diagnosis and treatment, especially in the area of cancer and cardiac disease.

References

1. Caby, M-P, Lankar D, Vincendeau-Scherrer C, Raposo G, Bonnerot C (2005) Exosomal-like vesicles are present in human plasma, *Int. Immunol.*, **17**(7), 879–887.
2. Owens A, Mackman N (2011) Microparticles in hemostasis and thrombosis, *Circ. Res.*, **109**, 1284–1297.
3. Aaron F, Orozco A, Lewis D (2010) Flow cytometric analysis of circulating microparticles in plasma, *Cytometry A*, **77A**(6), 502–514
4. Zitvogel L, Regnault A, Lozier A, Wolfers J, Flament C, Tenza D, Ricciardi-Castagnoli P, Raposo G, Amigorena S (1998) Eradication of established murine tumors using a novel cell-free vaccine: dendritic cell derived exosomes, *Nat. Med.*, **4**, 594 – 600.
5. Laulagnier K, Motta C, Hamdi S, Roy S, Fauvelle F, Pageaux J-F, Kobayashi T, Salles J-P, Perret B, Bonnerot C, Record M (2004) Mast cell- and dendritic cell-derived exosomes display a specific lipid composition and an unusual membrane organization, *Biochem. J.*, **380**(1), 161–171.
6. Gemmell C, Sefton M, Yeos E (1993) Platelet derived microparticle formation involved glycoprotein IIb IIIa, *JBC*, **268**(20), 14586–14589.
7. Berckmans R, Nieuwland R, Böing A, Romijn F, Hack C, Sturk A (2001) Cell-derived microparticles circulate in healthy humans and support low grade thrombin generation, *Thromb. Haemost.*, **85**(4), 639–646.
8. Berckmans J, Nieuwland R, Tak P, Böing N, Romijn P, Kraan C, Breedveld C, Hack E, Sturk A (2002) Cell-derived microparticles in synovial fluid from inflamed arthritic joints support coagulation exclusively via a factor VII-dependent mechanism, *Arthritis Rheum.*, **46**(11), 2857–2866.
9. Chargaff E, West R (1946) The biological significance of the blood thromboplastic protein of blood, *JBC*, **165**, 189–197.
10. Wolf P (1967) The nature and significance of platelet products in human plasma, *BJH*, **13**, 269–288.

11. Van Creveld S, Paulssen M (1952) Isolation and properties of the third clotting factor in blood-platelets, *Lancet*, **259**(6697), 23–25.
12. Exner T, Joseph J, Low J, Connor D, Ma D (2003) A new activated factor X-based clotting method with improved specificity for procoagulant phospholipids, *Blood Coagul. Fibrinolysis*, **14**, 773–779.
13. Owen, B, Xue A, Heit J, Owen G (2011) Procoagulant activity, but not number, of microparticles increases with age and in individuals after a single venous thromboembolism, *Thromb. Res.*, **127**(1), 39–46.
14. Giesen P, Rauch U, Bohrmann B, Kling D, Roque M, Fallon J, Badimon J, Himber J, Riederer M, Nemerson Y (1999) Blood-borne tissue factor: another view of thrombosis, *PNAS*, **96**, 2311–2315.
15. Butenas S, Bouchard B, Brummel-Ziedins K, Parhami-Seren B, Mann K (2005) Tissue factor activity in whole blood, *Blood*, **105**(7), 2764–2770.
16. Drake T, Ruf W, Morrissey J, Edgington T (1989) Functional tissue factor is entirely cell surface expressed on lipopolysaccharide-stimulated human blood monocytes and a constitutively tissue factor-producing neoplastic cell line, *JBC*, **109**, 389–395.
17. Bogdanov V, Balasubramanian V, Hathcock J, Vele O, Lieb M, Nemerson Y (2003) Alternatively spliced human tissue factor: a circulating, soluble, thrombogenic protein, *Nat. Med.*, **9**(4), 458–462.
18. Rak J, Milsom C, May L, Klement P, Yu J (2006) Tissue factor in cancer and angiogenesis: the molecular link between genetic tumor progression, tumor neovascularization, cancer coagulopathy, *Semin. Thromb. Haemost.*, **32**(1), 54–70.
19. Panes O, Matus V, Sáez CG, Quiroga T, Pereira J, Mezzano D (2007) Human platelets synthesize and express functional tissue, *Blood*, **109**, 5242–5250
20. Bouchard BA, Mann KG, Butenas S (2010) No evidence for tissue factor on platelets. *Blood*, **116**, 845–846
21. Osterud B, Olsen J, Bjorklid E (2009) What is blood borne tissue factor? *Thromb. Res.*, **124**(5), 640–641.
22. Sandberg H, Andersson L-O (1979) A highly sensitive assay of platelet factor 3 using a chromogenic substrate, *Thromb. Res.*, **14**(1), 113–124.
23. Hardisty R, Hutton R (1965) The kaolin clotting time of platelet-rich plasma: a test of platelet factor-3 availability, *BJH*, **11**, 258–268.
24. Spaet T, Cintron T (1965) Studies on PF3 availability, *BJH*, **11**, 269–275.

25. Boffa, M-C (1980) Evaluation of the phospholipid-related procoagulant activity in plasma. A new clue for the detecting tendency of thrombosis? *Thromb. Res.*, **17**, 567–572.

26. Wagenvoord R, Hendrix H, Kai H, Hemker H (1994) A chromogenic test to determine the procoagulant phospholipids in platelet-rich plasma and whole blood, *Thromb. Haemost.*, **72**(4), 582–7.

27. Van Dreden P, Rousseau A, Fontaine S, Woodhams B, Exner T (2009) Clinical evaluation of a new functional test for detection of plasma procoagulant phospholipids. *Blood Coagul. Fibrinolysis*, **20**(7), 494–502.

28. Exner T, Low J (2004) Detection of procoagulant phospholipid interfering in tests for lupus anticoagulant, *Thromb. Res.*, **114**(5–6), 547–552.

29. Exner T, Joseph J, Low J, Connor D, Ma D (2003) Procoagulant phospholipid in blood normally associates mainly with cells or aggregates larger than microparticles or platelets, *JTH*, (Suppl 1), Abstract No. P1078.

30. Yuana Y, Bertina R, Osanto S (2011) Pre-analytical and analytical issues in the analysis of blood microparticles. *Thromb. Haemost.*, **105**(3), 396–408.

31. International Society on Thrombosis and Haemostasis (2011) Vascular biology SSC of the ISTH, http://isth.org/default/index.cfm/ssc1/2011-ssc-subcommittee-minutes/2011-vascular-biology-minutes/.

32. Gerotziafas G, Chaari M, Van Dreden P, Woodhams B, Lionnet F, Girot R, Galea V, Elalamy I (2011) Erythrocyte derived microparticles and procoagulant phospholipids induce acceleration of thrombin generation in patients with steady state sickle cell anemia, *JTH*, **9**(Suppl 2), Abstract No. P-WE-429.

33. van Beers J, Schaap M, Berckmans R, Nieuwland R, Sturk A, van Doormaal F, Meijers J, Biemond B, CURAMA study group (2009) Circulating erythrocyte-derived microparticles are associated with coagulation activation in sickle cell disease, *Haematologica*, **94**(11), 1513–1519.

34. Connor D, Exner T, Ma D, Joseph J (2009) Detection of the procoagulant activity of microparticle associated phosphatidylserine using XACT, *Blood Coagul. Fibrinolysis*, **20**(7), 558–564.

35. Aupeix K, Hugel B, Martin T, Bischoff P, Lill H, Pasquali J, Freyssinet J (1997) The significance of shed membrane particles during programmed cell death *in vitro*, and *in vivo*, in HIV-1 infection, *JCI*, **99**(7), 1546–1554.

36. Vissac A, Peyrafitte M, Amiral J (2007) Microparticles procoagulant activity in citrated plasma, *JTH*, **5**(1), Abstract No. PS513.

37. Exner, T, Joseph J, Connor D, Low J, Ma, D (2005) Increased procoagulant phospholipid activity in blood from patients with suspected acute coronary syndromes: a pilot study, *Blood Coagul. Fibrinolysis*, **16**(5), 375–380.

38. Van Dreden P, Rousseau A, Savoure A, Lenormand B, Fontaine S, Vasse M (2009) Plasma thrombomodulin activity, tissue factor activity and high levels of circulating procoagulant phospholipid as prognostic factors for acute myocardial infarction. *Blood Coagul. Fibrinolysis*, **20**(8), 635–641.

39. Gerotziafas G, Van Dreden P, Chaari M, Khaterchi A, Lionnet F, Stankovic-Stojanovic K, Woodhams B, Meyer M, Hatmi M, Girot R, Elalamy I (2011) Gerotziafas GT, Van Dreden P, Chaari M, Galea V, Khaterchi A, Lionnet F, Stankovic-Stojanovic K, Blanc-Brude O, Woodhams B, Maier-Redelsperger M, Girot R, Hatmi M, Elalamy I (2012) The acceleration of the propagation phase of thrombin generation in patients with steady-state sickle cell disease is associated with circulating erythrocyte-derived microparticles, *Thromb. Haemost.*, **107**(6):1044–1052.

40. Schneider P, Van Dreden P, Rousseau A, Marie-Cardine A, Houivet E, Vannier J, Vasse M (2011) Decreased activity of soluble thrombomodulin and plasma procoagulant phospholipids in childhood bone marrow transplantation with severe complications, *Thromb. Res.*, **128**(3), 261–267.

41. Tartari C, Russo L, Rambaldi A, Finazzi G, Barbui T, Woodhams B, Falanga A (2010) Microparticle-associated thrombin generation and procoagulant activity is increased in patients with essential thrombocythemia, *Blood*, Abstract No. 1985.

42. Versteeg H, Ruf W (2007) Tissue factor coagulant function is enhanced by protein-disulfide isomerase independent of oxidoreductase activity, *JBC*, **282**, 25416–25424.

43. Popescu N, Lupu C, Lupu F (2010) Role of PDI in regulating tissue factor: FVIIa activity, *Thromb. Res.*, **125**(S1), S38–S31.

44. Usha R. Pendurthi U, Ghosh S, Mandal S, Rao L (2007) Tissue factor activation: is disulfide bond switching a regulatory mechanism? *Blood*, **110**(12), 3900–3908.

45. Bach R (2006) Tissue factor encryption, *Arterioscler. Thromb. Vasc. Biol.*, **26**, 456–461.

46. Lee R, Barcel D, Williams J, Wang J, Boles J, Manly D, Key N, Mackman N (2011) Pre-analytical and analytical variables affecting the measurement of plasma-derived microparticle tissue factor activity, *Thromb. Res.*, Jul. 5 (Epub ahead of print).

47. Macfarlane R, Biggs R (1953) A thrombin generation test; the application in haemophilia and thrombocytopenia. *J. Clin. Pathol.*, **6**, 3–8.

48. Pitney W, Dacie J (1953) A simple method of studying the generation of thrombin in recalcified plasma; application in the investigation of haemophilia, *J. Clin. Pathol.*, **6**, 9–14.

49. Hemker H, Willems G, Beguin S (1986) A computer assisted method to obtain the prothrombin activation velocity in whole plasma independent of thrombin decay processes, *Thromb. Haemost.*, **56**, 9–17.

50. Hemker H, Wielders S, Kessels H, Beguin S (1993) Continuous registration of thrombin generation in plasma, its use for the determination of the thrombin potential. *Thromb. Haemost.*, **70**(4), 617–624.

51. Hemker H, Giesen P, AlDieri R, Regnault V, de Smed E, Wagenvoord R, Lecompte T, Beguin S (2002) The calibrated automated thrombogram (CAT): a universal routine test for hyper- and hypocoagulability, *Pathophysiol. Haemost. Thromb.*, **32**, 249–253.

52. Gerotziafas G, Depasse F, Busson J, Leflem L, Elalamy I, Samama M (2005) Towards a standardization of thrombin generation assessment: the influence of tissue factor, platelets and phospholipids concentration on the normal values of thrombogram-thrombinoscope assay, *Thromb. J.*, **3**, 16.

53. Castoldi E, Rosing J (2011) Thrombin generation tests, *Thromb. Res.*, **127**(Suppl 3), S21–S25.

54. Dargaud Y, Luddington R, Gray E, Lecompte T, Siegemund T, Baglin T, Hogwood J, Regnault V, Siegemund A, Negrier C (2010) Standardisation of thrombin generation test—which reference plasma for TGT? An international multicentre study, *Thromb. Res.*, **125**(4), 353–356.

55. Laroche M, Peyrafitte M, Vissac A, Amiral J (2009) Measurement of procoagulant potential of blood microparticles carrying tissue factor, *Hamostaseologie*, **29**, A72, Abstract No. PP5.3–PP5.5.

56. Bogdanov V, Cimmino G, Tardos J, Tunstead J, Badimon I (2009) Assessment of plasma tissue factor activity in patients presenting with

coronary artery disease: limitations of a commercial assay, *JTH*, **7**(5), 894–897.

57. Tesselaar M, Romijn F, Van Der Linden I, Prins F, Bertina R, Osanto S (2007) Microparticle-associated tissue factor activity: a link between cancer and thrombosis? *JTH*, **5**(3), 520–527.

58. Levy G, Edgington T (1980) Lymphocyte cooperation is required for amplification of macrophage procoagulant activity, *J. Exp. Med.*, **151**(5), 1232–1244.

59. Key N, Slungaard A, Dandelet, L., Nelson S, Moertel C, Styles L, Kuypers F, Bach R (1998) Whole blood tissue factor procoagulant activity is elevated in patients with sickle cell disease, *Blood*, **91**(11), 4216–4223.

60. Jiang W, Kou J, Yuan S, Sun L, Yu B (2011) A simplified and high-throughput chromogenic assay for testing tissue factor-dependent procoagulant activity, *J. Biomol. Screening*, **16**(3), 295–302.

61. Delluc A, Rousseau A, Delluc C, Le Moigne E, Le Gal G, Mottier D, Van Dreden P, Lacut K (2011) Venous thromboembolism in patients with pancreatic cancer: implications of circulating tissue factor, *Blood Coagul. Fibrinolysis*, **22**(4), 295–300.

62. Khorana A, Francis C, Menzies K, Wang J, Hyrien O, Hathcock J, Mackman N, Taubman M (2008) Plasma tissue factor may be predictive of venous thromboembolism in pancreatic cancer, *JTH*, **6**(11), 1983–1985.

63. Jenny R, Butenas S (2009) Assay for tissue factor and factor VIIa, United States Patent 7, 604, 932.

64. Arthur C (2011) Tissue factor, blood coagulation, and beyond: an overview, Int. J. Inflammation, **2011**, Article ID 367284.

65. Van Dreden P, Woodhams B, Rousseau A, Favier M, Favier R (2011) Comparative evaluation of tissue factor and thrombomodulin activity changes in early and late foetal loss: the cause of hypercoagulability? *Thromb. Res.*, Aug. 2 (Epub ahead of print).

66. Gardiner C, Tannetta DS, Simms CA, Harrison P, Redman CWG, Sargent IL (2011) Syncytiotrophoblast microvesicles released from pre-eclampsia placentae exhibit increased tissue factor activity, *PLOS ONE*, **6**, e26313.

Chapter 7

Capture-Based Assays for Extracellular Vesicles within the Blood

Jean Amiral
Hyphen BioMed, 95000, Neuville sur Oise, France
jamiral@hyphen-biomed.com

7.1 Introduction

Activation or destruction of blood cells can generate circulating extracellular vesicles (EVs) comprising both microvesicles (MVs) and nanovesicles (exosomes). EVs therefore not only provide novel biomarkers of disease and its progression but also can be involved in pathogenesis and complications of the disease process.[1–4] Figure 7.1 shows the origin of blood cell EVs, mainly released from platelets, and their potential contribution to disease progression. Although EVs can originate from all blood cells, including endothelial cells, erythrocytes, monocytes, and leukocytes, the majority are platelet derived. Platelet-derived EVs are also highly procoagulant, although not all of them express phosphatidylserine (PS).[5] EVs can exhibit either homeostatic or pathological activities,[6,7] the latter being dependent on their mechanism of generation.[8–10] EVs are very

Extracellular Vesicles in Health and Disease
Edited by Paul Harrison, Christopher Gardiner, and Ian L. Sargent

ISBN 978-981-4411-98-1 (Hardcover), 978-981-4411-99-8 (eBook)
www.panstanford.com

heterogeneous in size (from 0.1 μm to 1.0 μm) and structure,[11] with exposure of different cell surface proteins according to their origin and the type of pathology.[12–14] Various assays are now available for their identification and measurement in blood or plasma. Flow cytometry allows the measurement of the number and phenotype of EVs, by characterizing their cellular origin using well-defined cluster of differentiation (CD)-specific antibodies.[15–17] Use of the high-affinity calcium-dependent annexin V (A5) and other probes can also characterize the exposure of PS thought to mediate procoagulant activity.[18]

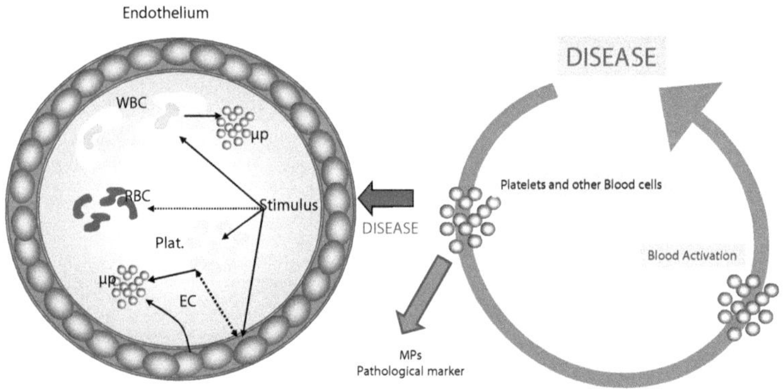

Figure 7.1 MVs generated in blood circulation can originate from various blood cells, although the majority are produced by platelets; they can be markers of pathology, but they also contribute to the progression of disease.

Capture-based assays, as first reported by Aupeix et al.,[19] also use surface-coated A5 for capturing these MVs in the presence of calcium and then simply measuring their procoagulant potential through thrombin generation, in the presence of factors Xa, Va, and prothrombin. Capture-based assays unlike flow cytometry measure the total procoagulant activity associated with all sizes of EVs that express PS. Theoretically EVs with the lowest size will also support a much higher procoagulant potential than larger vesicles because of the increase in surface area. Capture-based assays can also be designed for specifically measuring EVs derived from defined blood cells by using specific monoclonal antibodies to appropriate cell surface antigens (CDs). Of specific interest are EVs exposing tissue factor (TF), as they also associate the PS-dependent procoagulant activity with the blood coagulation–triggering effect of TF.[20–23]

A third class of assays has also been described for measuring the plasma procoagulant potential of cell-released phospholipids, mainly microparticles (MPs), using homogeneous clotting assays.[24] Although these assays are nonspecific unlike capture methods they can still measure intrinsic phospholipid-based procoagulant potential in plasma. In this chapter, the generation of EVs in health and disease and the technical characteristics of capture-based assays are described, as well as the preanalytical variables required.

7.2 Occurrence of EVs in Health and Disease

Micro- and nanoparticles can be generated in many pathological conditions, originating from many cell types, including platelets, monocytes, leukocytes, erythrocytes, or endothelial cells. More particularly, EVs exposing TF have been reported in cancer patients.[25] They are also associated with the high rate of thrombotic complications observed in malignancy[26] and also in infectious diseases[27] or in sepsis.[28] Many papers have been published that show the generation of various EV types in pathology and their possible contribution to disease progression and to thrombo-embolic complications. In physiological conditions, procoagulant activity increases with age (although the total number looks to remain stable). For example, in women, a significant change is observed at menopause, especially in metabolic syndrome.[29,30] However, development of pathological conditions can have important repercussions on the generation of various types of EVs and on their thrombogenicity.[31,9] Smoking habits can promote their release from monocytes/macrophages.[32] Abnormal levels of procoagulant activity and cell interactions have been described in obesity and metabolic syndrome.[33,34] As expected, these procoagulant vesicles and their thrombogenic effect are elevated in diabetic patients.[35–37] The presence of EVs from different cell origins can also stimulate some interactions and synergy between these cells.[38,39] Other clinical conditions, where EVs contribute to symptoms, have been reported, including Crohn disease[40] and injection of activated coagulation factors, such as factor VIIa.[41] In this latter case, the EVs can not only contribute to the hemostatic efficacy of factor VIIa but also may cause thrombotic complications. Similarly, their contribution to the development of thrombosis in patients with heparin-induced thrombocytopenia has also been suggested.[42] Conversely, in Scott syndrome, there is a defect in the procoagulant activity of platelets with failure of PS exposure and EV generation, resulting in a lifelong bleeding diathesis.[43] In cardiovascular diseases (myocardial infarction, atrial fibrillation) they can remain a predictor of possible recurrence and of poorer prognosis,[44–46] and those exposing TF are associated with poorer recovery.[47,48] In patients with cancer, blood activation and inflammation are important components, which contribute to pathological complications, and EVs are intimately involved in these processes. They can be derived from various cells and can expose TF in many circumstances.[26,49,21,25] This process can be potentiated by treatments implemented for cancer, such as chemotherapy. Therapeutic drugs used can also release EVs through the destruction of malignant cells, or they can directly activate blood cells.[50] However, the contribution of the membrane environment seems to be more important than TF expression for promoting thrombin formation.[51] EVs can also participate in a hypercoagulable state by allowing increased survival of the generated factor Va, as factor Va bound to EVs is protected from inactivation by activated protein C.[52] The participation of MPs in disease evolution is also supported by a reduction of their concentration and/or activity following initiation of targeted therapies, which potentially improve patient outcome.[53–57] This effect is also supported by laboratory studies that show a reduction in EVs or the use of PS-targeted proteins (such as A5) to reduce their hemostatic potential.[58–60]

7.3 Capture-Based Assays for Procoagulant EVs

Capture-based assays exhibit many advantages for the specific measurement of the procoagulant potential of EVs. The basic capture-based assay used for EV procoagulant activity is derived from the method described by Aupeix et al.[19] and is reported here next.

7.3.1 Assay Principle

Capture-based assays are designed using a specific PS capture protein (e.g., A5) or a monoclonal antibody (MoAb) coated on the solid phase for binding EVs within the tested specimens. Initially an assay of general use was developed for measuring procoagulant EVs. Basically, A5 is coated on a solid surface, usually a micro–

enzyme-linked immunosorbent assay (ELISA) plate (Maxisorb, Nunc, Roskilde, Denmark), and then stabilized. A diluted, tested plasma or specimen, a calibration standard of EVs, and controls are introduced into the A5-coated wells within a buffer diluent containing calcium. Any EVs within the samples therefore specifically bind with high affinity onto the A5 in the presence of calcium. Any unbound material is washed away with a washing solution also containing calcium in order to preserve the interaction of A5 with the EVs. Bound EVs are subsequently measured through thrombin generation. For this step, a solution containing factor Xa, factor Va, and calcium is added into the washed wells, and then a constant concentration of prothrombin is introduced. Prothrombinase forms on the EV surface, and its activity is dependent not only on the EV concentration but also on the exposure of PS. The amount of thrombin generated is proportional to the EV procoagulant activity, which is directly related to the PS concentration on the EV surface. The thrombin generated is measured through its activity on a specific chromogenic substrate. Results are expressed in nanomoles (nM) of PS equivalent. Calibration is performed using liposomes with a defined homogeneous size (e.g., 0.1 μm) and with a well-defined PS content (expressed in nanomoles). A direct linear relationship is obtained between EV concentration (expressed in nanometer PS) and the amount of procoagulant EVs. This assay principle and a typical calibration curve obtained are represented in Fig. 7.2.

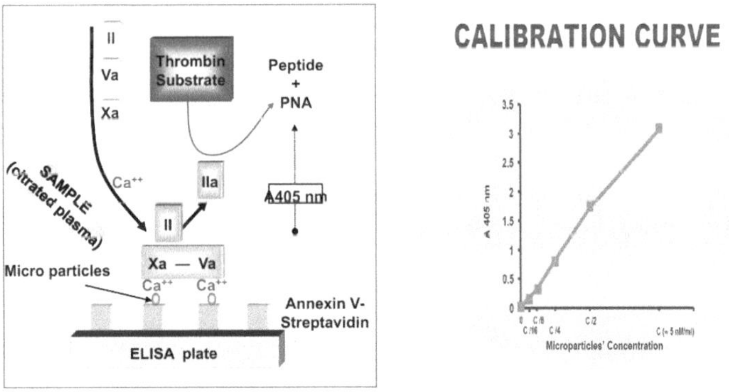

Figure 7.2 Scheme depicting an MV assay using A5 for capturing MVs and thrombin generation for measuring their associated coagulant activity, and the dose–response curve obtained with standardized liposome phospholipids spiked in plasma (concentrations expressed in nanomole of PS equivalent).

7.3.2 Preanalytical Variables

The most important variable is that any potential artifactual generation of EVs during blood sampling and plasma preparation must be avoided. For testing EVs in plasma, blood must be collected within a standard citrate anticoagulant (9 parts blood per 1 part 0.109 M citrate) and processed within 2 h at room temperature (18–25°C). Alternatively, a specific anticoagulant preventing platelet activation can also be used (citrate, adenosine, theophylline, dipyridamole [CTAD]), and this allows safer processing of the plasma preparation. Any activation during blood collection must be avoided, and no tourniquet must be used. Blood must be processed at room temperature and centrifuged at 2,500 g at room temperature, for 15 min, and the plasma supernatant must be collected, avoiding pipetting platelets present at the buffy coat interface. This plasma must then be centrifuged again for 2 min at 13,000 g at room temperature (which can be easily reached using an Eppendorf-type bench centrifuge, or equivalent) in order to remove any residual platelets. This platelet-free plasma is then ready for testing EVs and must be used within 4 h or stored frozen at –80°C for at least 6 months and thawed just before use (at least 15 min incubation at 37°C). Adhering to these recommendations for plasma preparation is very important to avoid any interference from contaminating platelets and preventing *ex vivo* EV release. These recommendations remain valid for all the assays designed for measuring EV procoagulant activity or EVs released from a specific cell line or EVs exposing a specific protein such as TF (EV-TF).

7.4 Capture-Based Assays for EVs Exposing Tissue Factor

Additional capture-specific assays are available for measuring EV-TF. These EVs potentially combine the procoagulant effect of negatively charged phospholipids (PS) with that of the coagulation trigger protein TF. The capture assay for EV-TF is designed using a solid phase, which is an ELISA microplate coated with a MoAb specific for the extracytoplasmic domain of TF but without inhibiting its activity, (e.g., 10H10 MoAb). Assayed specimens are then incubated in the coated microwells and EV-TF are captured by the MoAb, while maintaining their procoagulant activity. Following a washing step, FVIIa (Novoseven®) and FX (highly purified from human plasma and containing no factor Xa) are added to the reaction mixture. On the immobilized phospholipid surface, TF-FVIIa complexes form and activate FX into FXa in the presence of Ca^{++}. FXa generation is then dependent on TF and the concentration of anionic phospholipids (i.e., PS). Then a FXa-specific substrate is added, which reacts with the generated FXa, releasing pNA, and the absorbance is recorded at 405 nm. A lyophilized calibrator, containing recombinant (human) TF, relipidated with synthetic phospholipids and spiked in plasma, is used for calibration and standardization of the assay. The assay is highly sensitive and allows measurements in the very low range (pg/mL range). Calibration is from 0 pg/mL to 5 pg/mL of MPs exposing tissue factor, or MP-TF (expressed as a tissue factor (TF) equivalent), and a ratio of 0.1 nM PS/1 pg TF is used for preparing EV-TF.

Complementary to the measurement of the procoagulant activity of EVs, and EV-TF, ELISAs were also designed for measuring full-length TF (FL-TF) and total soluble TF (T-TF). FL-TF ELISA uses a capture murine MoAb specific for an extracellular TF epitope, unexposed on alternatively spliced TF (as-TF) and a second murine MoAb reactive with an extracytoplasmic epitope on TF, labeled with biotin. Streptavidin coupled to horse radish peroxidase (HRP) is then added and binds to biotinylated antibodies. This development system increases the assay sensitivity. Activity is revealed with the use of the tetramethyl benzindine/hydrogen peroxide (TMB/H_2O_2) peroxidase substrate. Tested specimens are assayed in a diluent, which enhances the assay reactivity and suppresses nonspecific interactions (heterophilic antibodies). T-TF ELISA is designed with a similar approach, but the capture MoAb is targeted to an extracytoplasmic epitope available on all TF forms, and a second biotinylated MoAb, reacting with the whole of TF, is used for detecting captured TF material. The T-TF assay is only reactive with truncated forms of TF (1-219). In both cases, assays range from 0 pg/mL to 500 pg/mL TF (for FL-TF or T-TF) in plasma or in the assayed milieu. A highly purified human TF is used for calibrating both assays. The assay principles for MP-TF, FL-TF, and T-TF are shown in Fig. 7.3.

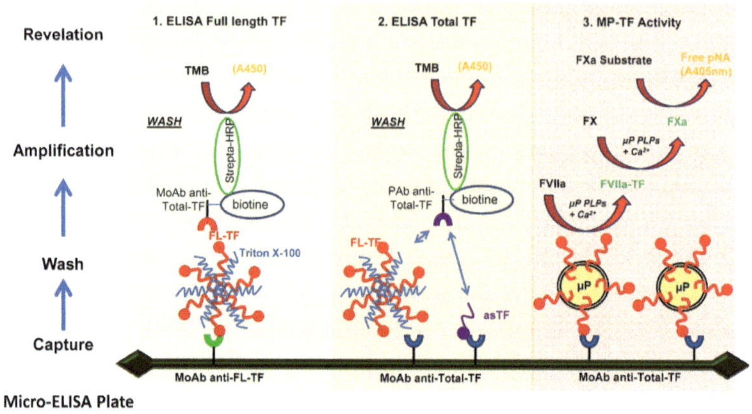

Figure 7.3 Principle of the various assays designed for measuring MVs exposing TF (MV-TF), full-length TF (FL-TF), or total TF (T-TF) and using a micro-ELISA plate. MV-TF is measured through factor Xa generation, whilst FL-TF and T-TF are measured through an amplified ELISA technique and tagged with specific antibodies.

7.5 Laboratory Results with Capture-Based Assays for EVs and EV-TF

7.5.1 MP Activity

For the EV activity assay (Zymuphen MP Activity, Hyphen Biomed, France), plasmas are assayed at 1:20 dilution in the assay diluent with calcium. The assay range in the tested dilution is from about 0.1 nM to about 2.5 nM PS equivalent, that is, from 2.0 nM to about 50.0 nM in the assayed plasma. The assay has good reproducibility and repeatability (interassay coefficient of variation [cv] of 5–10% and intraassay cv of 3–8%) and good recovery of synthetic MVs and liposomes exposing PS spiked in plasma. The total assay time is less than 90 min. When normal plasmas were tested, EV activity ranged from <1.0 nM to 9.6 nM. The normal range is then established for EV activity <10 nM. Forty-six pathological plasmas with various pathologies were tested, and the mean EV activity and ranges are reported in Table 7.1. Results obtained with the assay here reported were compared with the original Freyssinet method published in

1997 (Aupeix et al.[19]), used for the same specimens. Results obtained with both methods are very similar. Increased concentrations of EV activity are associated with a higher risk of thrombotic diseases. This assay can offer utility not only for diagnosing patients at risk for developing thrombotic diseases but also for following the efficacy of the antithrombotic therapies used.

Table 7.1 Microparticle activity in normals and various diseases associated with blood activation, as measured with the Zymuphen MP Activity assay or the original method reported by Aupeix et al. in 1997

	Number of patients	**nM of MPs (mean of two lots) Zymuphen MP Activity**	**nM of MPs with Freyssinet's method[19]**
Normals (<10 nM)	8	4.3	6.0
Patients with a cardiologic risk factor in primary prevention	6	20.5	23.53
Cardiac patients transplanted in suspicion of acute or chronic rejection	12	16.2	14.9
Patients treated for auricular fibrillation	13	6.7	6.0
Transplanted patients: receivers of small islets of Langerhans (unstable diabetes)	4	6.7	5.2
Patient in brain death	1	10.3	6.2
Indetermined patients	2	6.2	8.7
Total mean (nM)	46	11.22	10.7
SD (nM)	46	10.9	10.7

Source: Adapted from Ref. 19

Note: Very similar results are obtained with both methods.

7.5.2 EV-TF and TF

The specific assay for EV-TF needs to be sensitive enough for measuring the very low concentrations expected. It is sensitized by using an overnight incubation of the assayed plasma in the ELISA microwell coated with the MoAb specific for TF and by using the amplification system through factor Xa generation. EV-TF concentrations in normals are very low and usually <0.2 pg/mL, while two pathological plasmas with high EV activity concentrations (9.8 and >63 nM PS equivalent), as measured with Zymuphen MP Activity, were found at 0.95 pg/mL and >5 pg/mL for MP-TF concentration using Zymuphen MP-TF (expressed as TF equivalent). EV-TF can also be generated from normal blood by stimulating blood cells with lipopolysaccharide (LPS) as a positive control. Plasmas from LPS-stimulated whole blood following 6 h incubation at 37°C had significantly higher EV-TF concentrations than baseline values with EV-TF generation increasing from <1 ng/mL to values in the range of 10 pg/mL to 22 pg/mL. Further incubation of LPS-stimulated whole blood with polyclonal antibodies specific for human TF also suppressed the increased values measured. Conversely, spiking whole blood with synthetic phospholipid liposomes did not increase the concentration of EV-TF, demonstrating the specificity of the assay for EV-TF. Results are presented in Fig. 7.4. Recombinant truncated TF (1-219), even though mixed at 100 pg/mL with synthetic saturating phospholipids (870 nM PS), was measured at <0.2 pg/mL, confirming the specificity for EV-TF and the absence of reactivity with truncated forms of human TF (1-219).

ELISAs for TF, in its different possible presentations, have also been used for generating specific biological data. Measurement of the concentrations of the different TF forms, FL-TF and T-TF, in normal plasmas were as follows: FL-TF is below the detection limit of the Zymutest FL-TF assay (<10 pg/mL); when assayed with the Zymutest Total TF assay, this analyte is at a mean concentration of about 50 pg/mL in normal plasmas. These data show that no free FL-TF is measured, at least in normal plasmas, or that it is below the detection limit. This is consistent with the very low concentration of EV-TF measured in these normal plasmas. Stimulation with LPS of whole blood induced an increase of EV-TF but with only a minor increase in FL-TF. When truncated recombinant human TF (1-219) is spiked in normal plasma at 100 pg/mL, it has only 1 % reactivity

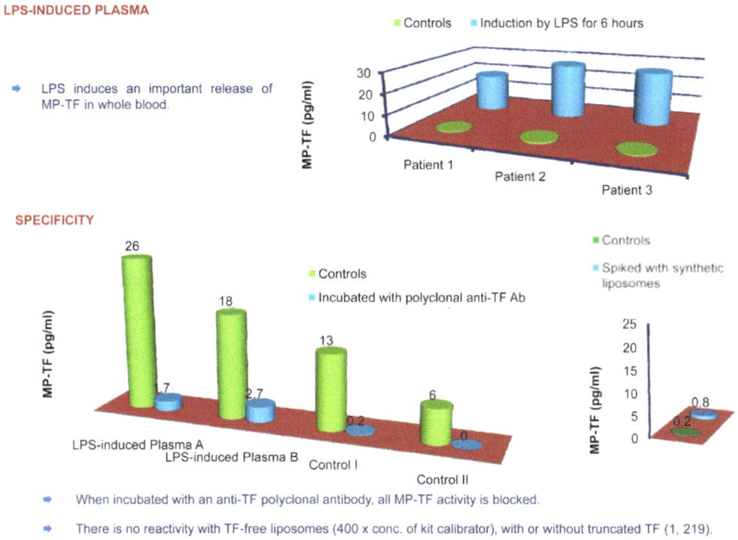

Figure 7.4 Specificity of a MV-TF assay. Incubation of whole blood with LPS induces an important increase of MV-TF concentration, and this elevation is totally abolished if blood is incubated with a polyclonal antibody specific for TF. Conversely, addition of a phospholipid in liposomes to blood has no effect on MV-TF measurements.

in the FL-TF (1-263) ELISA and 60 % reactivity in the T-TF ELISA. Incubation of human blood with LPS increased MP-TF significantly but FL-TF and T-TF only slightly. Results are presented in Fig. 7.5, which reports the various TF-associated activities (MP-TF, FL-TF, and T-TF) measured in normal and one pathological plasma with the various TF assays described above. Stimulation with LPS slightly increases T-TF but induces a very significant increase of EV-TF in normal blood, and concentrations then become similar to those measured in pathological plasmas.

7.6 Discussion and Conclusions

These studies show the potential utility of measuring EVs as a biological marker or a prognostic indicator in various diseases associated with blood activation and increased risk of thromboembolic events. Approaches with flow cytometry (see Chapter 8) also allow meas-

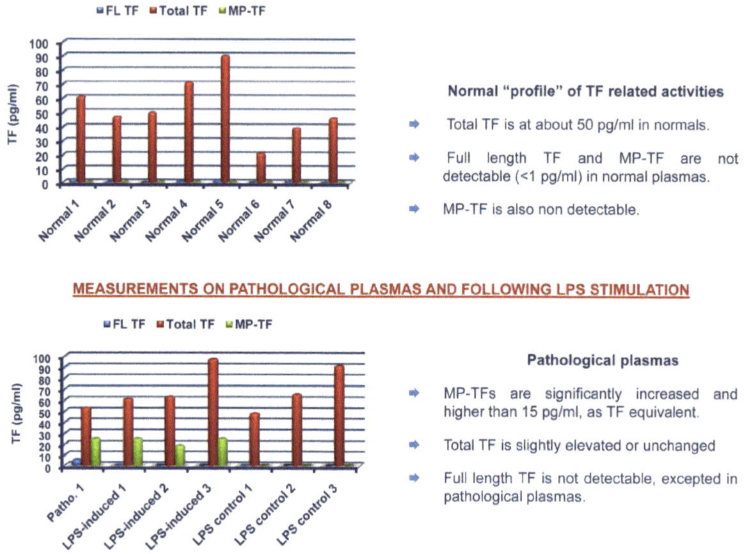

Figure 7.5 Measurement of the various TF activities, MV-TF, FL-TF, and T-TF in normal individuals, in normal blood stimulated with polysaccharides (LPS), and in pathological plasma. Only T-TF is measurable in normal individuals, and MV-TF becomes detectable only following LPS stimulation. In pathological plasma both T-TF and MV-TF are present.

urement of the level and cellular origin of EVs. However, they only focus on those with a larger size (>0.5 μm or >0.25 μm, depending upon the type and generation of instrument). Limited information on their potential procoagulant activity is also available through their binding of A5. However, a better correlation with clinical conditions is obtained with capture-based assays, which do not discriminate on EV size and measure the whole of procoagulant potential by combining A5 capture with the generation of thrombin. A subpopulation of vesicles exposing TF is of special relevance, especially in thrombotic complications associated with infectious diseases, sepsis, and cancer. Although they only represent a very low proportion of the total EVs, they can have a highly significant pathological implication for and contribution to disease evolution. Technology now allows measuring all these different species in a reliable and reproducible method. Many studies and publications will hopefully become available for reporting the high value of these specific determinations.

There is still a lack of standardization for these methodologies, as each laboratory is developing its own calibration and quality control material. Use of phospholipid liposomes or of relipidated TF is a convenient approach, but it needs to be validated and documented for each assay and each application. Therefore the normal ranges and pathological cutoff values are assay specific at this time. No doubt that scientific committees and standardization organizations will eventually propose materials that contribute toward improved standardization of procoagulant and EV measurement. This should then contribute to extending the possible applications of these assays and perhaps to their introduction in routine practice. EVs and their subpopulations are good candidates for large-scale evaluation analytes as they offer all the expected requirements: long half-life with little variations with time, values associated with diagnostic and prognostic value, variations reflecting the efficacy of antithrombotic therapies in diseases, and good correlation with pathological diagnosis and disease progression. Simplification and automation of laboratory assays for capture-based EV assays can also be a key factor for extending their applications.

Acknowledgments

The author would like to thank Anne Marie Vissac for contributing to these studies and Maxime Laroche for his participation in the development and validation of TF-based ELISAs.

References

1. Ardoin SP, Shanahan JC, Pisetsky DS (2007) The role of microparticles in inflammation and thrombosis, *Scand. J. Immunol.*, **66**(2–3), 159–165.
2. Michelson AD, Furman MI (1999) Laboratory markers of platelet activation and their clinical significance, *Curr. Opin. Hematol.*, **6**(5), 342–348.
3. Owen III AP, Mackman N (2011) Microparticles in hemostasis and thrombosis, *Circ. Res.*, **108**, 1284–1297.
4. Puddu P, Puddu GM, Cravero E, Muscari S, Muscari A (2010) The involvement of circulating microparticles in inflammation, coagulation and cardiovascular diseases, *Can. J. Cardiol.*, **26**(4), 140–145.

5. Connor DE, Exner T, Ma DD, Joseph JE (2010) The majority of circulating platelet-derived microparticles fail to bind annexin V, lack phospholipid-dependent procoagulant activity and demonstrate greater expression of glycoprotein Ib, *Thromb. Haemost.*, **103**(5), 1044–1052.
6. Freyssinet JM (2003) Cellular microparticles: what are they bad or good for? *J. Thromb. Haemost.*, **1**(7), 1655–1662.
7. Tans G, Rosing J, Thomassen MC, Heeb MJ, Zwaal RF, Friffin JH (1991) Comparison of anticoagulant and prociagulant activities of stimulated platelets and platelet-derived microparticles, *Blood*, **77**(12), 2641–4648.
8. Bernimoulin M, Waters EK, Foy M, Steele BM, Sullivan M, Falet H, Walsh MT, Barteneva N, Geng JG, Hartwig JH, Maguire PB, Wagner DD (2009) Differential stimulation of monocytic cells results in distinct populations of microparticle, *J. Thromb. Haemost.*, **7**(6), 1019–1028.
9. Morel O, Jesel L, Freyssinet JM, Toti F (2011) Cellular mechanisms underlying the formation of circulating microparticles, *Arterioscler. Thromb. Vasc. Biol.*, **31**, 15–26.
10. Perez-Pujol S, Marker PH, Key NS (2007) Platelet microparticles are heterogeneous and highly dependent on the activation mechanism: studies using a new digital flow cytometer, Cytometry A, **71**, 38–45.
11. Lawrie AS, Albanyan A, Cardigan RA, Mackie IJ, Harrison P (2009) Microparticle sizing by dynamic light scattering in fresh-frozen plasma, *Vox Sang.*, **96**(3), 206–212.
12. Goon PKY, Lip GYH, Boos CJ, Stonelake PS, Blann AD (2006) Circulating endothelial cells, endothelial progenitor cells, and endothelial microparticles in cancer, *Neoplasia*, **8**(2), 79–88.
13. Trappenburg MC, SchilfgaardeMv, Marchetti M, Spronk HM, Cate Ht, Leyte A, Terpstra WE, Falanga A (2009) Elevated procoagulant microparticles expressing endothelial and platelet markers in essential thrombocythemia, *Haematologica*, **94**(7), 911–918.
14. van Beers EJ, Schaap MCL, Berckmans RJ, Nieuwland R, Sturk A, DoormaalFFv, Meijers JCM, Biemond BJ (2009) Circulating erythrocyte-derived microparticles are associated with coagulation activation in sickle cell disease, *Haematologica*, **94**(11), 1513–1519.
15. Michelson AD, Barnard MR, Krueger LA, Frelinger AL 3rd, Furman MI (2000) Evaluation of platelet function by flow cytometry, *Methods*, **21**(3), 259–270.

16. Morel O, Toti F, Hugel B, Freyssinet JM (2004a) Cellular microparticles: a disseminated storage pool of bioactive vascular effectors, *Curr. Opin. Hematol.*, **11**, 156–164.

17. Morel O, Hugel B, Jesel L, Zobairi F, Bareiss P, Freyssinet JM, Toti F (2004b) Procoagulant membranous microparticles and atherothrombotic complications in diabetics, *Arch. Mal. Coeur Vaiss*, **97**(10), 1006–1012.

18. Fink K, Feldbrügge L, Schwarz M, Bourgeois N, Helbing T, Bode C, Schwab T, Busch HJ (2011) Circulating annexin V positive microparticles in patients after successful cardiopulmonary resuscitation, *Crit. Care*, **15**, R251.

19. Aupeix K, Hugel B, Martin T, Bischoff P, Lill H, Pasqualli JL, Freyssinet JM (1997) The significance of shed membrane particles during programmed cell death *in vitro*, and *in vivo*, in HIV-1 infection, *J. Clin. Invest.*, **99**(7), 1546–1554.

20. Davila M, Amirkhosravi A, Coll E, Desai H, Robles L, Colon J, Baker CH, Francis JL (2008) Tissue factor-bearing microparticles derived from tumor cells: impact on coagulation activation, *J. Thromb. Haemost.*, **6**(9), 1517–1524.

21. Key NS (2010) Analysis of tissue factor positive microparticles, *Thromb. Res.*, **125**(Suppl 1), S42–S45.

22. Thaler J, Ay C, Mackman N, Bertina RM, Kaider A, Marosi C, Key NS, Barcel DA, Scheithauer W, Kornek G, Zielinski C, Pabinger I (2012) Microparticle-associated tissue factor activity, venous thromboembolism and mortality in pancreatic, gastric, colorectal and brain cancer patients, *J. Thromb. Haemost.*, **10**(7), 1363–1370.

23. Wang JG, Manly D, Kirchhofer D, Rawlinski R, Mackman N (2009) Levels of microparticle tissue factor activity correlate with coagulation activation in endotoxemic mice, *J. Thromb. Haemost.*, **7**(7), 1092–1098.

24. Connor DE, Exner T, Ma DD, Joseph JE (2009) Detection of the procoagulant activity of microparticle-associated phosphatidylserine using XACT, *Blood Coagul. Fibrinolysis*, **20**(7), 558–564.

25. Thaler J, Ay C, Weinstabl H, Dunker D Simanek R Vormittag R, Freyssinet JM, Zielinski C, Pabinger I (2011) Circulating procoagulant microparticles in cancer patients, *Ann. Hematol.*, **90**(4), 447–453.

26. Amin C, Mackman N, Key NS (2008) Microparticles and cancer, *Pathophysiol. Haemost. Thromb.*, **36**(3–4), 177–183.

27. Lopez-Bezerra LM, Filler SG (2003) Endothelias cells, tissue factor and infectious diseases, *Braz. J. Med. Biol. Res.* **36**(8), 987–991.

28. Meziani F, Delabranche X, Asfar P, Toti F (2010) Bench-to-bedside review: circulating microparticles—a new player in sepsis? *Crit. Care*, **14**, 236.

29. Jayachandran M, Litwiller RD, Lahr BD, Bailey KR, Owen WG, Mulvagh SL Heit JA, Hodis HN, Harman SM, Miller VM (2011) Alterations in platelet function and cell-derived microvesicles in recently menopausal women: relationship to metabolic syndrome and atherogenic risk, *J. Cardiovasc. Trans. Res.*, **4**, 811–822.

30. Owen BAL, Xue A, Heit JA, Owen WG (2011) Procoagulant activity, but not number, of microparticles increases with age and in individuals after a single venous thromboembolism, *Thromb. Res.*, **127**(1), 39–46.

31. Esmon CT (2009) Basic mechanisms and pathogenesis of venous thrombosis, *Blood Rev.*, **23**(5), 225–229.

32. Li M, Yu D, Williams KJ, Liu ML (2010) Tobacco smoke induces the generation of procoagulant microvesicles from human monocytes/macrophages, *Arterioscler. Thromb. Vasc. Biol.*, **30**, 1818–1824.

33. Agouni A, Lagrue-Lak-Hal AH, Ducluzeau PH, Mostefai HA, Draunet-Busson C, Leftheriotis G, Heymes C, Martinez MC, Andriantsitohaina R (2008) Endothelial dysfunction caused by circulating microparticles from patients with metabolic syndrome, *Am. J. Pathol.*, **173**(4) 1210–1219.

34. Goichot B, Grunebaum L, Desprez D, Vinzio S, Meyer L, Schlienger JL, Lessard M, Simon C (2006) Circulating procoagulant microparticles in obesity, *Diabetes Metab.*, **32**(1), 82–85.

35. Curtis AM, Zhang L, Mendenilla E, Gui M, Wilkinson PF, Hu E, Giri J, Doraiswamy V, Gunda S, Burgert ME, Moor JS, Edelberg JM, Mohler III ER (2010) Relationship of microparticles to progenitor cells as a measure of vascular heath in a diabetic population, *Cytometry B*, **78**(5), 329–337.

36. Leroyer AS, Tedgui A, Boulanger CM (2008) Microparticles and type 2 diabetes, *Diabetes Metab.*, **34**(S1), 27–32.

37. Morel O, Hugel B, Jesel L, Lanza F, Douchet MP, Zupan M, Chauvin M, Cazenave JP, Freyssinet JM, Toti F (2004c) Sustained elevated amounts of circulating procoagulant membrane microparticles and soluble GPV after acute myocardial infarction in diabetes mellitus, *Thromb. Haemost.*, **91**(2), 345–353.

38. Sabatier F, Roux V, Anfosso F, Camoin L, Sampol J, Dignat-George F (2002) Interaction of endothelial microparticles with monocytic cells *in vitro* induces tissue factor-dependent procoagulant activity, *Blood*, **99**(11), 3962–3970.

39. Satta N, Toti F, Feugeas O, Bohbot A, Dachary-Prigent J, Eschwège V, Hedman H, Freyssinet JM (1994) Monocyte vesiculation is a possible mechanism for dissemination of membrane-associated procoagulant activities and adhesion molecules after stimulation by lipopolysaccharide, *J. Immunol.*, **153**(7), 3245–3255.

40. Chamouard P, Desprez D, Hugel B, Kunzelmann C, Gidon-jeangirard C, Lessard M, Baumann R, Freyssiner JM, Grunebaum L (2005) Circulating cell-derived microparticles in Crohn's disease, *Dig. Dis. Sci.*, **50**(3), 574–580.

41. Proulle V, Hugel B, Guillet B, Trichet C, Rafowicz A, Lambert T, Freyssinet JM, Dreyfus M (2004) Injection of recombinant activated factor VII can induce transient increase in circulating procoagulant microparticles, *Thromb. Haemost.*, **91**(5), 873–878.

42. Warkentin TE, Hayward CPM, Boshkov KL, Santos AV, Sheppard JAI, Bode AP, Kelton JG (1994) Sera from patients with heparin-induced thrombocytopenia generate platelet-derived microparticles with procoagulant activity: an explanation for the thrombotic complications of heparin-induced thrombocytopenia, *Blood*, **84**(11), 3691–3699.

43. Satta N, Toti F, Fressinaud E, Meyer D, Freyssinet JM (1997) Scott syndrome: an inherited defect of the procoagulant activity of platelets, *Platelets*, **8**(2–3), 117–124.

44. Choudhury A, Chung I, Blan AD, Lip GYH (2007) Elevated platelet microparticle levels in nonvalvular atrial fibrillation, *Chest*, **131**(3), 809–815.

45. Ederhy S, Di Angelantonio E, Mallat Z, Hugel B, Janower S, Meuleman C, Boccara F, Freyssinet JM, Tedgui A, Cohen A (2007) Levels of circulating procoagulant microparticles in nonvalvular atrial fibrillation, *Am. J. Cardiol.*, **100**(6), 989–994.

46. Michelsen AE, Brodin E, Brosstad F, Hansen JB (2008) Increased level of platelet microparticles in survivors of myocardial infarction, *Scand. J. Clin. Lab. Invest.*, **68**(5), 386–392.

47. Huisse MG, Ajzenberg N, Feldman L, Guillin MC, Steg PG (2009) Microparticle-linked tissue factor activity and increased thrombin acitivity play a potential role in fibrinolysis failure in ST-segment elevation myocardial infarction, *Thromb. Haemost.*, **101**(4), 734–740.

48. Morel O, Pereira B, Averous G, Faure A, Jesel L, Germain P, Grunebaum L, Ohlmann, Freyssinet JM, Bareiss P, Toti F (2009) Increased levels of procoagulant tissue factor-bearing microparticles within the occluded coronary artery of patients with ST-segment elevation myocardial

infarction: role of endothelial damage and leukocyte activation, *Atherosclerosis*, **204**(2), 636–641.

49. Kanazawa S, Nomura S, Kuwana M, Muramatsu M, Yamaguchi K, Fukuhara Q (2003) Monocyte-derived microparticles may be a sign of vascular complication in patients with lung cancer, *Lung Cancer*, **39**(2), 145–149.

50. Fu Y, Zhou J, Li H, Cao F, Su Y, Fan S, Li Y, Wang S, Li L, Gilbert GE, Shi J (2010) Daunorubicin induces procoagulant activity of cultured endothelial cells through phosphatidylserine exposure and microparticles release, *Thromb. Haemost.*, **104**(6), 1235–1241.

51. Stampfuss JJ, Censarek P, Bein D, Schrör K, Grandoch M, Naber C, Weber AA (2008) Membrane environment rather than tissue factor expression determines thrombin formation triggered by monocytic cells apoptosis, *J. Leukocyte Biol.*, **83**, 1379–1381.

52. Magdeleyns EJ, Keuren JF, Curvers J (2007) Factor Va, bound to microparticles released during platelet storage, is resistant to inactivation by activated protein C, *Transfusion*, **47**(10), 1880–1888.

53. Bouchama A, Kunzelmann C, Fehbi M, Kwaasi A, Eldali A, Zobairi F, Freyssinet JM De Prost D (2008) Recombinant activated Protein C attenuates endothelial injury and inhibits procoagulant microparticles release in baboon heatstroke, *Arterioscler. Thromb. Vasc. Biol.*, **28**, 1318–1325.

54. Del Turco S, Basta G, Lazzerini G, Evangelista M, Rainaldi G, Tanganelli P, Christensen JH, Schmidt EB, De Caterina R (2008) Effect of the administration of n-3 polyunsaturated fatty acids on circulating levels of microparticles in patients with a previous myocardial infarction, *Haematologica*, **93**(6), 892–899.

55. Lubsczyk B, Kollars M, Hron G, Kyrle PA, Weltermann A, Gartner V (2010) Low dose acetylsalicylic acid and shedding of microparticles *in vivo* in humans, *Eur. J. Clin. Invest.*, **40**(6), 477–482.

56. Morel O, Jesel L, Hugel B, Douchet MP, Zupan M, Chauvin M, Freyssinet JM, Toti F (2003) Protective effect of vitamin C on endothelium damage and platelet activation during myocardial infarction in patients with sustained generation of circulating microparticles, *J. Thromb. Haemost.*, **1**, 171–177.

57. Rössig L, Haendeler J, Mallat Z, Hugel B, Freyssinet JM, Tedgui A, Dimmeler S, Zeiher AM (2000) Congestive heart failure induces endothelial cell apoptosis: protective role of carvedilol, *J. Am. Coll. Cardiol.*, **36**(7), 2081–2089.

58. Chen HH, Vicente CP, He L, Tollefsen DM, Wun TC (2005) Fusion proteins comprising annexin V and kunits protease inhibitors are highly potent thrombogenic site-directed anticoagulants, *Blood*, **105**(10), 3902–3909.

59. Matijevic N, Wang YWW, Kostousov V, Wade CE, Vijayan KV, Holcomb JB (2011) Decline in platelet microparticles contributes to reduced hemostatic potential of stored plasma, *Thromb. Res.*, **128**(1), 35–41.

60. Morel O, Toti F, Hugel B, Bakouboula B, Camoin-Jau L, Dignat-George F, Freyssinet JM (2006) Procoagulant microparticles: disrupting the vascular homeostasis equation? *Arterioscler. Thromb. Vasc. Biol.*, **26**, 2594–2604.

Chapter 8

Flow Cytometry

Romaric Lacroix,[a,b] **Stephane Robert,**[a] **Florence Sabatier,**[a,b] **and Francoise Dignat-George**[a,b]
[a]*VRCM, UMR-1076, UFR de Pharmacie, INSERM/Aix-Marseille University, Marseille, France*
[b]*Hematology and Vascular Biology Department, CHU La Conception, AP-HM, Marseille, France*
Romaric.Lacroix@univ-amu.fr

Microvesicles (MVs) are submicron-size vesicles released from cell membranes in response to activation or apoptosis.[1] They are generally defined as 0.1 μm to 1 μm membrane fragments exposing the anionic phospholipid phosphatidylserine (PS) and membrane antigens representative of their cellular origin. MVs originating from several cell sources have been described in human plasma. Among them, platelet-derived MVs (PMVs) are believed to account for the majority of circulating MVs in healthy subjects, followed by erythrocyte-derived MVs (Ery-MVs).[2–4] It is now well recognized that MVs behave as vectors of bioactive molecules, playing a role in blood coagulation, inflammation, cell activation, and cancer spreading.[5–7] In clinical practice, circulating MVs originating from blood and vascular cells are elevated in a variety of prothrombotic

Extracellular Vesicles in Health and Disease
Edited by Paul Harrison, Christopher Gardiner, and Ian L. Sargent

ISBN 978-981-4411-98-1 (Hardcover), 978-981-4411-99-8 (eBook)
www.panstanford.com

and inflammatory disorders; cardiovascular diseases; autoimmune, infectious diseases; and cancers.[8-10] In these clinical settings, MVs may give information about nonaccessible tissues (tumors, endothelium, placenta, etc.), correlate with disease activity, have a prognostic value to identify patients with thrombotic or vascular risk, and help in treatment monitoring.[11-14] Among the different methodologies available to measure MVs in biological samples, flow cytometry (FCM) is the most commonly used technique.[15] The goal of this chapter is to review the recent progress in terms of standardization, size resolution, and probing to overcome the current limitations of this methodology.

8.1 Advantages and Limitations of FCM

8.1.1 Advantages

Flow cytometry (FCM) analyzes cells and particles by suspending them in a stream of fluid and passing them by a flow cell interacting with a laser beam. Scattering of light and fluorescence emissions' detection allow simultaneous multiparametric analysis of the physical and multiple antigen characteristics of up to thousands of particles per second. Thus, microvesicle (MV) measurement by FCM relies mainly on the combination of their scatter light properties and the expression of phosphatidylserine (PS) and specific antigens detected by fluorescence. FCM features give a high potential for MV analysis; indeed, FCM has the major advantage to identify and quantify MV subpopulations, because each particle is interrogated individually. The multiparameter analysis improves the specificity of MV detection. The speed of this technique contrasts with the sensitive but time-consuming microscopy techniques. Absolute counting is also possible in a wide dynamic range using generally counting beads of a known concentration.[16] Finally, this technique is available in many research and clinical laboratories.

8.1.2 Limitations

Most FCM limitations are due to the fact that MVs are very small and heterogeneous in size, ranging from 0.1 μm to 1 μm according to electron microscopy. Currently available flow cytometers (FCMrs) are unable to count all MVs except the larger ones.[17] However, the size of the smallest MV an FCMr can detect is a difficult question to answer. As discussed later in this chapter, sizes of calibration beads are not perfectly equivalent to the size of MVs, meaning that if an FCMr can resolve two small latex beads, this does not imply that it can resolve two MVs of the same size range. In addition, performances of the instruments vary between various types of cytometers and even among individual instruments of the same type and also evolve with time on the same apparatus.[18] However, we can estimate that the smallest detected MV is around 400 nm for a standard FCMr (sd-FCMr) and 200 nm for a high-sensitivity FCMr (hs-FCMr) using scatter light parameters. Depending on the antigen density on the MVs and the signal/noise ratio of the probe, events smaller than 200 nm can also sometimes be detected by fluorescence.[19]

Limitations to count all MVs are technologically driven, being dependent both on the fluorescence sensitivity of the instrument and on the intrinsic resolution limit of instruments for the size-related parameter used for the initial discrimination of interesting events from background noise. It is often quoted that the most commonly used laser wavelength (488 nm) is a major determinant of the limit of detection,[20,21] but despite its impact in the physics of light scattering, numerous experiments have proven this statement is not entirely accurate.[22,23] Among potential sources of instrument noise that impede MV detection, optical noise has generally the greatest impact over electronic and fluidic noise. Thus, optimizing settings can greatly improve results in MV enumeration.[17] Because forward scatter (FS/FSC) has to be measured in a narrow scattering angle range close to a very intense light source (the laser beam itself), more precise alignment is required than for wide-angle scattering or fluorescence measurements.[24–26] The position of the obscuration bar, which blocks direct laser light, is also critical. Therefore, fine optical alignment by an experienced technician may be highly beneficial for MV analysis, more than for any other normal FCM application. This optical noise can also come from dust particles and/or optical coupling gel stuck on the side of the flow cell. This noise increases slowly with time. It may be prevented by a protected atmosphere in the cytometer's lab or reduced by cleaning the external side of the flow cell with optical grade alcohol (e.g., isopropanol) during regular technical service.[27] In conclusion, laser alignment and careful optical cleaning by service technicians are critically needed to improve MV detection by an FCMr.

To overcome these limitations, two main strategies have been developed: (a) standardization of the MV window of analysis in a defined but reduced size range in order to try and obtain reproducible results in clinical studies and (b) improvement of the detection of small-size MVs with the introduction of the hs-FCMr.

8.2 Standardization of the MV Gate

8.2.1 Gating Strategies

To measure MVs by FCM, two main strategies have been proposed. The first selects MVs according to their size, using scatter parameters (scatter-based threshold or discriminator) or impedance, and then analyzes fluorescent signals only on the populations preselected on size. The second preselects MVs on the basis of fluorescence staining first (using a fluorescence threshold or discriminator) and then eventually restricts the population(s) of interest via scatter gating on size. Each option has advantages and drawbacks. Fluorescence triggering may offer the greatest sensitivity for MV detection but at the expense of specificity.[28] Because the event analysis is not restricted to any size region, all kinds of elements (specific MVs, free dyes, antibody aggregates, etc.), giving a fluorescence signal above a limit usually defined by an isotype control, will be counted as events of interest. This methodology is therefore more sensitive than the scatter strategy to small differences between isotype controls and specific antibodies. Other limitations in the fluorescence strategy are mainly the low signal-to-background ratio and variability in staining intensity. Moreover, standardized strategies to reproducibly position the fluorescence threshold are in their infancy. In the scatter-gating strategy, strongly labeled, small events can be missed but standardization of the MV measurement is also possible. Interestingly with the recent improvement of the scatter resolution on modern instruments, setting of the fluorescence detection limit is becoming more critical and standardization of fluorescence now also seems to be required with this strategy.

8.2.2 Calibration Beads

MV populations are generally defined by using size calibration beads. Beads of 1 μm diameter were first used to draw the upper limit of the MV population and to discriminate the MV population from larger events such as platelets.[29,30] However, although this facilitated detection of MVs below 1 μm, the lower limit was not well defined as a consequence of being present within a moving zone of the instrument noise. A few years ago, a blend of commercial fluorescent beads, called Megamix, was developed by Biocytex (Marseille, France) in an attempt to better standardize both upper and lower limits of detection. It provides size-calibrated fluorescent beads of, respectively, 0.5 μm and 0.9 μm purposely formulated with a 2:1 numerical ratio. The 0.9 μm beads help to place the upper limit of the MV-gated region; the 0.5 μm beads help to set up the FS threshold/discriminator and lower MV limit at a standardized level corresponding to its median value. This is easily achieved on the basis of the intrinsic 2:1 ratio of 0.5 μm to 0.9 μm beads in order to always include the same percentage of the 0.5 μm beads in the analysis.[16,31] This calibrated-bead strategy allowed full, long-term control of the FCM-based protocol and reproducible PMV counts over time on a Cytomics FC500 (Beckman Coulter [BC]). Optimal settings were easily transferred between three instruments of the same type using Megamix as a stable template. Similar PMV counts were therefore obtained using the three instruments.[16]

Three International Society on Thrombosis and Haemostasis (ISTH) scientific and standardization subcommittees (SSC Vascular Biology, Disseminated Intravascular Coagulation [DIC], and Haemostasis and Malignancy) have initiated a project aimed at standardizing the enumeration of cellular MVs by FCM. Using the Megamix beads strategy, a first collaborative workshop was set up to, first, establish the resolution and level of background noise of the FCMrs currently used in laboratories with respect to the strategy requirements, and, second, define the interinstrument reproducibility of PMV enumeration in human plasma. The study included 40 laboratories and a total of 59 FCMrs.[17] As a result, Megamix beads proved to be a useful standard that allows reliable assessment of an instrument's ability to detect MVs within this standardized region. However, they do not constitute a universal biological standard for PMV enumeration. Indeed, PMV concentrations appeared consistent among instruments that measure the FS parameter with a relatively wide solid angle (1–19°, BC). However, this strategy was inapplicable without substantial modifications among instruments with FS measured with a lower solid angle (1–8° Becton–Dickinson [BD]).[17] Thus, this study indicates that standardization of PMV enumeration by FCM was feasible but was dependent on intrinsic characteristics of both the FCMr and the calibration strategy.

8.2.3 Value of Calibration Beads for MV Sizing

Light scattered at low solid angles along the axis of the laser beam (FS/FSC) tends to be more related to the size of the particles than light scattered at large solid angles (side scatter [SS/SSC]), which is highly influenced by the cells' internal structure (granularity and nucleocytoplasmic ratio).[25] This is why FS/FSC is generally used to define the analysis gate for MVs. Despite the fact that Megamix beads were not intended to be used as a size calibrator but rather as a tool to monitor MV enumeration in a standardized manner, a debate began on the value of calibrated polystyrene beads for sizing MVs by FCM. The use of polystyrene beads remains an imperfect model for general size calibration in FCM. Indeed, many factors other than event size influence FS/FSC, among which are the refractive index, surface roughness, and the presence of any absorptive material.[25] The FS/FSC signal is also strongly affected by the laser wavelength and by the range of solid angles over which scattered light is collected. Consequently, it was showed on Apogee A40 (Apogee system) that the difference in refractive index between polystyrene beads (Megamix) and biological MVs is so drastic that the MV gate, defined between the 0.5 μm and the 0.9 μm Megamix beads, includes, in fact, a large proportion of the platelet population.[32] Accordingly, it was deduced by theoretical calculations and experiments with Apogee A40 that setting the gate with 0.5 μm and 0.9 μm Megamix beads results in the measurement of equivalent biological vesicles between the diameters of 1.4 μm and 2.7 μm, respectively. These size values were far outside the expected size range for MVs and suggested that all previously published PMV counts, using this methodology, were in fact counts of remaining platelets. However, several laboratories did not reproduce such overlap between the relative positions of beads and the platelet population on other instruments.[33,34] We also did not find a dramatic difference between the nominal diameter of the beads and the size of giant viruses on the Gallios instrument (BC), demonstrating that the difference in refractive index between beads and biological particles does not seem to impact extensively on FS-/FSC-based sizing on this cytometer.[34] However, an important difference was observed when using a large angle (closer than SS/SSC) between beads and the MV population.[33]

As a potential explanation, the FS/FSC signal is strongly affected by the range of solid angles over which scattered light is collected. Because manufacturers of FCMrs do not use the same optical design for FS/FSC measurement, it is unlikely that exactly the same results could be obtained by measuring the same events within different instruments.[17] The surprising results found on Apogee A40/A50 are explained by the fact that a larger angle than on usual FS/FSC was used to collect scattered light. Polystyrene beads therefore still behave as potential suitable tools with which to define windows for MV analysis, provided that the intrinsic technical characteristics of each FCMr are understood and taken into account.

8.2.4 Standardization Perspectives

Different behaviors of biological MVs and beads on different FCMrs show that the current ISTH protocol developed on BC instruments is not fully transposable to all types and brands of FCMrs. Moreover, beads or microorganisms from a nominal diameter close to the wavelength of the laser also highly scatter light at large angles (SS/SSC). In some cases, the resolution obtained is even better than when FS/FSC is used, especially on BD instruments.[35] Thus, some instruments can easily discriminate two beads from 100 nm and 300 nm using SS/SSC. Therefore, taking into account the impact of the large angle on the differential sizing of beads and MVs as discussed in the previous paragraph, the SS/SSC parameter can be also be chosen to alternatively act as a trigger for measuring MV populations.[36] Philippe Poncelet (BioCytex, France) therefore recently proposed a comparison of forward- versus side-scatter strategies to standardize MV counts by FCM.[37] This strategy aimed to provide scatter-based reference levels for threshold and/or gating, permitting general standardization of MV counts across platforms. Comparisons of PMV counts were therefore measured on the same plasma samples on an FCMr of each group (FS/FSC and

SS/SSC optimized). Delineation of small PMVs was made in FS on Gallios (BC) using 0.3 µm and 0.5 µm beads; large PMVs were found between 0.5 µm and 0.9 µm beads. The boundaries of the same two subsets were found on the SSC scale of a BD FCMr to between 0.17 µm and 0.22 µm bead equivalents (small PMVs) and between 0.22 µm and over 0.5 µm bead equivalents (large PMVs). Similar counts were obtained on these size-defined PMV subsets using different platforms, demonstrating the ability of this standardization strategy to monitor interinstrument reproducibility. This strategy is going to be evaluated in a future ISTH SSC international workshop.

In the near future, biological standards such as homogeneous suspensions of lipid vesicles, bacteria, or giant viruses may be very helpful to qualify homogeneous preparations of polystyrene beads and could be circulated as standards within the scientific community.

8.3 Small-Size MV Detection by FCM

Recently a new generation of instruments became available with technical characteristics in the fluidic or optical system theoretically more suitable for MV detection. Indeed, these hs-FCMrs show greatly improved scatter performance in terms of resolution and background. Improvement of FS resolution on Gallios is illustrated on Fig. 8.1 using a blend of fluorescent beads of 0.1, 0.3, 0.5, and 0.9 µm. This is also the case, for example, with instruments such as LSRII Fortessa, the BD Influx,[19,23] and Apogee A40/50.[32,33] More progress may rely on specific optical designs or amplification strategies to detect smaller vesicles. Most of these instruments are fitted with a high-performance photomultiplier instead of a photodiode for detection of FS/FSC signals. Gallios takes advantage of a differential amplification of signal between internal (1–8°) and external (8–19°) angles measured by the FS detector (W2 strategy) (Fig. 8.1).[22] The Influx cell sorter uses also wide-angle FSC instead of conventional FSC to improve small-particle detection. It is also optionally equipped with a pinhole system for detection of small-size particles. Remarkably, the optimized BD Influx FCMr can detect particles with sizes as small as 100 nm and with sufficient resolution to easily distinguish between 100 nm and 200 nm particles on the basis of light scatter.[19,23]

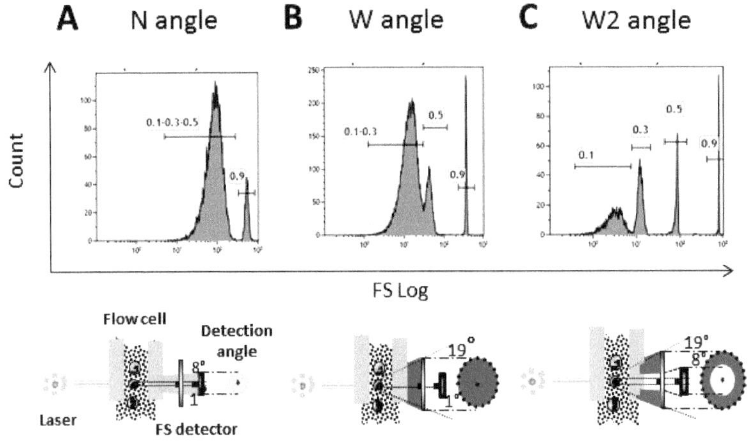

Figure 8.1 FS resolution according to the collection angle option on Gallios. FS resolution was tested on a mix of four beads (0.9, 0.5, 0.3, and 0.1 μm beads). Their distributions are shown as monoparametric FS histograms. Configurations shown in A, B, and C show acquisition with an FS normal collection angle *N* (1–8°), with an FS wide angle *W* (1–19°) and with an FS wide angle plus a scatter signal treatment amplifying the signal at the external angle *W*2 (8–19°), respectively.

Consequently, a clear improvement in MV counting was observed. When PMV profiles between the sd-FCMr and the hs-FCMr were compared at their respective optimal cutoff, a new but unknown subset of less intense, smaller PMVs appeared in the analysis, coupled with a marked increase in PMV counts. A new standardization protocol was then proposed using calibrated beads to reproducibly measure both large and small MVs using Megamix plus beads (BioCytex) (Fig. 8.2).[22] On Apogee A50, PMV quantification in plasma from a healthy subject sampled in ideal preanalytical conditions increase the annexin V–positive PMV count from 100 μl to 1,370 μl.[33] This improvement is also observed for other MV subpopulations, including Ery-MVs, leukocyte-derived MVs (LMVs), and endothelial-derived MVs (EMVs).[22,38] With the defined adaptations of BD Influx and an optimized protocol for vesicle labeling and measurement using a fluorescent threshold, this method is also superior in the detection and characterization of nanosized cell-derived vesicles from *in vitro* cultures.[19]

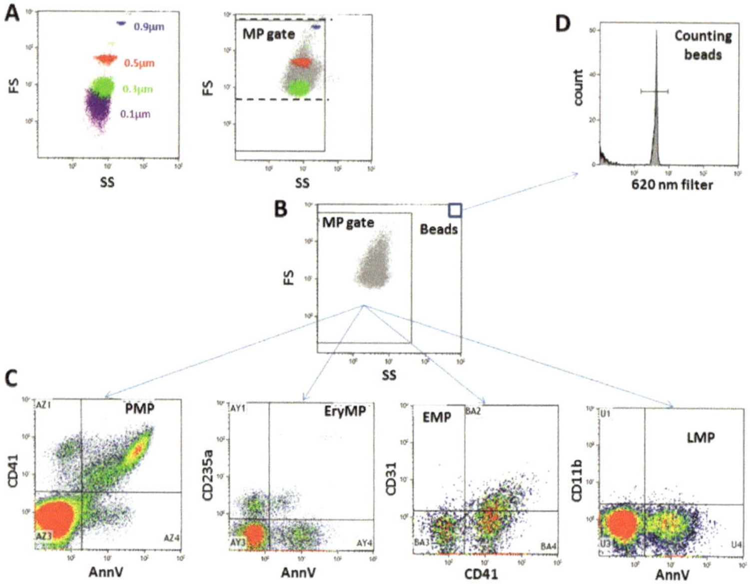

Figure 8.2 Standardized flow cytometry protocol for MV subpopulation analysis on representative samples from a healthy donor. (A) A standardized MV gate is defined by Megamix plus beads. Beads were first analyzed using a fluorescence threshold. Each subset is characterized by colored dots: 3 μm (pink), 0.9 μm (blue), 0.5 μm (red), 0.3 μm (green), and 0.1 μm (violet) (left panel). Then, to move to MV analysis conditions, the threshold was switched to FS, leading to the appearance of a background (gray dots) and the disappearance of all 0.1 μm beads as well as 50% of 0.3 μm beads (right panel). A reproducible FS threshold can be set up by adjusting the FS discriminator value and the PMT voltage to move from a 0.3 μm/0.5 μm bead ratio of 2 (66% vs. 34%) originally to a ratio of 1 (50% of each) at the requested 0.3 μm bead equivalent. (B) On a representative plasma sample, events included in the previously defined MV gate are selected. (C) Among MV gate events, MV subpopulations are defined as PMVs (annexin V/CD41), Ery-MVs (annexin V/CD235a), EMVs (annexin V/CD31+/CD41−), and LMVs (annexin V/CD11b). (D) Absolute enumeration of MV subpopulations is performed using fluorescent counting beads.

Interestingly, access to the measurement of the MV subset of a small size gives access to new information: applied to coronary patient samples, resulting in 8- to 20-fold increases in MV counts as compared with an sd-FCMr. Interestingly, the ratio between small and large MVs varied according to clinical status but also depended on the MV subset.[22] Moreover, significant differences in LMVs between patients with stable and unstable atherosclerotic plaques were only detected using the new standardized protocol (unpublished data). Little is known so far about the differential nature and properties of the large and small MVs detected by FCM. Several hypotheses have been formulated for large PMVs, including small platelets, aggregates, or even swarm detection of small MVs simultaneously illuminated by the laser beam and counted as a single event signal.[39] There are several arguments against the swarm hypothesis, including (1) filtration experiments can remove all large MVs, (2) analysis of a mixed population of PMVs labeled with different fluorochromes does not result in the detection of double-labeled events, and (3) design of a separate population by simultaneous detection of small events has poor probability. In any case, the recent detection of small MVs by FCM results in many new questions. Among the new technical issues is the poor fluorescence intensity of the smallest MV events, suggesting our current labeling strategies need to be improved and optimized.

8.4 Labeling Strategies

MVs are usually identified and enumerated by measuring specific antigens at their surface. Because of the MV size, however, the amount of any surface marker is drastically reduced in comparison to intact cells. It is therefore possible that some smaller MVs expressing specific antigens could remain below the limit of detection even if labeling is optimally performed. This is suggested by the lack of true separation between events considered as positive and the fluorescent noise. This stresses the relevance of calibrating the fluorescence axis in addition to the scatter axis by developing new reference standards such as fluorescent liposomes.[40]

8.4.1 Labeling Controls

Better controls in MV measurement are needed to differentiate true MVs from other cell debris and especially those generated during preanalytical steps. To further improve specificity in PMV measurement, Mobarrez et al. have proposed the use of phalloidin.[41] Phalloidin is a cyclic peptide that binds to f-actin with high affinity.

Intact platelets do not have exposed f-actin, but if the cell membrane becomes disrupted, for example, by freezing, phalloidin will bind to exposed f-actin. Thus, phalloidin binding may indicate the presence of platelet fragments, which otherwise may falsely be counted as PMVs. Calcein AM, a nonfluorescent marker that becomes fluorescent upon cleavage by cytosolic esterases only stains intact MVs/cells and is not reactive with cellular fragments.[42,43] The use of detergents has also been suggested by Gyorgy et al. to help differentiate MVs from potential immune complexes.[44]

In the absence of consensus and standardized recommended antibodies and fluorochrome panel, the quality of obtained results may be improved by following simple rules.[27] High-g centrifugation of antibodies before use may help to reduce interference with aggregates. Titration of antibodies must be performed on appropriate MV samples before use to avoid use of excess free antibody and improve resolution of specific events compared to the noise.[45] More recently, Orozco et al. have stressed the need to study and quantify the analyzed population before any antibody labeling.[46]

Currently most groups quantifying MVs by FCM are using commercially available isotypic controls to set limits for the discrimination of antigen-positive and antigen-negative events. Due to low fluorescence intensity, small variations in determining the antibody background binding by isotypic controls may have significant consequences on the quantification of antigen-positive MVs. Variation in staining of the same MV population by different isotypic controls has been well demonstrated by Trummer et al.[47] As described, their variability regarding the background level can be mainly attributed to (1) noncompliance of isotype with antibody subgroup, (2) conjugation variation resulting in differences in the fluorochrome-to-protein ratio (F:P ratio), and (3) isotypic control concentration not exactly matching antibody concentration. In our hands, a useful strategy consists of checking the overlapping of histograms between a conjugated specific antibody and its paired isotypic control on MV populations definitely known to be nonreactive with the specific antibody or by use of an MV-free plasma. Alternatively, the use of antigen-negative MVs to adjust instrument settings has been suggested with the use of an Fc receptor (FcR)-blocking reagent prior to antibody labeling.[47] When identifying a new antigen on MVs by FCM, it is recommended to remove the signal using a high concentration of an unlabeled specific antibody to saturate the antigen in order to prove the specificity of the detection signal.

8.4.2 Fluorescent Probes

8.4.2.1 General probes

Annexin V binding has been extensively used as evidence of the PS-rich surface.[48] This placental protein with high affinity and a strict specificity for aminophospholipids, however, requires the presence of physiologic calcium concentrations.[49] A negative control is provided by annexin V in a buffer that lacks bivalent cations or use of ethylenediaminetetraacetic acid (EDTA) to prevent any specific binding. The addition of calcium to diluted plasma samples may also cause problems with fibrin formation and subsequent analysis depending upon the final dilution factors utilized. This can be controlled by using agents that inhibit thrombin generation of fibrin cross-linking. It is controversial whether staining for phospholipids with annexin V should be performed. Some groups define MVs only by size and surface antigen staining without testing for annexin V positivity.[8,50] In addition, the existence of an annexin V–negative MV population has also been suggested. It has been reported that the concentration of MVs (without the use of annexin staining) is 30 times higher than the concentration of MVs that are detected by annexin V.[51] However, it has been demonstrated that simply analyzing unfiltered buffers results in a significant amount of background noise, cell debris, and precipitates, which often have the same size range as MVs.[52] This could influence and disturb the analysis and quantify false-positive MVs. Therefore, it seems preferable that positive staining for annexin V be applied to be able to distinguish true events from cell debris or precipitates. The binding of annexin V to MVs is influenced by the calcium concentration and the membrane PS content. Consequently, the concentration of calcium in the buffer should be titrated for optimal staining of MVs.

Alternatively to annexin V, other probes were proposed to detect whole circulating MVs. Lactadherin is an opsonin released by stimulated macrophages and characterized by a PS-bonding motif and an integrin-binding motif. Lactadherin bridges PS-exposing apoptotic cells to macrophages, facilitating their engulfment.[53,54] This protein has demonstrated higher affinity than annexin V for PS and has the potential advantage of non-calcium-dependent binding to PS.[55] It is also more expensive. BODIPY-maleimide has been identified as a potential generic stain for detection and quantification of MVs because of its property to attach to biological membranes via cysteine residues and thiol groups in proteins.[56–58]

On the basis of recent findings that MVs may also contains nucleic acids, nucleic acid–staining dyes have also been tested in the detection of MVs generated *in vitro* from cell lines.[59] Ullal et al. utilized the cell-permeant dye SYTO 13, which binds ribonucleic acid (RNA) to detect MVs.[60] For all these potential generic markers for total enumeration of MVs, the determination of a limit between positive and negative events is often hazardous due to a poor signal/noise ratio, high fluorescent noise in plasma, and the lack of an appropriate negative control. Their practical advantages need to be evaluated further in terms of specificity and performance in plasma samples.

8.4.2.2 Specific probes

FCM uses fluorescently labeled specific antibodies to identify the cellular origin of MVs. MVs have an antigenic composition originating, in theory, from their parent cells; however, they also differ regarding some crucial characteristics important for their detection, such as antigenic density, loss of antigen by shedding or during formation, size, and membrane curvature. Because of these differences, we cannot predict the ability of an antibody to efficiently detect specific MVs by FCM only on the basis of cellular models. Over the past years, some antibodies have been selected to distinguish MVs derived from platelets, leucocytes, erythrocytes, and endothelial cells (Table 8.1).[58,61] In addition, antibodies have been used for the detection of MVs derived from malignant cells, such as MUC1 for breast and pancreatic cancer cells, Fas ligand (FasL) for melanoma cells, epidermal growth factor receptor variant III (EGFRvIII) for glioblastoma cells, and CXCR-4 for leukemic cells.[62] Table 8.1 summarizes the most widely used specificities with regard to their recognition of MV subsets and to the relative intensity of the fluorescent signal obtained with *in vitro* or *in vivo* MVs. Depending on the population studied and its relative abundance in human plasma, the difficulties to identify them can vary. Platelet- and erythrocyte-derived MVs are the most abundant populations and benefit from specific antigens well represented at their surface, such as CD41 and CD42 for PMVs and CD235a for Ery-MVs. LMVs are physiologically less represented. CD45 is present on all leukocytes and is often used to identify LMVs. For leukocyte subsets, the following antibodies are used: CD14/CD11b (monocytes) and CD66b (granulocytes). Most of the leukocyte antigens are weakly expressed, and it is possible that a large fraction of these MVs may be also lost during analysis. The challenge of rare event detection is even more difficult when this is associated with the measurement of only weakly expressed antigens as is the case with EMVs. CD105 (endoglin) is not only expressed on endothelial cells but also weakly expressed on leukocytes. EMVs are therefore CD105+ and CD45−. CD144 (VE-cadherin) is also used as a marker for EMVs, but its expression is very weak. CD54 (ICAM1), CD62E (E-selectin), CD62P (P-selectin), and CD106 (VCAM1) can be markedly up-regulated on endothelial cells by various agonists and therefore may provide a measure of whether subsets of EMVs are derived from activated endothelial cells.[27,63]

Table 8.1 Main antigens used in measuring cell-derived MV subpopulations

	Antigens	**Specificity**	**Sensitivity (fluorescence intensity in vitro)**
Pan-MV	PS	++	+++
PMV	CD41	+++	+++
	CD42 (a or b)	+++	+++
	CD61	++	++
Ery-MV	CD235a	+++	+++
LMV	CD45	+++	+
	CD11b[a]	+++	++
	CD14[b]	+++	+
	CD66b[c]	+++	+
	CD15[c]	+++	++
EMV	CD144	+++	+/−
	CD146	++	+++
	CD105-LMV	++	++
	CD31-PMV	+	+++
	CD62E	+++	+

Source: Adapted from Lacroix et al., *Semin. Thromb. Haemost.*, 2010

Abbreviations: MV, Microvesicles; PMV, Platelet-MV; EryMV, Erythrocyte-MV; LMV, Leukocytes-MV; EMV, Endothelium-MV; PS, Phosphatidylserine.

[a]Myeloid and NK cell–derived MVs

[b]Monocyte-derived MVs

[c]Granulocyte-derived MVs

Better sensitivity of the hs-FCMr allows developing multilabeling strategies that improve the efficiency and specificity of MV subpopulations detection by FCM,[41] but new MV-dedicated reagents are also required.

8.5 Other Flow Cytometry Methods

Over the past years, other FCM-based methods have been used to analyze MVs, such as impedance FCM and, more recently, imaging FCM.

Impedance FCM is based on the Coulter principle. It determines the electronic volume of a particle, which is directly proportional to the change of impedance associated with the displacement of electrolyte in a flow cell orifice by the particle of interest. Analysis has been performed on a modified version of the Quanta MPL instrument (BC, Miami, FL). Measurement depends on the diameter of the flow cell aperture, and smaller apertures with diameters of 25 μm or 40 μm are required in order to analyze small particles and vesicles.[31,64] Although this system allowed accurate sizing and excellent discrimination between the MV population and platelets, its full potential use remained limited by a high electrical background.

Amnis has also developed an imaging FCMr (ImageStream), which may overcome some of the problems associated with flow cytometric assessment of MVs. The instrument provides an actual measurement of size rather than FS. The signal/noise ratio for small particles is markedly better by using imaging cytometry. Because signals arising from small objects occupy a minimal number of pixels, there is very little pixel-associated noise, and thus ImageStream is very sensitive to MV fluorescence. This is especially important for the analysis of MVs that exhibit low levels of markers on their surface and hence low fluorescence intensity when compared with the parent cell. Furthermore, obvious coincident events can be flagged and potentially included in the analysis. However, the limitation of ImageStream is the accurate measurement of size for small particles and vesicles.[63]

8.6 Conclusions

The standardization process to measure MVs by FCM remains a challenge, especially for multicenter studies. The diversity of scatter measurements between manufacturers certainly impedes standardization. However, provided that the intrinsic behavior of any instrument for size-related measurements has been taken into account, polystyrene beads remain as useful tools to standardize windows for MV analysis. The new-generation hs-FCMrs also maintain the lead of FCM in the MV analysis field, with potential access to new biological information. Another challenge will be to develop new tools to improve the fluorescence signals on small vesicles. Validation of all of these advances in large series of pathological settings will be mandatory to evaluate their true impact upon clinical practice.

References

1. Freyssinet JM, Dignat-George F (2005) More on: measuring circulating cell-derived microparticles, *J. Thromb. Haemost.*, **3**(3), 613–614.
2. Heijnen HF, Schiel AE, Fijnheer R, Geuze HJ, Sixma JJ (1999) Activated platelets release two types of membrane vesicles: microvesicles by surface shedding and exosomes derived from exocytosis of multivesicular bodies and alpha-granules, *Blood*, **94**(11), 3791–3799.
3. Lynch SF, Ludlam CA (2007) Plasma microparticles and vascular disorders, *BJH*, **137**(1), 36–48.
4. Piccin A., Murphy WG, Smith OP (2007) Circulating microparticles: pathophysiology and clinical implications, *Blood Rev.*, **21**(3), 157–171.
5. Al-Nedawi K, Meehan B, Rak J (2009) Microvesicles: messengers and mediators of tumor progression, *Cell Cycle*, **8**(13), 2014–2018.
6. Ardoin SP, Shanahan JC, Pisetsky DS (2007) The role of microparticles in inflammation and thrombosis, *Scand. J. Immunol.*, **66**(2–3), 159–165.
7. Davizon P, Lopez JA (2009) Microparticles and thrombotic disease, *Curr. Opin. Hematol.*, **16**(5), 334–341.
8. Boulanger CM, Amabile N, Tedgui A (2006) Circulating microparticles: a potential prognostic marker for atherosclerotic vascular disease, *Hypertension*, **48**(2), 180–186.
9. Diamant M, Tushuizen ME, Sturk A, Nieuwland R (2004) Cellular microparticles: new players in the field of vascular disease? *Eur. J. Clin. Invest.*, **34**(6), 392–401.
10. Zwicker JI (2008) Tissue factor-bearing microparticles and cancer, *Semin. Thromb. Haemost.*, **34**(2), 195–198.

11. Amabile N, Heiss C, Chang V, Angeli FS, Damon L, Rame EJ, Mcglothlin D, Grossman W, De Marco T, Yeghiazarians Y (2009) Increased CD62e(+) endothelial microparticle levels predict poor outcome in pulmonary hypertension patients, *J. Heart Lung Transplant.*, **28**(10), 1081–1086.

12. Koga H, Sugiyama S, Kugiyama K, Watanabe K, Fukushima H, Tanaka T, Sakamoto T, Yoshimura M, Jinnouchi H, Ogawa H (2005) Elevated levels of VE-cadherin-positive endothelial microparticles in patients with type 2 diabetes mellitus and coronary artery disease, *J. Am. Coll. Cardiol.*, **45**(10), 1622–1630.

13. Nomura S, Inami N, Kimura Y, Omoto S, Shouzu A, Nishikawa M, Iwasaka T (2007) Effect of nifedipine on adiponectin in hypertensive patients with type 2 diabetes mellitus, *J. Hum. Hypertens.*, **21**(1), 38–44.

14. Sabatier F, Camoin-Jau L, Anfosso F, Sampol J, Dignat-George F (2009) Circulating endothelial cells, microparticles and progenitors: key players towards the definition of vascular competence, *J. Cell Mol. Med.*, **13**(3), 454–471.

15. Dignat-George F, Freyssinet JM (2005) *Questionnaire on Microparticle Detection and Characterization : A Retrospective Analysis*, Sydney: Vascular Biology Working Group, SSC ISTH.

16. Robert S, Poncelet P, Lacroix R, Arnaud L, Giraudo L, Hauchard A, Sampol J, Dignat-George F (2009) Standardization of platelet-derived microparticle counting using calibrated beads and a Cytomics FC500 routine flow cytometer: a first step towards multicenter studies? *J. Thromb. Haemost.*, **7**(1), 190–197.

17. Lacroix R, Robert S, Poncelet P, Kasthuri RS, Key NS, Dignat-George F (2010) Standardization of platelet-derived microparticle enumeration by flow cytometry using calibrated beads: results of ISTH SSC collaborative workshop, *J. Thromb. Haemost.*, **8**(11), 2571–2574.

18. Lacroix R, Robert S, Poncelet P, Kasthuri RS, Key NS, Dignat-George F (2010) Standardization of platelet-derived microparticle enumeration by flow cytometry with calibrated beads: results of the International Society on Thrombosis and Haemostasis SSC collaborative workshop, *J. Thromb. Haemost.*, **8**(11), 2571–2574.

19. Hoen EN, Van Der Vlist EJ, Aalberts M, Mertens HC, Bosch BJ, Bartelink W, Mastrobattista E, Van Gaal EV, Stoorvogel W, Arkesteijn GJ, Wauben MH (2011) Quantitative and qualitative flow cytometric analysis of nanosized cell-derived membrane vesicles, *Nanomedicine*, **8**(5), 712–720.

20. Furie B, Furie BC (2006) Cancer-associated thrombosis, *Blood Cells Mol. Dis.*, **36**(2), 177–181.

21. Yuana Y, Oosterkamp TH, Bahatyrova S, Ashcroft B, Garcia Rodriguez P, Bertina RM, Osanto S (2010) Atomic force microscopy: a novel approach to the detection of nanosized blood microparticles, *J. Thromb. Haemost.*, **8**(2), 315–323.

22. Robert S, Lacroix R, Poncelet P, Harhouri K, Bouriche T, Judicone C, Wischhusen J, Arnaud L, Dignat-George F (2012) High-sensitivity flow cytometry provides access to standardized measurement of small-size microparticles—brief report, *Arterioscler. Thromb. Vasc. Biol.*, **32**(4), 1054–1058.

23. Van Der Vlist EJ, Nolte-'t Hoen EN, Stoorvogel W, Arkesteijn GJ, Wauben MH (2012) Fluorescent labeling of nano-sized vesicles released by cells and subsequent quantitative and qualitative analysis by high-resolution flow cytometry, *Nat. Prot.*, **7**(7), 1311–1326.

24. Nebe-Von-Caron G (2009) Standardization in microbial cytometry, *Cytometry A*, **75**(2), 86–89.

25. Shapiro H (2003) *Practical Flow Cytometry*, Wiley-Liss.

26. Yao B, Luo GA, Feng X, Wang W, Chen LX, Wang YM (2004) A microfluidic device based on gravity and electric force driving for flow cytometry and fluorescence activated cell sorting, *Lab Chip*, **4**(6), 603–607.

27. Lacroix R, Robert S, Poncelet P, Dignat-George F (2010) Overcoming limitations of microparticle measurement by flow cytometry, *Semin. Thromb. Haemost.*, **36**(8), 807–818.

28. Mooberry M (2012) *ISTH SSC-Vascular Biology*, Liverpool, SSC minutes.

29. Aras O, Shet A, Bach RR, Hysjulien JL, Slungaard A, Hebbel RP, Escolar G, Jilma B, Key NS (2004) Induction of microparticle- and cell-associated intravascular tissue factor in human endotoxemia, *Blood*, **103**(12), 4545–4553.

30. Kim HK, Song KS, Lee ES, Lee YJ, Park YS, Lee KR, Lee SN (2002) Optimized flow cytometric assay for the measurement of platelet microparticles in plasma: pre-analytic and analytic considerations, *Blood Coagul. Fibrinolysis*, **13**(5), 393–397.

31. Zwicker JI, Lacroix R, Dignat-George F, Furie BC, Furie B (2012) Measurement of platelet microparticles, *Methods Mol. Biol.*, **788**(127–39.

32. Chandler WL, Yeung W, Tait JF (2011) A new microparticle size calibration standard for use in measuring smaller microparticles using a new flow cytometer, *J. Thromb. Haemost.*, **9**(6), 1216–1224.

33. Mullier F, Bailly N, Chatelain C, Dogne JM, Chatelain B (2011) More on: calibration for the measurement of microparticles: needs, interests,

and limitations of calibrated polystyrene beads for flow cytometry-based quantification of biological microparticles, *J. Thromb. Haemost.*, **9**(8), 1679–1681; author reply 81–82.

34. Robert S, Poncelet P, Lacroix R, Raoult D, Dignat-George F (2011) More on: calibration for the measurement of microparticles: value of calibrated polystyrene beads for flow cytometry-based sizing of biological microparticles, *J. Thromb. Haemost.*, **9**(8), 1676–1678; author reply 81–82.
35. Tocchetti EV, Flower RL, Lloyd JV (2001) Assessment of *in vitro*-generated platelet microparticles using a modified flow cytometric strategy, *Thromb. Res.*, **103**(1), 47–55.
36. Bode AP, Orton SM, Frye MJ, Udis BJ (1991) Vesiculation of platelets during *in vitro* aging, *Blood*, **77**(4), 887–895.
37. Poncelet P (2012) *ISTH SSC-Vascular Biology*, Liverpool, SSC minutes.
38. Pelletier F, Garnache-Ottou F, Angelot F, Biichle S, Vidal C, Humbert P, Saas P, Seilles E, Aubin F (2011) Increased Levels of Circulating Endothelial-Derived Microparticles and Small-Size Platelet-Derived Microparticles in Psoriasis, J Invest Dermatol,
39. Van Der Pol E, Van Gemert MJ, Sturk A, Nieuwland R, Van Leeuwen T. G (2012) Single vs. swarm detection of microparticles and exosomes by flow cytometry, *J. Thromb. Haemost.*, **10**(5), 919–930.
40. Nolan J (2012) ISTH SSC-Vascular Biology, Liverpool, SSC minutes,
41. Mobarrez F, Antovic J, Egberg N, Hansson M, Jorneskog G, Hultenby K, Wallen H (2010) A multicolor flow cytometric assay for measurement of platelet-derived microparticles, *Thromb. Res.*, **125**(3), 110–116.
42. Bernimoulin M, Waters EK, Foy M, Steele BM, Sullivan M, Falet H, Walsh MT, Barteneva N, Geng JG, Hartwig JH, Maguire PB, Wagner DD (2009) Differential stimulation of monocytic cells results in distinct populations of microparticles, *J. Thromb. Haemost.*, **7**(6), 1019–1028.
43. Bratosin D, Mitrofan L, Palii C, Estaquier J, Montreuil J (2005) Novel fluorescence assay using calcein-AM for the determination of human erythrocyte viability and aging, *Cytometry A*, **66**(1), 78–84.
44. Gyorgy B, Modos K, Pallinger E, Paloczi K, Pasztoi M, Misjak P, Deli MA, Sipos A, Szalai A, Voszka I, Polgar A, Toth K, Csete M, Nagy G, Gay S, Falus A, Kittel A, Buzas EI (2010) Detection and isolation of cell-derived microparticles are compromised by protein complexes resulting from shared biophysical parameters, *Blood*, **117**(4), e39–e48.
45. Biro E, Nieuwland R, Sturk A (2004) Measuring circulating cell-derived microparticles, *J. Thromb. Haemost.*, **2**(10), 1842–1851.

46. Orozco AF, Lewis DE (2010) Flow cytometric analysis of circulating microparticles in plasma, *Cytometry A*, **77**(6), 502–514.
47. Trummer A, De Rop C, Tiede A, Ganser A, Eisert R (2008) Isotype controls in phenotyping and quantification of microparticles: a major source of error and how to evade it, *Thromb. Res.*, **122**(5), 691–700.
48. Tait JF, Gibson D (1992) Phospholipid binding of annexin V: effects of calcium and membrane phosphatidylserine content, *Arch. Biochem. Biophys.*, **298**(1), 187–191.
49. Falati S, Liu Q, Gross P, Merrill-Skoloff G, Chou J, Vandendries E, Celi A, Croce K, Furie BC, Furie B (2003) Accumulation of tissue factor into developing thrombi *in vivo* is dependent upon microparticle P-selectin glycoprotein ligand 1 and platelet P-selectin, *J. Exp. Med.*, **197**(11), 1585–1598.
50. Shet AS, Aras O, Gupta K, Hass MJ, Rausch DJ, Saba N, Koopmeiners L, Key NS, Hebbel RP (2003) Sickle blood contains tissue factor-positive microparticles derived from endothelial cells and monocytes, *Blood*, **102**(7), 2678–2683.
51. Horstman LL, Jy W, Jimenez JJ, Bidot C, Ahn YS (2004) New horizons in the analysis of circulating cell-derived microparticles, *Keio J. Med.*, **53**(4), 210–230.
52. Dey-Hazra E, Hertel B, Kirsch T, Woywodt A, Lovric S, Haller H, Haubitz M, Erdbruegger U (2010) Detection of circulating microparticles by flow cytometry: influence of centrifugation, filtration of buffer, and freezing, *Vasc. Health Risk Manag.*, **6**, 1125–1133.
53. Andersen MH, Graversen H, Fedosov SN, Petersen TE, Rasmussen JT (2000) Functional analyses of two cellular binding domains of bovine lactadherin, *Biochemistry*, **39**(20), 6200–6206.
54. Hanayama R, Tanaka M, Miwa K, Shinohara A, Iwamatsu A, Nagata S (2002) Identification of a factor that links apoptotic cells to phagocytes, *Nature*, **417**(6885), 182–187.
55. Shi J, Heegaard CW, Rasmussen JT, Gilbert GE (2004) Lactadherin binds selectively to membranes containing phosphatidyl-L-serine and increased curvature, *Biochim. Biophys. Acta*, **1667**(1), 82–90.
56. Dale GL, Remenyi G, Friese P (2005) Quantitation of microparticles released from coated-platelets, *J. Thromb. Haemost.*, **3**(9), 2081–2088.
57. Dasgupta SK, Abdel-Monem H, Niravath P, Le A, Bellera RV, Langlois K, Nagata S, Rumbaut RE, Thiagarajan P (2009) Lactadherin and clearance of platelet-derived microvesicles, *Blood*, **113**(6), 1332–1339.

58. Enjeti AK, Lincz LF, Seldon M (2007) Detection and measurement of microparticles: an evolving research tool for vascular biology, *Semin. Thromb. Haemost.*, **33**(8), 771–779.

59. Reich CF, 3rd, Pisetsky DS (2009) The content of DNA and RNA in microparticles released by Jurkat and HL-60 cells undergoing *in vitro* apoptosis, *Exp. Cell Res.*, **315**(5), 760–768.

60. Ullal AJ, Pisetsky DS, Reich CF, 3rd (2010) Use of SYTO 13, a fluorescent dye binding nucleic acids, for the detection of microparticles in *in vitro* systems, *Cytometry A*, **77**(3), 294–301.

61. Burnier L, Fontana P, Kwak BR, Angelillo-Scherrer A (2009) Cell-derived microparticles in haemostasis and vascular medicine, *Thromb. Haemost.*, **101**(3), 439–451.

62. Yuana Y, Bertina RM, Osanto S (2010) Pre-analytical and analytical issues in the analysis of blood microparticles, *Thromb. Haemost.*, **105**(3), 396–408.

63. Van Der Heyde HC, Gramaglia I, Combes V, George TC, Grau GE (2011) Flow cytometric analysis of microparticles, *Methods Mol. Biol.*, **699**, 337–354.

64. Zwicker JI, Liebman HA, Neuberg D, Lacroix R, Bauer KA, Furie BC, Furie B (2009) Tumor-derived tissue factor-bearing microparticles are associated with venous thromboembolic events in malignancy, *Clin. Cancer Res.*, **15**(22), 6830–6840.

Chapter 9

Atomic Force Microscopy Measurement of Extracellular Vesicles Derived from Plasma

Yuana Yuana and Susanne Osanto
Clinical Oncology Department, Leiden University Medical Centre, Albinusdreef 2, Leiden, 2333ZA, the Netherlands
milis_yuana@yahoo.com

The remarkable feature of atomic force microscopy (AFM) is its ability to "view" details at the atomic and molecular levels. This makes AFM a suitable method to detect and quantify extracellular vesicles (EVs). AFM operated in fluid tapping mode allows the detection of EVs, while applying a minimal force, thereby preserving their natural state. Monoclonal antibodies immobilized on a modified mica surface enable capturing subsets of EVs. AFM allows nanoscale measurements of individual EVs and simultaneously measures the 3D size of the EVs. The numbers of EVs attached to the antibody-coated surface can be quantified by using image processing software. For the first time we have shown that AFM detects 1,000-fold more CD41-positive EVs than flow cytometry (FCM) does. These EVs have sizes ranging between 10 nm and 475 nm with a peak at 67.5 nm,

Extracellular Vesicles in Health and Disease
Edited by Paul Harrison, Christopher Gardiner, and Ian L. Sargent

ISBN 978-981-4411-98-1 (Hardcover), 978-981-4411-99-8 (eBook)
www.panstanford.com

which is clearly below the detection limit of conventional FCM. This AFM method was also used to detect EVs bearing tissue factor (TF) antigen. We also demonstrated the feasibility of the AFM method combined with a microfluidic method to detect and quantify CD41-positive EVs directly in plasma, reducing time between venepuncture and EV measurement and also preventing EV loss during the isolation procedure. Future research will focus on further ways to improve and standardize the AFM method for EV measurement, its use to optimize preanalytical variables of EV preparation, and quantification of EV subsets in clinical samples. Ultimately, the accurate measurement of EVs may contribute to the development of EVs as a diagnostic tool and possible prognostic and/or predictive (bio) markers in various diseases.

9.1 Introduction: Atomic Force Microscopy

Invented in 1986 by Binnig, Quate, and Gerber,[1] the atomic force microscope is one of a family of scanning probe microscopes that produces three-dimensional (3D) images. The atomic force microscope, with its ability to produce images of exquisitely high resolution (down to the Ångstrom range), was originally developed for its use in the physical sciences. An atomic force microscope measures the forces between a nanostylus (fine tip) and a sample immobilized on an atomically flat substrate, usually mica or gold.[2,3] The tip is attached to the free end of a microcantilever and is brought very close to the surface of the sample. Attractive or repulsive forces resulting from interactions between the tip and the surface will cause positive or negative bending of the cantilever. A laser beam is reflected on the back of the cantilever, and the upward and downward deviations of the cantilever are read by a sensitive photosensor monitor.[4] Figure 9.1 shows the basic concept of an atomic force microscope.

The piezoelectric ceramic transducer, probe, and photodiode are the three major components in atomic force microscopy (AFM).[4] The piezoelectric ceramic transducer expands or contracts in the presence of a voltage gradient, and conversely, it develops an electrical potential in response to mechanical pressure, enabling movements in x-, y-, and z-directions. The probe represents a microcantilever bearing a sharp tip at one end. Cantilevers are usually made from silicon (Si) or silicon nitride (Si_3N_4). Different cantilever lengths, materials, and shapes allow for varied spring constants and resonant frequencies. A photodiode collects the signal of the laser beam deflection from the back of the cantilever.

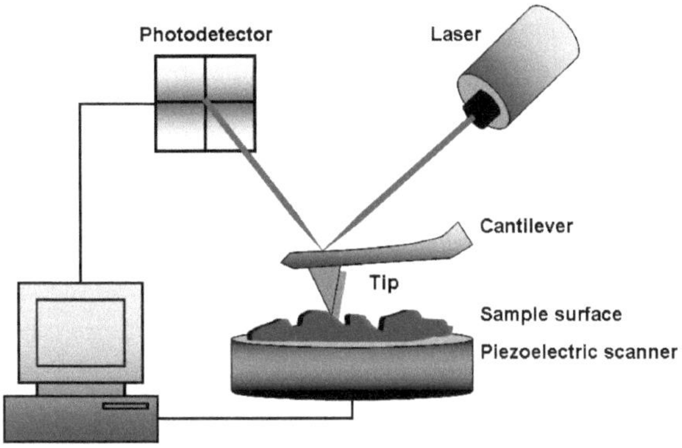

Figure 9.1 Drawing of the basic principle of an atomic force microscope. A cantilever bearing a very small tip (probe) moves along the sample surface attached to a piezoelectric scanner. The tip experiences attractive or repulsive forces. A laser and a photodiode are used to measure these forces.

AFM imaging relies on van der Waals (vdW) forces created from the interaction between the probe and the sample.[4] During contact with the sample (<0.5 nm probe–surface separation), the probe predominantly experiences repulsive vdW forces (contact mode). This leads to deflection of the tip. By maintaining a constant cantilever deflection the force between the probe and the sample remains constant and an image of the surface is obtained. As the tip moves further away from the surface (0.1–10 nm probe–surface separation) the attractive vdW forces are dominant (noncontact mode). The probe does not contact the sample surface but oscillates above the surface during scanning. When the cantilever is oscillated at its resonant frequency, the probe lightly "taps" on the sample surface (0.5–2 nm probe–surface separation) during scanning (tapping mode). By maintaining a constant oscillation amplitude a constant tip–sample interaction is maintained and an image of the surface is obtained. Tapping mode is applied when a minimal force between the tip and the sample is needed, especially during imaging of biological samples.[5]

One of the strengths of the AFM technique is the possibility to operate and measure objects within aqueous fluids. This enables imaging of a wide variety of biological samples ranging from single molecules, such as nucleic acids and proteins, to macromolecular assemblies and even whole cells. AFM also enables measurement of interaction forces between tip and specimen or to exert forces on a specimen. This means that AFM can also be used to measure the forces required to produce conformational changes in proteins. Furthermore, atomic force microscope tips can be functionalized with ligands that interact with receptors and thus be used to map the distribution of the (ligand-defined) receptors on cell surfaces.[6]

9.2 Measurement of Extracellular Vesicles

Extracellular vesicles (EVs) have important physiological and pathological roles in blood coagulation, inflammation, and tumor progression.[7] In recent years EVs also have been recognized to participate in important biological processes, such as in signaling and in the horizontal transfer of their membrane and/or cargo molecules, which are enriched in specific proteins and messenger ribonucleic acids (mRNAs).[8] Since EVs may vary in size within the submicron meter range (50–1,000 nm), studies have been hampered by the lack of methods for the sensitive detection and accurate quantification of EVs of all sizes.

Flow cytometry (FCM) has been used extensively to quantify EVs in biological samples because this method is high throughput and allows identification of the parental origin of the EVs by using combinations of several fluorescent antibodies. Conventional FCM instruments, however, employ laser light, which excites at a wavelength of 488 nm. Thus, EVs with a size smaller than the cutoff limit of 0.5 μm cannot be measured accurately, often resulting in an underestimation of EV counts.[9,10] Although there are advances made in developing new-generation FCM instruments, these have not been tested extensively for routine measurements of EVs.[11] Recently, the Vascular Biology Group of the Scientific and Standardization Committee (SSC) of the International Society on Thrombosis and Haemostasis (ISTH) proposed to standardize the enumeration of platelet-derived microparticles (PMPs) larger than 0.5 μm by using size-calibrated standard beads (Megamix beads) to adjust the instrument setting and increase the resolution of FCM.[12]

AFM is an uncommon method for the measurement of EVs, but while taking into account the detection limit of FCM instruments,

AFM is currently one of the most suitable methods to detect nanosized EVs. Siedlecki et al.[13] were the first to demonstrate the possibility to image EVs derived from activated platelets by using AFM in fluid/liquid tapping mode. They generated EVs by using surface-activated and thrombin-activated platelets deposited on glass. Their results show that EVs generated from contact-activated platelets are clustered in close proximity to adherent platelets and in some cases appear to originate from platelet pseudopodia.

We reported the development of a novel method to measure EVs bearing CD41 antigen by using AFM in fluid tapping mode.[14] These EVs were isolated from double-centrifuged, fresh platelet-poor plasma (PPP) and were captured on the surface of amine-modified mica by using anti-CD41 monoclonal antibody (MoAb). The advantage of this method compared to Siedlecki's method is that other subsets of EVs can also be potentially captured by using different antigen-specific antibodies to coat the surface of modified mica. We also used muscovite mica, a nonconducting layered mineral that has an atomically flat surface, to distinguish single EVs from the roughness of the supporting surface and to functionalize the surface for protein binding.[15]

In this study, we observed that the binding of EVs to anti-CD41-coated mica was specific, as a markedly lower number of EVs bound to immunoglobulin G1 (IgG1) isotype control MoAb-coated mica. Also, a linear relationship was obtained between the number of CD41-positive EVs per 100 μm^2 surface and the EV concentration, which saturated at a concentration of ~500 per 100 μm^2 surface. We also quantified the number of CD41-positive EVs isolated from fresh PPP of seven blood donors and three cancer patients, both by AFM and by FCM. We found that the numbers of CD41-positive EVs measured by AFM were 1,000-fold higher than those measured simultaneously by FCM (3-702 × 10^9/L plasma vs. 11-626 × 10^6/L plasma).

When EVs are bound to a surface, they are flat. Siedlecki et al. reported that EVs derived from surface-activated platelets are 125 ± 21 nm in the *x*-/*y*-dimension (width) and 5.2 ± 3.6 nm in the *z*-dimension (height), comparable with results obtained from thrombin-activated EVs (125 ± 22 nm width and 4.1 ± 1.6 nm height). The height of these EVs is only about 4% of their width. We assumed that unbound EVs are spherical. Thus, we calculated the spherical diameter (d_{sph}) of EVs from their volume (Fig. 9.2). This volume was obtained by analyzing the AFM images using SPMEDIATOR version 6.1 that evaluates the center of mass of particles in comparison with the surface background. The EVs bearing CD41 antigen isolated from fresh PPP have diameters (d_{sph}) ranging from 10-475 nm with a peak at 67.5 nm, which is clearly far below the detection limit of conventional FCM. Moreover, after filtration of isolated EVs through a 0.22 μm filter, CD41-positive EVs were still detected in the filtrate by AFM (mean d_{sph}: 37.2 ± 11.6 nm) but not by FCM. This demonstrates that AFM is much more sensitive for the detection of EVs compared to FCM.

We also generated EVs by stimulating isolated platelets with thrombin. Similar to Siedlecki's results, AFM analysis showed that these contained CD41-positive EVs with a width of 125.3 ± 11.7 nm and a height of 4.3 ± 1.9 nm. The d_{sph} of these EVs was 44.1 ± 13.7 nm (range 4–189 nm) corresponding to the d_{sph} observed for CD41-positive EVs from fresh PPP.

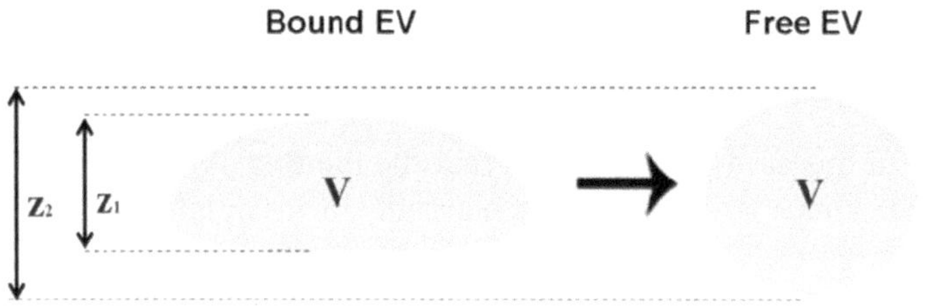

$$d_{sph} = \sqrt[3]{6V/\pi}$$

Figure 9.2 Bound vs. unbound state of an EV. The height of a bound EV (z_1) is lower than the height of a free EV (z_2), but the measured volume (V) remains the same. The spherical diameter (d_{sph}) is calculated from the volume (V) of the EV.

We have also used the AFM method to detect tissue factor (TF)-bearing EVs by coating modified mica with a fluorescein isothiocyanate (FITC)-labeled anti–human TF MoAb (Fig. 9.3A) to capture EVs isolated from the supernatant of a cultured human breast adenocarcinoma cell line, MDA-MB231 cells, which were stimulated with a calcium ionophore (Fig. 9.3C). An FITC-labeled mouse IgG1 isotype control MoAb was used as a negative control (Fig. 9.3B). TF-positive EVs with diameters (d_{sph}) ranging from 6-296 nm (median: 40.4 nm; mean: 46.4 ± 26.0 nm) were successfully captured on an FITC-labeled anti-TF-coated surface (Fig. 9.4). Only a few EVs were captured on the mica surface coated with an FITC-labeled mouse IgG1 isotype control MoAb (Fig. 9.3D). We also found that the numbers of TF-positive EVs measured by AFM were 2,000-fold higher than those measured by FCM in the same sample (8×10^6 per μL supernatant vs. 4×10^3 per μL supernatant).

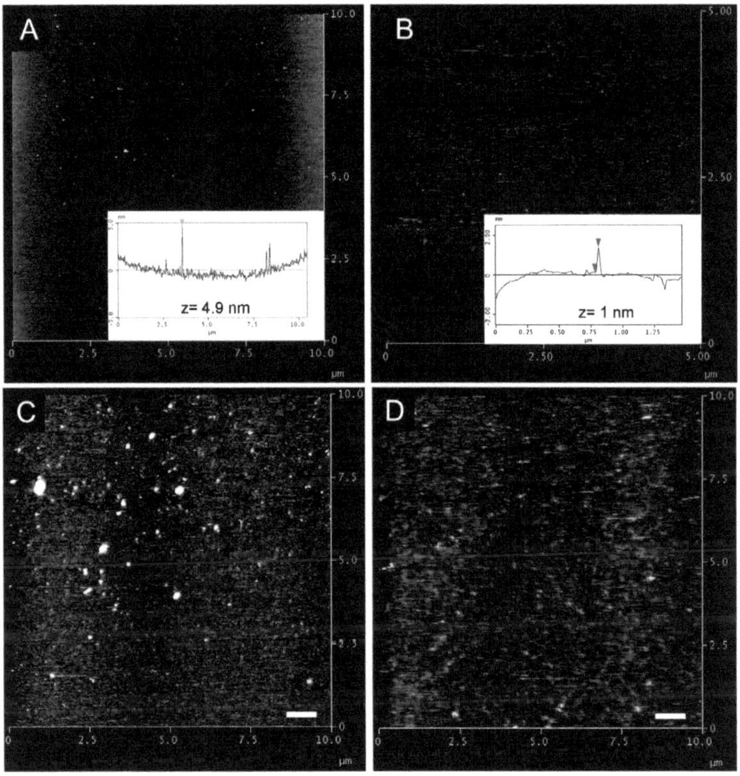

Figure 9.3 AFM images showing the 3D surface topography of (A) FITC-labeled anti-TF MoAb-coated mica (0.1 mg/mL) and (B) FITC-labeled IgG1 isotype control MoAb-coated mica (0.1 mg/mL). EVs bound to (C) FITC-labeled anti-TF-coated mica and FITC-labeled isotype control (IgG1)-coated mica were used as a negative control (D). Inserts show the results of the height analysis (*z*) of the surfaces. The scale bar in these images is 1 μm.

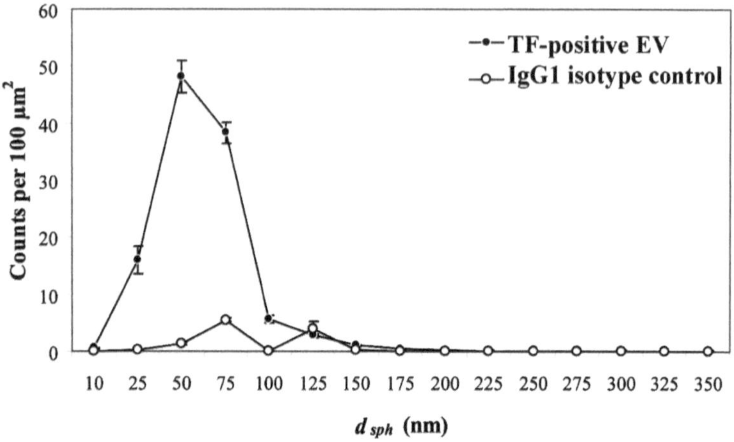

Figure 9.4 Size distribution (d_{sph}) and counts of TF-positive EVs per 100 μm^2 surface area.

In conclusion, AFM operated in fluid tapping mode can sensitively detect and accurately quantify subsets of EVs by using different antigen-specific MoAbs.

9.3 Important Factors for AFM Measurement of EVs

Preanalytical procedures such as blood collection, plasma/EV preparation, and storage are some of the important factors that significantly influence the number, morphology, and characteristics of EVs.[16] Generally, sodium citrate is used to collect blood for preparing plasma EVs. However, the effect of anticoagulant on EV characteristics is still relatively unknown. Centrifugation speed to isolate plasma or EVs may also affect the yield of EVs and their characteristics. Unfortunately, until now there is no ideal standardized protocol to prepare plasma/EsV, although the Vascular Biology Group of the SSC of the ISTH is working toward defining this.[17]

In multicenter studies and prospective trials it is often inevitable to freeze and store the plasma samples obtained over time before simultaneous analysis of serially collected samples. For our AFM measurement, we used freshly collected, citrated plasma to avoid changes in the EV number and characteristics that may be caused by storage and freezing-thawing procedures of plasma.[14] We observed no significant difference in size distribution of CD41-positive EVs after freezing-thawing of plasma obtained from two of three donors by using AFM. However, the number of CD41-positive EVs from these two donors decreased about two- to seven-fold after freezing-thawing.[14]

Direct detection of EVs in plasma is preferred because isolation procedures will inevitably cause the loss of some EVs, especially when multiple washing steps are included. Importantly, it is not known which centrifugation speed is optimal to pellet all EVs from plasma.[16] For that reason, we have explored a method that combines the use of microfluidics and AFM to detect EVs directly in plasma.[18] Some advantages of isolating EVs are that samples will be more concentrated and have less interference with protein complexes, especially insoluble immune complexes, and lipoprotein particles.[19,20] Because of their size lipoprotein particles, such as very low-density lipoprotein (VLDL) and low-density lipoprotein (LDL), with sizes between 25 nm and 90 nm[21–23] might also interfere with the measurement of EVs. As the AFM method we developed relies on the use of a specific and high-affinity antibody, EVs can be distinguished from these lipoprotein particles. AFM imaging is also performed in the presence of a physiological buffer to enable the detection of EVs with minimal force and to prevent drying of the samples. Other groups who detected EVs by using AFM in tapping mode without the presence of fluid observed EVs mainly with sizes around 10 nm,[24] much smaller than what we observed in our study.

9.4 Future Research and Discussion

The current AFM method is unfortunately still low throughput and labor intensive. Furthermore, to assess the precise number of EVs in a given sample, the availability of an external standard (e.g., EVs with a known concentration) will be essential. However, until now there are no suitable reference materials (standards) to calibrate and adjust the settings of the methods for EV analysis.[25] Thus, improvements are needed, for example, by using automated sample-handling systems, standardized reagents, reference materials for calibration and settings, and an external standard to obtain absolute numbers of EVs by using AFM.

9.5 Conclusion

AFM operated in fluid tapping mode provides a novel method enabling the sensitive detection of defined subsets of EVs in the nanosize range, far below the lower limit of what can be measured by conventional FCM, by using specific MoAbs. In combination with microfluidics EVs can also be directly measured in plasma, thus reducing the time between venepuncture and EV measurement and also preventing their loss because of washing steps in the isolation procedure. A standardized AFM method will also be very useful for optimizing preanalytical variables of EV preparation to allow their optimal measurements in various clinical samples.

Acknowledgments

The authors thank Prof. Dr. Rogier M. Bertina and Prof. Dr. Tjerk H. Oosterkamp for their fruitful discussion and Dr. Brian A. Aschroft, Dr. Marten van Es, Dr. Frederica Galli, and Nicole de Groot for their contribution in the AFM measurements. This work was supported by the Dutch Cancer Society (KWF UL 2006-3618).

References

1. Binnig G, Quate CF, Gerber C (1986) Atomic force microscope, *Phys. Rev. Lett.*, **56**(9), 930–933.
2. Muller DJ, Amrein M, Engel A (1997) Adsorption of biological molecules to a solid support for scanning probe microscopy, *J. Struct. Biol*, **119**(2), 172–188.
3. Maslova MV, Gerasimova LG, Forsling W (2004) Surface properties of cleaved mica, *Colloid J.*, **66**, 322–328.
4. Ricci D and Braga PC (2004) How the atomic force microscopy works, in *Methods in Molecular Biology: Biomedical Methods and Applications* (ed. Braga PC, Ricci, D), USA: Humana Press, 3–12.
5. Radmacher, M, Tillmann RW, Fritz M, Gaub HE (1992) From molecules to cells—imaging soft samples with the atomic force microscope, *Science*, **257**(5078), 1900–1905.

6. Parot P, Dufrene YF, Hinterdorfer P, Le Grimellec C, Navajas D, Pellequer JL, Scheuring S (2007) Past, present and future of atomic force microscopy in life sciences and medicine, *J. Mol. Recognit.*, **20**(6), 418–431.
7. Cocucci E, Racchetti G, Meldolesi J (2009) Shedding microvesicles: artefacts no more, *Trends Cell Biol.*, **19**(2), 43–51.
8. Pap E, Pallinger E, Pasztoi M, Falus A (2009) Highlights of a new type of intercellular communication: microvesicle-based information transfer, *Inflamm. Res.*, **58**(1), 1–8.
9. Furie B, Furie BC (2006) Cancer-associated thrombosis, *Blood Cells Mol. Dis.*, **36**(2), 177–181.
10. Van Der Pol E, Hoekstra AG, Sturk A, Otto C, Van Leeuwen TG, Nieuwland R (2010) Optical and non-optical methods for detection and characterisation of microparticles and exosomes, *J. Thromb. Haemost.*, **8**(12), 2596–2607.
11. Lacroix R, Robert S, Poncelet P, Dignat-George F (2010) Overcoming limitations of microparticle measurement by flow cytometry, *Semin. Thromb. Haemost.*, **36**(8), 807–818.
12. Lacroix R, Robert S, Poncelet P, Kasthuri RS, Key NS, Dignat-George F (2010) Standardization of platelet-derived microparticle enumeration by flow cytometry using calibrated beads: results of ISTH SSC collaborative workshop, *J. Thromb. Haemost.*, **8**(11), 2571–2574.
13. Siedlecki CA, Wang IW, Higashi JM, Kottke-Marchant K, Marchant RE (1999) Platelet-derived microparticles on synthetic surfaces observed by atomic force microscopy and fluorescence microscopy, *Biomaterials*, **20**(16), 1521–1529.
14. Yuana Y, Oosterkamp TH, Bahatyrova S, Ashcroft B, Garcia RP, Bertina RM, Osanto S (2010) Atomic force microscopy: a novel approach to the detection of nanosized blood microparticles, *J. Thromb. Haemost.*, **8**(2), 315–323.
15. Klein DC, Stroh CM, Jensenius H, van Es M, Kamruzzahan AS, Stamouli, A, Gruber HJ, Oosterkamp TH, Hinterdorfer P (2003) Covalent immobilization of single proteins on mica for molecular recognition force microscopy, *ChemPhysChem*, **4**(12), 1367–1371.
16. Yuana Y, Bertina RM, Osanto S (2010) Pre-analytical and analytical issues in the analysis of blood microparticles, *Thromb. Haemost.*, **105**(3), 396–408.
17. Lacroix R, Judicone C, Poncelet P, Robert S, Arnaud L, Sampol J, Dignat-George F (2012) Impact of pre-analytical parameters on the

measurement of circulating microparticles: towards standardization of protocol, *J. Thromb. Haemost.*, **10**(3), 437–446.

18. Aschroft BA, de Sonneville J, Yuana Y, Osanto S, Bertina RM, Kuil ME, Oosterkamp TH (2012) Determination of the size distribution of blood microparticles directly in plasma using atomic force microscopy and microfluidics, *Biomed. Microdevices*, **14**(4), 641–649.
19. György B, Módos K, Pállinger E, Pálóczi K, Pásztói M, Misják P, Deli MA, Sipos A, Szalai A, Voszka I, Polgár A, Tóth K, Csete M, Nagy G, Gay S, Falus A, Kittel A, Buzás EI (2010) Detection and isolation of cell-derived microparticles are compromised by protein complexes resulting from shared biophysical parameters, *Blood*, **117**(4), e39–e48.
20. Dragovic RA, Gardiner C, Brooks AS, Tannetta DS, Ferguson DJP., Hole P Carr B, Redman CWG, Harris AL, Dobson PJ, Harrison P, Sargent IL (2011) Sizing and phenotyping of cellular vesicles using nanoparticle tracking analysis, *Nanomedicine*, **7**(6), 780–788.
21. van Antwerpen R, La Belle M, Navratilova E, Krauss RM (1999) Structural heterogeneity of apoB-containing serum lipoproteins visualized using cryo-electron microscopy, *J. Lipid Res.*, **40**(10), 1827–1836.
22. Orlova EV, Sherman MB, Chiu W, Mowri H, Smith LC, Gotto AM, Jr. (1999) Three-dimensional structure of low density lipoproteins by electron cryomicroscopy, *Proc. Natl. Acad. Sci. U. S. A.*, **96**(15), 8420–8425.
23. van Antwerpen R (2004) Preferred orientations of LDL in vitreous ice indicate a discoid shape of the lipoprotein particle, *Arch. Biochem. Biophys.*, **432**(1), 122–127.
24. Baran J, Baj-Krzyworzeka M, Weglarczyk K, Szatanek R, Zembala M, Barbasz J, Czupryna A, Szczepanik A, Zembala M (2010) Circulating tumour-derived microvesicles in plasma of gastric cancer patients, *Cancer Immunol. Immunother.*, **59**(6), 841–850.
25. Chandler WL, Yeung W, Tait J.F (2011) A new microparticle size calibration standard for use in measuring smaller microparticles using a new flow cytometer, *J. Thromb. Haemost.*, **9**(6), 1216–1224.

Chapter 10

Light-Scattering Methods to Characterize Extracellular Vesicles

Don A. Gabriel,[a,b] **Sam Glover,**[b] **Regina Bowling,**[b] **and Nigel Key**[a]
[a]*University of North Carolina School of Medicine, Division Hematology/Oncology, Bone Marrow Transplant Program, Chapel Hill, NC 27599, USA*
[b]*Invitrox Inc., PO Box 13319, Durham, NC 27709, USA*
dgabriel@invitrox.com

10.1 Introduction

10.1.1 Background

Both circulating cells and the cellular components of the vascular wall, under certain circumstances, can shed fragments of their cell wall into circulating blood. These tiny chards of cellular material are called extracellular vesicles (EVs). Microvesicles (MVs) are thought to range in size from 0.1 μm to 1 μm and are released during cell activation, apoptosis, or high mechanical shear. MVs also express surface antigens or epitopes as well as some cytoplasmic content that present at that region of the cell during their formation. In contrast, exosomes are true nanovesicles released via exocytosis of preformed vesicles or multivesicular bodies within the endosome.

Extracellular Vesicles in Health and Disease
Edited by Paul Harrison, Christopher Gardiner, and Ian L. Sargent

ISBN 978-981-4411-98-1 (Hardcover), 978-981-4411-99-8 (eBook)
www.panstanford.com

Under resting conditions a dynamic asymmetric steady state between the outer and inner membrane lipid leaflets is maintained via enzyme activity. The exact mechanism(s) for MV generation is not fully known, but membrane enzymes such as scramblase, gelsolin (present only in platelets), aminophospholipidtranslocase, calpain, and floppase are thought to be involved and result in the expression of negatively charged phospholipids on the cell surface, coupled with membrane blebbing.[1–4]

In a clinical setting, EVs may have significant potential for earlier and more sensitive diagnostics and therapeutic monitoring.[5,6] EVs could potentially serve as unique biomarkers that describe the current metabolic state of their host cell at the time of their formation. Although platelet MVs are among the most widely studied, they can also originate from many different cell types, including leukocytes, endothelial cells, monocytes, tumor cells, and erythrocytes.

There now exists a better understanding of the metabolic function of EVs and their possible diagnostic utility.[1,6–8] Although EVs can be found circulating in low levels in healthy individuals, several investigators have documented elevated levels of EVs in diseases with vascular involvement and hypercoagulable states such as disseminated intravascular coagulation (DIC), acute coronary syndrome, systemic inflammatory disease, and kidney disease.[6–14] EVs have also been associated with inflammation, cellular activation and dysfunction, angiogenesis, and transport of cellular receptors.[15–18]

The major impediment to advancements in the clinical diagnostic exploitation of EVs has been the inability to simultaneously and accurately determine their, size, number, phenotype, and other characteristics such as charge. Although flow cytometry is currently considered the technique of choice for EV characterization, conventional flow cytometers cannot routinely examine EVs below 0.5 μm, including the exosome and small MV populations. This is particularly relevant since many groups have reported large numbers of EVs in the 0.05–0.3 μm range using alternative methodologies (e.g., electron microscopy). However, recent advances in flow cytometry with the advent of modern digital instruments now facilitate analysis of EVs in the 0.2–0.5 μm range. This coupled with other types of instrumentation suggests that there is significant biological information and EVs below the traditional cutoff of cytometers. Two instruments capable of analyzing smaller EVs will be discussed herein.

10.2 Theoretical Considerations

10.2.1 Principle of Quasi Elastic Light Scattering

Quasi elastic light scattering (QELS) is also referred to as dynamic light scattering (DLS) and photon correlation spectroscopy (PCS). In QELS, incident coherent light from a laser is scattered from a moving particle. When this occurs, a small amount of energy from the incident light is exchanged with the scattering particle so that the scattered light can have a slightly lower frequency than the incident light. The detection of the scattered light is possible because of random local fluctuations in the refractive index caused by concentration fluctuations of the scattering particles. The fluctuations are a result of the Brownian movement or diffusion of the particles. QELS is a method that detects particle motion and permits the determination of a *z*-averaged diffusion coefficient for the particle. Particle dimensions are then calculated from the diffusion coefficient, assuming a spherical shape for the particle.

10.2.2 Diffusion Coefficients

The diffusion coefficient of a particle is important because it can be related to particle size. For example, in a dynamic system such as the aggregation of microtubules, fibrin monomers, or sickle cell hemoglobin, each state of aggregation will have a slightly different diffusion coefficient.[19,20,22] In principle, knowledge of the diffusion coefficient can give information on the extent and dynamics of aggregation. A significant problem with this technique is resolution between the diffusion coefficients for each species, since the differences between the diffusion coefficients for an ensemble of particles the size of EVs or intact blood cells may be extremely small. Thus, the result often appears as a broad envelope without resolution between discrete EVs of different sizes, and only an average diffusion coefficient for the entire ensemble of vesicle sizes is possible.[20]

The calculation of the diffusion coefficient to determine size should be approached with care.[20–23] Additionally, the type of diffusion being assessed should be clearly understood. Tracer, or self-diffusion coefficient, D_S, is described as particles that exchange without mass accumulation in a system of uniform chemical composition.[24] D_S is the usual quantity measured in heterodyne light-scattering experiments. In mutual diffusion the flux is driven by particle concentration gradients. Homodyne light-scattering experiments measure mutual diffusion.

For spheres D_S is given by the well-known Stokes–Einstein equation:

$$Ds = \frac{kT}{6\pi\eta\alpha}$$

where k is the Boltzmann constant, T the absolute temperature, η the solvent viscosity, and a the radius of the sphere. Since the magnitude of the diffusion coefficient is dependent on the concentrations of the diffusing particles, the diffusion coefficient is usually defined in the limit of infinite dilution. Figure 10.1 illustrates this important point by showing the diffusion coefficient for highly purified human hemoglobin as a function of hemoglobin concentration.[25] The top curve is the result for a homodyne QELS experiment that yields the mutual diffusion coefficient for hemoglobin. The bottom curve is the result for the tracer or self-diffusion coefficient. Notice two important features. First, the mutual and self-diffusion coefficients are equal only in the limit of infinite dilution. The second feature is that the diffusion coefficient changes significantly as the concentration of hemoglobin is varied. The relationship between the diffusion coefficient, D, and the measured quantity, the self-diffusion coefficient, D_S, is given by:[20]

$$D = D_S(1 - \upsilon c)$$

where D is the diffusion coefficient, D_S is the measured quantity (i.e., the self-diffusion coefficient), υ is the specific volume, and c is the mass concentration of particle, typically of a macromolecule. We have been unable to find published results to show the effect of particles as large as EVs and intact cells on D. Other factors that influence the value of the diffusion coefficient include the particle volume (mass), solution viscosity, temperature, and potential particle–particle interactions.

10.2.3 Electrophoretic Quasi Elastic Light Scattering

A more powerful application of QELS adds an intense electric field across the scattering sample. Electrophoretic quasi elastic light scattering (EQELS®) does not measure the diffusion coefficient, but the electrophoretic mobility and zeta potential for particles in

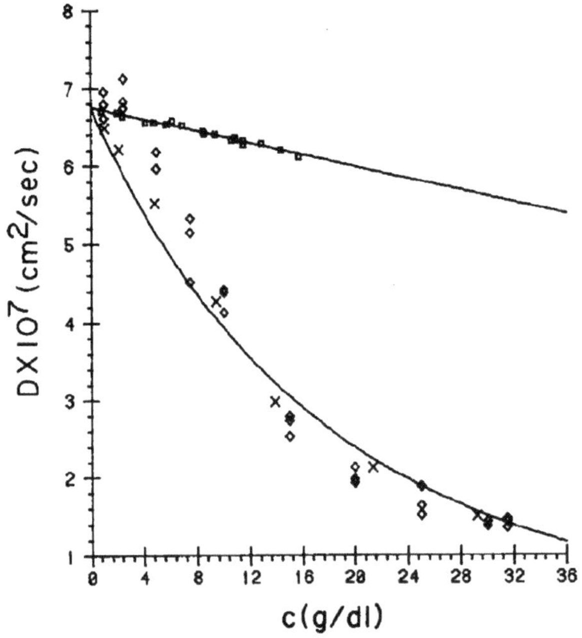

Figure 10.1 Effect of concentration on the value of diffusion coefficients. Mutual diffusion coefficients, *D*, for hemoglobin (squares) measured by QELS and self-diffusion coefficients, measured by the diaphragm method (diamonds and Xs), for met-hemoglobin versus concentration are shown. Two important results can be seen. Mutual and self-diffusion coefficients may not be equal except in the limit of infinite dilution. Second, the effect of concentration is demonstrated for both types of diffusion.

solution. The movement of large particles like blood cells and EVs in this application depends on the particles' surface charge density and the strength of the electric field. The electrophoretic mobility for particles the size of EVs and blood cells is not influenced by the particle shape but by only its surface charge density.[26,27] It should be emphasized that there is no physical separation of particles in an EQELS® analysis. Only the differences in the particle rates of movement in the applied electric field are measured.

The actual measured quantity is the time dependence of the autocorrelation of the intensity fluctuations of the scattered light measured at a specific scattering angle for the entire ensemble of scattering particles. The scattered intensity (I_S) from a moving particle at a fixed scattering angle (θ_S) is observed as an oscillating intensity described in the heterodyne experiment as a second-order field autocorrelation function,[28-32]

$$G^2_{\text{Lhet}}(\tau) = I_L^2 + 2I_L < I_s > \cos(\mathbf{K} \cdot \upsilon_d)e^{-Dk^2\tau}$$

where τ is the time increment, I_L is the intensity of the reference beam (local oscillator), I_S is the average intensity of the scattered light, and D is the diffusion coefficient. **K** is the scattering vector defined by Eq. 10.4:

$$K = \frac{4\pi n \sin\left(\frac{\theta s}{2}\right)}{\lambda}$$

where n is the refractive index, λ is the wavelength of the incident light, and υ_d is the velocity of the scattering particle. The important quantity in this expression is $\mathbf{K}\cdot\upsilon_d$, the Doppler shift of the signal resulting from the particle motion.

The Fourier transform of the measured autocorrelation function (Eq. 10.3) gives the power spectrum from which the diffusion coefficient, in the case of QELS, or the electrophoretic mobility, in the case of EQELS,® of the scattering particles is calculated.[26-29] The Doppler shift can then be related to the electrophoretic mobility by Eq. 10.5,

$$\delta \upsilon = \frac{n \upsilon_d \sin(\theta_S)}{\lambda_o}$$

where $\delta\upsilon$ is the Doppler shift, λ_o is the frequency of the incident light, and θ_Ss the scattering angle. The electrophoretic mobility is related to the velocity of the scattering particle by the simple equation, $\upsilon_d = \mu E$, where μ is the electrophoretic mobility and E is the applied electric field.[26]

In the case of EQELS,® resolution can be improved by going to lower scattering angles or increasing the electric field strength (see Eq. 10.6).[28]

$$\text{Resolution} = \frac{\lambda_o u E_{dc}}{2\pi n D_T \theta_s}$$

Scattering angles less than 5 degrees are complicated by technical issues such as stray light that is also peaked in the forward direction, that is, at low angles. Caution must be used with high electric fields because damage to the electrode surface may occur and the potential for Joule heating, especially if the solution ionic strength is high, can result.

10.2.4 Debye–Hückel Considerations

In solution, charged particles attract solution counter ions of opposite charge in a layer over the particle surface called the Stern layer (see Fig. 10.2). As the distance from the particle surface increases, the ability to selectively attract oppositely charged counter ions decreases. At a distance defined as the electrical double layer, the particle surface charge ceases to influence solution counter ions. When the diameter of the particle is large compared to the Debye length, the mobility of the particle in the electric field is solely governed by the surface charge density and not by its shape.[26,27] For platelets the Debye length is estimated to be 8Å.[33,34] When the particle is induced to move, as when an electric field is superimposed, the attraction for counterions is not intense enough to move the entire cloud of counterions. The plane at which the surface potential is not strong enough to move counterions lying at a distance further out is called the shear plane and defines the zeta potential of the particle.

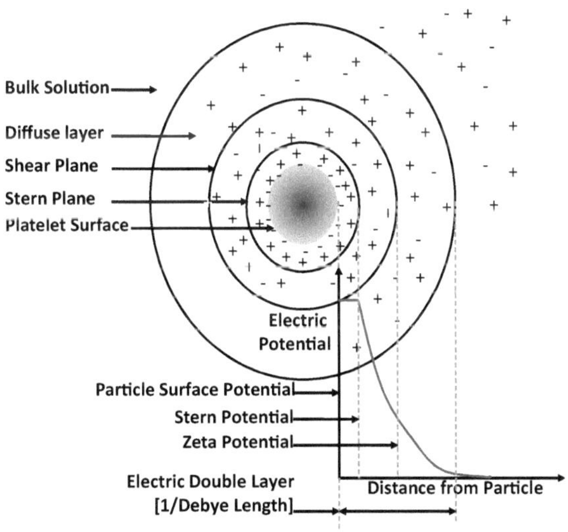

Figure 10.2 Debye–Hückle models. The influence of the particle surface potential on the surrounding counterions varies with the distance from the cell surface.

10.2.5 Experimental Setup

The electrophoretic effect is obtained by superimposing a uniform electric field across the sample. Temperature, ionic strength, pH, and conductivity all affect the electrophoretic mobility of the scattering particle and are therefore carefully controlled. Snell's law corrections are made for all scattering angles.

Experiments with EQELS® are typically run at low ionic strength for at least two reasons. First, low solution ionic strength provides low solution conductivity and then allows a high voltage drop across the sample at low current. The low ionic strength minimizes Joule heating, convection effects, and electrode reactions. Second, lower ionic strength minimizes counterion screening of the particle surface charge and maximizes electrophoretic mobility. Solution osmolarity is maintained by inert molecules such as sucrose.

10.2.6 Possible Artifacts in Dynamic and Electrophoretic Light Scattering

As with any analytic technique, care must be taken to avoid artifacts. Multiple scattering can arise from a sample that is too concentrated, and can be avoided by simple dilution. Excessive laser power can overheat the sample and induce convection in the sample, leading to errors in measurement. Thermal lensing is also avoided by controlling the incident laser power. Joule heating can create convection within the sample and is minimized by regulation of the intensity of the electric field and the pulse duration of the electric field and the pulse frequency of the electric field. The electric field pulse duration is typically 2 s, followed by 8 s off. Samples that absorb photons near the frequency of the laser output can cause serious errors due to loss of laser power. Contaminated particulate material can be a substantial issue in QELS but is usually not an issue with electrophoretic light-scattering measurements, since the difference in the scale of mobility from the contaminants compared to the sample is usually large. In the case of electrophoretic light scattering, contaminates that result from electrode reactions can be avoided by appropriate choice of buffers. It is also important that the field be pulsed and its polarity alternated in order to avoid mass accumulation.

10.2.7 Mie Scattering Methods

When the scattering particle is large compared to the wavelength of incident light, the standard Rayleigh conditions do not apply. Gustav Mie provided solutions to Maxwell's equations that permit analysis of particles in this size range. More than a century after Gustav Mie's original paper in 1908, Mie scattering is still an active area of scientific research. Mie's solution applies to scattering from homogeneous spherical particles larger than the wavelength of incident light. Basically, Mie's classical solution is described in terms of two parameters, refractive index and particle volume. The theoretical development for Mie scattering is beyond the scope of this review but can be found in other sources.[35]

10.3 Experimental Results

10.3.1 Results of QELS Diffusion Methods

10.3.1.1 ThromboLUX

Maurer-Spurej et al. have developed a QELS device known as ThromboLUX (LightIntegra, Vancouver, British Columbia, Canada) to assess platelet viability and function after storage in blood banks. Although the device has been targeted to assess platelet function, they also report that platelet-derived MVs can also be detected. In their approach, a scoring criterion is utilized to determine if a specific platelet product is acceptable for clinical use on the basis of the platelet size distribution, number of MVs, and the platelet response to changes in temperature during the measurement.[36–38] ThromboLUX measures the change in the average size of platelets and MVs via cycling of temperature within the sample holder.

10.3.1.2 N5 and Zetasizer

Harrison et al. have evaluated two different QELS spectrometers to determine the presence of MVs in human plasma.[39] The first device evaluated was the BC N5's (Beckman Coulter [BC], Hialeah, FL) ability to determine the size distribution of MVs present in fresh frozen plasma, both pre- and postfiltration and compared its performance to the second spectrometer, the Zetasizer Nano S (Malvern Instruments Ltd., Worcestershire, U.K.). N5 and Zetasizer are both dynamic light-scattering spectrometers that are relatively simple to use. The differences between the two platforms lie in the differences of both the hardware and the software design, with each device offering its own set of advantages based on the sample sources.[39] Data from both instruments can be used to determine particle size. N5 demonstrated a greater numerical difference between the pre- and postfiltration MV size and polydispersity as compared to Zetasizer. BC N5 can measure particles in the range of 3 nm to 3 µm, where the reported size range for Zetasizer Nano S is 0.6 nm to 6 µm. The benefit of these QELS devices is that for the first time, investigators can assess EV size and polydispersity in a nonsubjective manner. However, the primary drawback of these particular instruments is the inability to resolve EVs into discrete sizes or identify their cellular source.[40]

10.3.2 Results of Electrophoretic Methods

As previously discussed, the principal behind EQELS® is the measurement of the Doppler shift that occurs between the incident light and the scattered light. The magnitude of the Doppler shift can then be related to the electrophoretic mobility for a scattering particle, in our case a cell or an EV, moving in response to an applied electric field. The net surface charge density, or electrokinetic fingerprint, of the cell depends on the cell's metabolic state. When the cell surface charge density changes, the cell's electrophoretic mobility also changes. As noted, for scattering particles the size of blood cells and EVs the mobility is dependent on the particle surface charge density, that is, its electrokinetic fingerprint, and not the particle shape.[26,27] Thus, changes in mobility permit detection of differences in the activation state or changes in the cell surface caused by drug binding. Thus, measuring the change in mobility as a function of ligand concentration can yield a binding curve. Fitting the curve to a binding model allows calculation of the binding coefficient.

10.3.3 Cell Activation

The biological utility of EQELS® is its ability to exploit changes in the cell electrokinetic fingerprint as a function of the cell's metabolic state. When cells activate, undergo apoptosis, or bind to a ligand, their surface charge density changes through the activation of surface molecules, the translocation of other molecules to the cell surface, or the binding of a ligand to the cell surface. Thus, a different electrokinetic "fingerprint" is present for different metabolic states for the cell. The cell's electrophoretic mobility is highly sensitive to subtle changes in the surface charge and easily detected by EQELS.®

An early application of EQELS® was to determine the activation state of human platelets. Resting platelets have a negative surface charge (Fig. 10.3) that is derived from various sources such as sialic acid. A fully activated platelet has a substantial decrease in the magnitude of the negative charge that may eventually become positive (Fig. 10.3). This reduction in negativity of the cell's surface charge density occurs despite the translocation of negatively charged phosphatidylserine (PS) to the outer surface of the platelet membrane. The exact cause for the direction of the change in the surface charge density is unknown, but the release of calcium molecules to the platelet surface and the activation of numerous surface epitopes are speculated to be the source. These data are highly reproducible when an appropriate buffer system and a high uniform electric field are used.

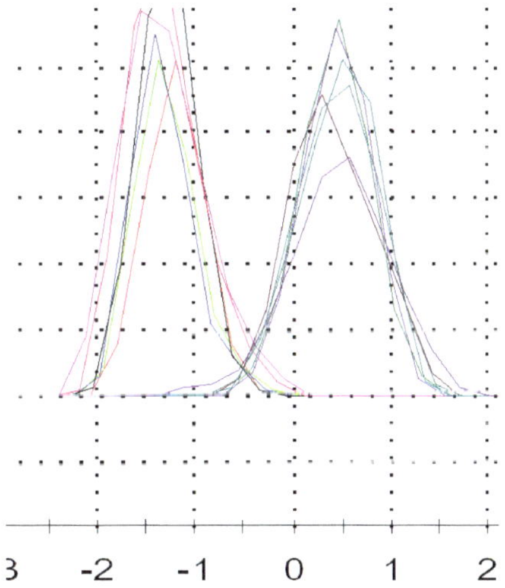

Figure 10.3 Effect of platelet activation on the electrophoretic mobility spectrum, as assessed by EQELS® The results for six different platelet donors are shown in this figure. Platelets were activated with thrombin. The activation state of platelets is shown to effect its electrokinetic fingerprint, as reflected in the difference in its electrophoretic mobility ($p = 0.00001$).

10.3.4 Ligand–Cellular Chemical Exchange

EQELS® technology has also been used to examine the chemical exchange of coagulation factor VIII (FVIII) in the inactive state between an activated platelet surface and the von Willebrand factor (VWF).[41] Because of the large negative charge present in the B-domain of coagulation FVIII, an increase in the electrophoretic mobility of a platelet occurs when FVIII is bound to its surface. The binding kinetics and binding coefficient can be accurately monitored. FVIII only binds to activated platelets (data not shown). In the absence of VWF, FVIII bound to activated platelets with a K_d of 10 nM (data not shown). If VWF was then titrated into a solution of platelet-bound FVIII, FVIII was pulled from the platelet surface and bound toVWF (Fig. 10.4b). This was due to the fact that binding of FVIII to VWF is tighter than FVIII to activated platelets. In this manner, a K_d of 1.5 nM for FVIII–VWF binding was found; this value is consistent with published values determined by other means. When FVIII is activated (to FVIIIa), no binding to VWF occurs, and a K_d of 2 nM for FVIIIa to activated platelets was observed (Fig. 10.4a), which is also consistent with previously published values. B-domainless FVIII was seen to bind to activated platelets with an intermediate K_d of 5 nM [41] (data not shown). The method is quite sensitive to extremely low concentrations of the ligand. For example, ligand concentrations as low as 10^{-17} M can be detected using EQELS®.[41]

10.3.5 Protein-Binding Site Mapping

EQELS® technology also allows for binding-site mapping. Binding curves (as described above) are obtained for a series of chimeric molecules that have had a specific region of the molecule modified by selectively exchanging specific amino acid side chains. A series of binding curves are then obtained and binding coefficients calculated. In this manner side chains critical to binding can be rapidly identified. These results then identify side chains to be modified in the protein that will then yield the desired attributes that are sought. For example, the amino acid side chains in coagulation factor IX (FIX) that are critical for platelet binding were determined.[42] In these experiments, a series of mutant molecules involving the "gla" region of FIX were engineered and their binding curves to activated platelets determined (data not shown). The result showed that lysine

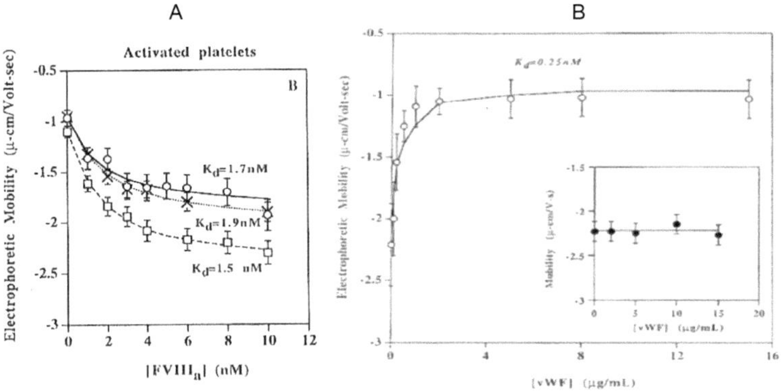

Figure 10.4 Chemical exchange of coagulation FVIII between activated human platelets and vWF as measured by EQELS®. Panel A shows the binding curves for 3 different sources of activated human coagulation FVIII binding to activated human platelets. Panel B shows activated human platelets that have been loaded with un-activated human recombinant FVIII. As purified vWF is titrated in, the mobility of the platelet changes reflecting loss of FVIII form its surface. vWF has a much higher binding affinity for FVIII than activated platelets. The insert is a control experiment in which VWF is blocked by a polyclonal anti-VWF. The blocked VWF does not bind FVIII.

in position 5 was critical for binding. Experiments and results such as these can aid molecular biologists to develop "super" molecules with either gain or loss of function, depending on the therapeutic goal.

10.3.6 EQELS® and MPs

Extension of EQELS® technology to the measurement of EVs requires no modification of the technology. Furthermore, experiments with EVs may actually be less demanding due to their greater stability compared to intact cells. An example of the electrophoretic mobility spectrum for MVs derived from platelets is seen in Fig. 10.5. As previously stated, sample handling is critical when generating EVs both *in vitro* and *ex vivo*. The activation state of the cell, method of activation, strength of the centrifugal field applied, buffer composition, and water quality all affect the results across all sizing and light-scattering methods. The electrokinetic fingerprint of

EVs and hence its electrophoretic mobility will depend on several factors. First is, as noted above, the cell's metabolic state—resting, activated, or apoptotic. The second factor depends on the region of the cell surface that the MV was derived from. Factors such as capping, lipid rafts, etc., could all potentially affect the composition of the MV surface and thus its electrokinetic fingerprint. The MVs shown in Fig. 10.5 were produced from calcium ionophore activation of human platelets. Platelet activation was carried out using human platelets in Tyrodes buffer that were activated with a calcium ionophore. The sample was exposed to a 1,500 g centrifugal field for 15 min, followed by careful resuspension of the activated platelets. One hundred microliters of the sample was then diluted with 4 mL of EQELS® buffer containing 280 mM sucrose and 2 mM 4-(2-hydroxyethyl)piperazine-1-ethanesulfonic acid (HEPES) and 5mM NaCl at pH 7.4. Resting platelets have a mobility value in the range of –1.2 mobility units (see Fig. 10.3), while activated platelets have a mobility in the range of –0.1 to +0.5 mobility units. The presence of a specific surface epitope, or phenotyping, can be easily probed for by addition of a second molecule that binds specifically to the epitope of interest; this in turn alters the surface charge density and is observed as a change in mobility, as shown above.

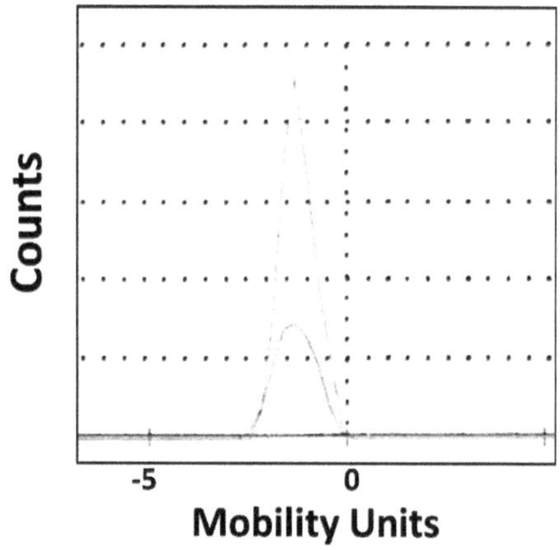

Figure 10.5 Electrophoretic mobility spectrum for human platelet microparticles obtained by activation of normal human PRP with a calcium ionophore. Two different preparations of platelet microparticles are shown. The zeta potential was –13.9 and the electrophoretic mobility –1.329 mobility units. *Abbreviation*: PRP, platelet-rich plasma.

10.3.7 ISADE Results

The Invitrox Surface Antigen Detector and Enumerator (ISADE®) is an innovative light-scattering device with powerful EV detection and analytic capabilities. In its most basic terms the device consists of a coherent light source that illuminates a sample contained in a flow cell. The sample is diluted so that single-particle counting occurs. Light scattered from the particle then passes through a spatial filter that defines a scattering angle, and is then collimated onto a second lens that focuses the scattered light onto a detector. The amount of light scattered by the particle is proportional to the size of the particle, and the voltage change in the detector is proportional to the number of photons scattered. Thus, a relationship between the particle size and the voltage output from the detector can be made, allowing for measurement of a sample's particle size distribution. As mentioned, ISADE® allows for detection and analysis of particles in the typical range of sizes for EVs and is able to measure particles that fall below the size range of flow cytometry.

10.3.8 Sample Preparation

Careful sample handling and preparation are critical for consistent and reliable results. Poor sample preparation can lead to significant variations in results from laboratory to laboratory and even between runs. A positive aspect of EVs is that samples stored up to 18 months at −80°C have shown the same size distribution when thawed and reanalyzed. A major determinant of the size distribution depends on the intensity of the centrifugal field applied during sample preparation. Figure 10.6 shows this effect on the size distribution for platelet MVs. Raw data for particle size distribution using ISADE are shown in Fig. 10.6A. Data from Fig. 10.6A were then extracted for incremental sizes to show the number of particles in a specific size range (Fig. 10.6B). Notice that as the intensity of the centrifugal field increases, the number of MVs of all sizes decreases. Further, high centrifugal fields and long centrifugation times promote MV aggregation that is not dispersed when resuspended. These large aggregates distort the sample size distribution and mislead the interpretation of the experimental results.

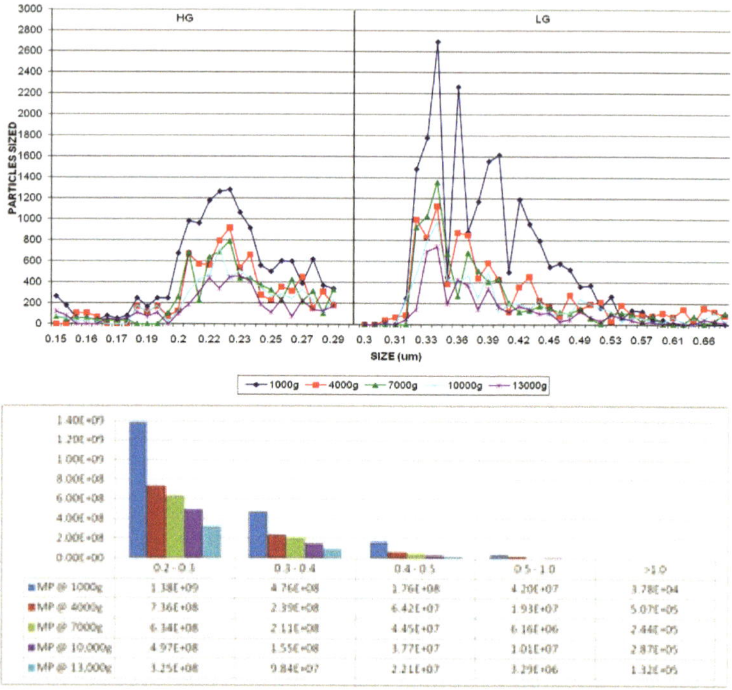

Figure 10.6 Effect of the centrifugal field strength on the resulting MV size distribution. The raw data are shown in the upper panel, and the number of particles/mL in a specific size range are shown in the bottom panel. As the centrifugal field was varied from 1 g to 13 g, more that 50% of the MVs were lost. As expected larger particles are affected more that smaller particles.

Another critical factor is the contribution from background particles present in solvents and buffers. Figure 10.7 shows the effect of the purity of the buffer on the size distribution of platelet-derived MVs. Even with high-purity water and buffers, MVs within the small size range are still detectable. Particulate and inert material present in reagents also contributes to the background particulate count. Therefore to have reproducible data, interlaboratory comparison of data and clinical diagnostics, it is crucial to control for all preanalytical variables.

The electron micrograph of MVs after a 13,000 g centrifugation step demonstrates that MV aggregation can occur during pelleting (Fig. 10.8). Note the aggregates seen in the sample centrifuged at

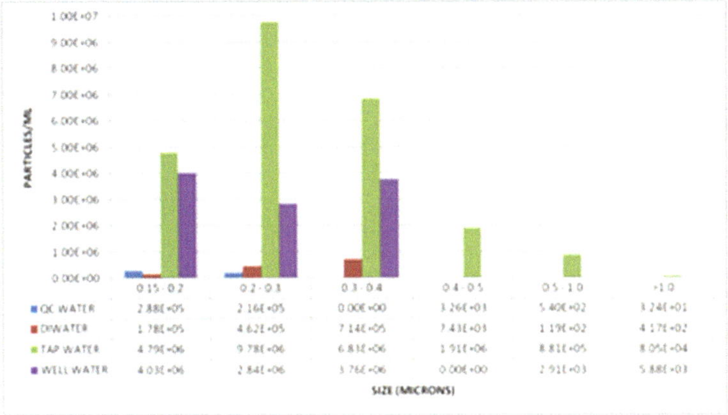

Figure 10.7 The importance of the purity of water and buffers with respect to particle content and contamination of the sample. Addition of extraneous particles from impure water and buffers can mislead results. Separate particle size analysis should be run for all solvents, as possible.

13,000 g. In our experience the aggregation is irreversible. Larger aggregates of MVs will of course lead to erroneous conclusions regarding the sample size distribution. Thus, care must be taken not to use conditions that cause MVs to aggregate. Aggregation of MVs will be observed as a large particle, so every attempt to provide a monodispersed sample should be made. The centrifugal force that is applied is also critical; thus, the use of high g centrifugation may remove important MV subsets from the sample and lead to MV aggregation. The electron micrographs also illustrate a second important factor, namely, contamination of the buffer with extraneous non-EV material.

10.3.9 Particle Size Resolution

Figure 10.9 demonstrates the power of ISADE by showing the ability to resolve a mixture of polystyrene beads with sizes of 0.2, 0.24, 0.3, 0.35, 0.4, and 0.5 µm. Both the size and the number of beads were accurately reported. Resolution between the bead sizes is sharp and well defined. The measurement requires approximately 2 min. The low counts seen on either side of each bead size result

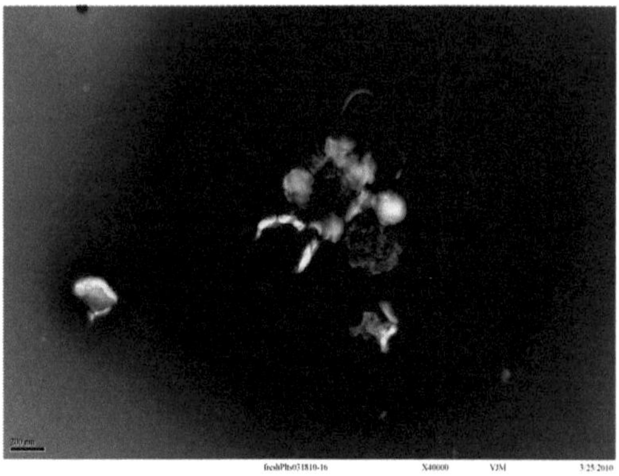

Figure 10.8 Electron micrographs of MVs after centrifugation at 13,000 g. MVs were produced from fresh PRP by activation with a calcium ionophore. Notice the clumps of MVs that remain even after multiple attempts to disperse and resuspend. Aggregates such as these can distort the size distribution.

from a small variation in the actual bead size that was confirmed by scanning electron microscopy and also seen in the ISADE analysis.

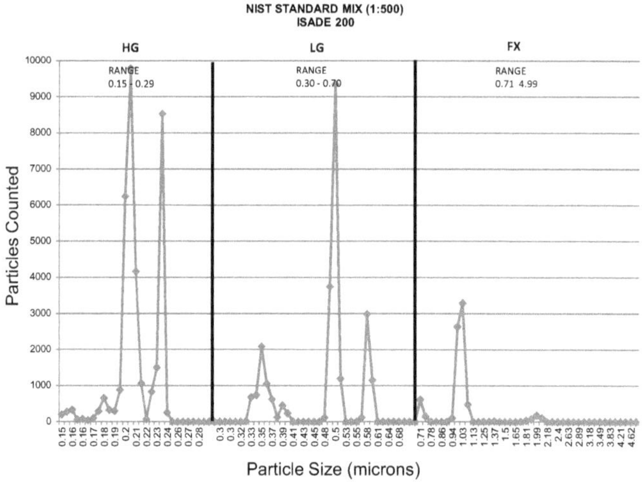

Figure 10.9 ISADE size distribution analysis of polystyrene beads of different sizes. A mixture of NIST standard polystyrene beads of known diameters was analyzed using ISADE. ISADE shows clear resolution of the mixture with respect to bead size and numbers. *Abbreviation*: NIST, National Institute of Standards and Technology.

10.3.10 Platelet Activation

When platelets undergo activation, MVs are often formed. The rate of activation and the extent of membrane fragmentation are determined by numerous factors, including the specific agonist used, the concentration of the agonist, and mechanical shear. Figure 10.10 illustrates the importance of these factors. Again, to facilitate comparison of results and to provide a quantitative result, the size distribution is divided into specific size increments and the number of particles in each size compartment is shown. In this manner changes in a specific MV size can be assessed. Using this data format to distinguish between a normal and a disease state or between a control and an activated or apoptotic state can be more easily seen. In this example resting platelets from a standard platelet-rich plasma (PRP) preparation are analyzed using ISADE. MVs are shown to be present in the PRP prior to platelet activation. When the platelets are activated with arachidonic acid in the absence of mechanical shear, a partial reduction in the number of intact platelets occurs, coupled with a small increase in the number of MVs. When the platelets are activated in the presence of mechanical shear, the majority of platelets appear to be activated and the number of MVs is nearly doubled. Further, activation by different agonists such as thrombin receptor agonist peptide (TRAP), thrombin, adenosine diphosphate (ADP), arachidonic acid, or epinephrine shows a variable extent of platelet activation and differences in the number of MVs produced (data not shown). In all cases the addition of mechanical shear increases the percentage of platelets activated and a significant increase in the number of MVs produced (data not shown).

10.3.11 Plasma MVs

Since platelet MVs are generated by platelet activation, they are also are thought to express PS on their surface. It is also well established that variable levels of circulating platelet-derived PS-positive (PS^+) MVs are present in healthy individuals. In contrast, the presence of tissue factor–positive (TF^+) PS^+ MVs is virtually undetectable in normal subjects. Conversely, levels of PS^+, TF^+ MVs are readily detected in a variety of diseases. *In vitro*, hypercoagulation models can be used to mimic disease states in regard to both cellular activation states and MV formation. Analysis of plasma from normal

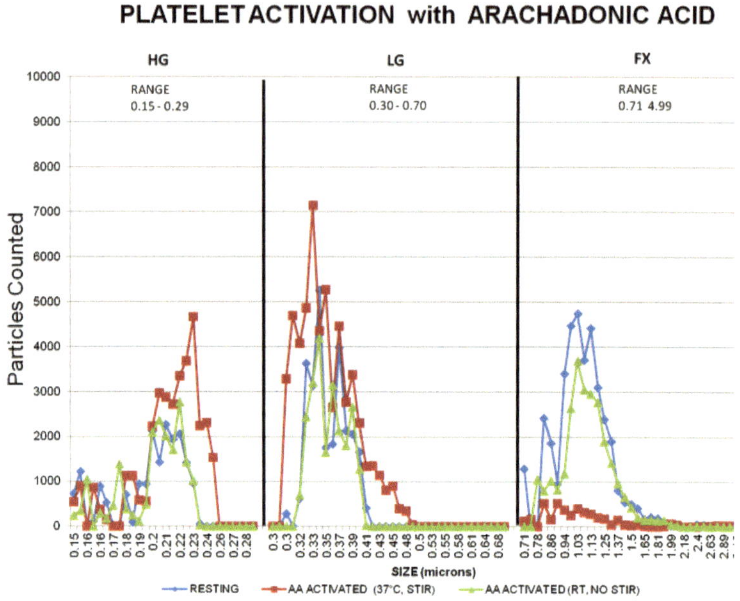

Figure 10.10 ISADE application to platelet activation assays with and without mechanical shear. One hundred microliters of resting human platelets obtained from PRP is diluted in 30 mL of buffer and analyzed using ISADE. MVs from the resting preparation (blue circles) are detected. The same preparation was then activated with arachidonic acid (green triangles) and analyzed, with very little change in the size distribution observed. When the platelets were activated and subjected to mechanical shear (red squares), a dramatic change is observed with a decrease in the intact platelet number and an increase in MVs.

subjects using ISADE to detect MVs reveals two main populations. One of these is a small population of residual platelets remaining after centrifugation of the sample that have a size distribution of 0.8–1.0 μm. A second population consists of MVs that have a size range from 0.15 μm to 0.5 μm. Both platelet and MV size ranges were also confirmed using electron microscopy (data not shown).

10.3.12 Cell Activation by LPS

Incubation of whole blood with lipopolysaccharide (LPS) at 10 μM for four to 6 h results in the activation of both monocytes (direct) and platelets (indirectly), in addition to MV formation. When analyzed using ISADE, the platelet population again decreases and the MV population grows twofold (Fig. 10.11). Time courses of the whole-blood activation were correlated with both flow cytometry and an MV-TF activity assay. Correlation with flow cytometry was weak, presumably due to the higher sensitivity and lower size range capabilities of ISADE as compared to flow cytometry. ISADE results, however, correlated well with the MV-TF activity assay.

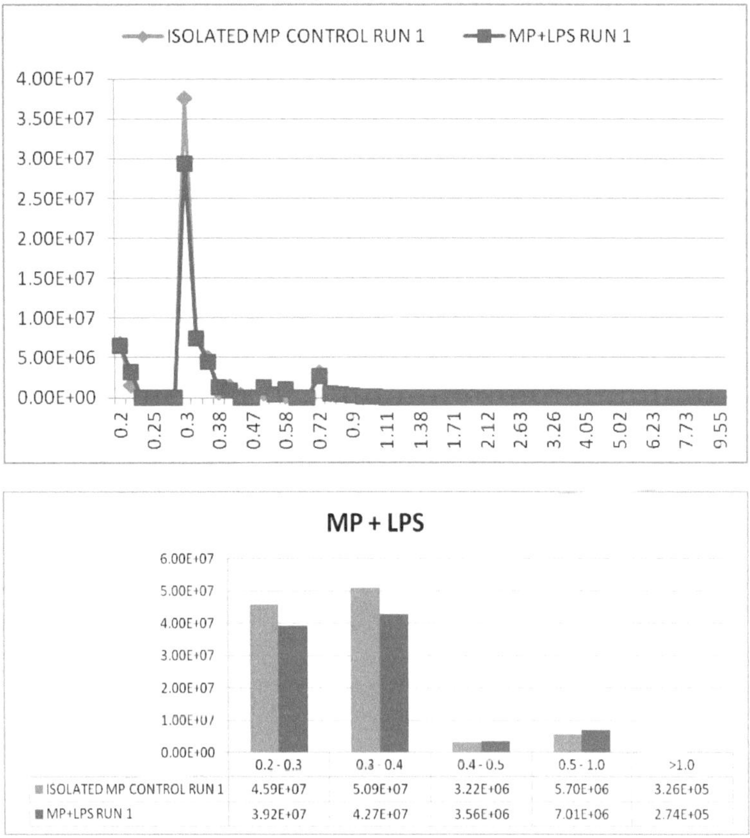

Figure 10.11 MVs produced after the activation of whole blood with LPS for 6 h.

Experiments similar to this highlight the importance of MV assays, specifically ISADE's unique ability to potentially serve as a diagnostic tool for clinical use. Examining the activation state of a clinical blood sample to determine the number and phenotype of MVs has potentially important clinical diagnostic utility. Further, the ability to monitor MV numbers and phenotype can be used to monitor efficacy of drug therapy. These assays also have the potential to be performed at both higher specificity and higher sensitivity.

10.3.13 EVs from Tumor Cell Lines

MVs produced from a leukemic cell line (THP-1) that was activated with LPS, as well as MVs that were constitutively produced from two different breast cancer cell lines (MCF-10CA1 and MBA-MB231), are shown in Fig. 10.12 using ISADE for sizing and counting. All results were confirmed using electron microscopy. Very similar size distribution was seen in both the ISADE results and the electron micrographs (data not shown).

10.4 Discussion

EV detection technology continues to evolve. The field has grown from the very basic enzyme-linked immunosorbent assay (ELISA) and flow cytometric approach to the more elegant methodologies utilizing a variety of advanced light-scattering technologies. Two different technologies are presented here. EQELS® offers a powerful tool to probe cell and MV surfaces. Determination of ligand binding, monitoring of cell surface reactions, and detection of the activation state of cells are examples. ISADE provides a simple, accurate, and rapid means to analyze the MV size distribution of a clinical sample. Further studies of MVs are necessary to determine their predictive value and to ultimately aid in the diagnosis and even prognosis of patients with various diseases. The challenge that lies ahead is in development of instrumentation and reagents that can further elucidate the therapeutic indications of MVs. Standardization of assays and methodology is critical for comparison of information between different laboratories. For successful entry of new EV measurement technology into the clinical arena, EV assays will have to provide accurate and reproducible size, numbers, and disease-specific changes in either charge and/or phenotype. The assays should be easy to run, be affordable, and have a rapid turnaround time.

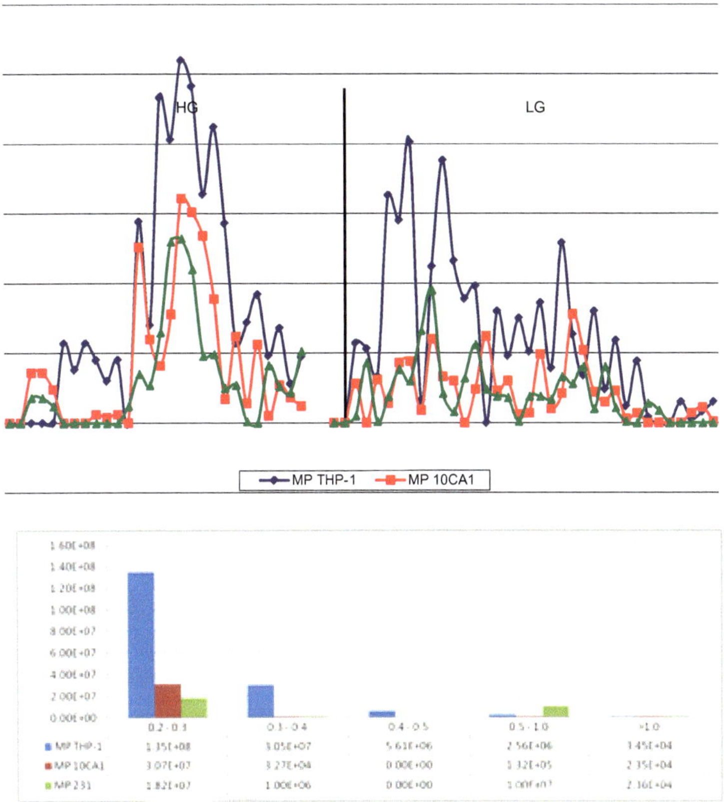

Figure 10.12 ISADE analysis of MVs from tumor cell lines. MVs from three different cell lines analyzed by ISADE are shown. MVs (blue diamonds) from THP-1 leukemia cells are produced by LPS activation. MVs from two different breast cancer cell lines, 10CA1 (red squares) and CA31 (green triangles), are constitutively produced.

Acknowledgments

The authors wish to thank Professor Charles Johnson, Jr., for his insightful enhancements to the theoretical section of this manuscript.

References

1. Piccin A, Murphy WG, Smith OP (2006) Circulating microparticles: pathophysiology and clinical implications, *Blood Rev.*, **21**, 157–171.

2. Fox JE, Austin CD, Boyles JK, Steffen PK (1990) Role of the membrane skeleton in preventing the shedding of procoagulant-rich microvesicles from the platelet plasma membrane, *J. Cell Biol.*, **111**(2), 483–493.
3. Connor J, Pak CH, Zwaal RFA, Schroit AJ (1992) Bidirectional transbilayer movement of phospholipid analogs in human red blood cells, *J. Biol.* Chem., **267**, 19412–19417.
4. Diaz C, Schroit AJ (1996) Role of translocases in the generation of phosphatidylserine asymmetry, *J. Membr. Biol.*, **151**, 1–9.
5. Zwaal RFA, Schroit AJ (1997) Pathophysiologic implications of membrane phospholipid asymmetry in blood cells, *J. Am. Soc. Hematol.*, **89**, 1121–1132
6. Nomura S, Ozaki Y, Ikeda Y (2008) Function and role of microparticles in various clinical settings, *Thromb. Res.*, **123**, 8–23.
7. George FD (2008) Microparticles in vascular diseases, *Thromb. Res.*, **122**(Suppl 1), S55–S59.
8. Chironi GN, Boulanger CM, Simon A, Dignat-George F, Freyssinet JM, Tedgui A (2009) Endothelial microparticles in disease, *Cell Tissue Res.*, **335**, 143–151.
9. Mallet Z, Benamer H, Hugel B, Benessiano J, Freyssinet J, Tedgui A (2000) Elevated levels of shed membrane microparticles with procoagulant potential in the peripheral circulating blood of patients with acute coronary syndromes, *Circulation*, **101**, 841–843.
10. Morel O, et al. (2004) Circulating procoagulant microparticles and soluble GPV in myocardial infarction treated by primary percutaneous transluminal coronary angioplasty. A possible role for GPIIb-IIIa antagonists, *J. Thromb. Haemost.*, **2**, 1118–1126.
11. Matsumoto N, Nomura S, Kamihata H, Kimura Y, Iwasaka T (2004) Increased level of oxidized LDL-dependent monocyte-derived microparticles in acute coronary syndrome, *Thromb. Haemost.*, **91**, 146–154.
12. Chirinos JA, et al. (2005) Elevation of endothelial microparticles, platelets and leukocyte activation in patients with venous thromboembolism, *J. Am. Coll. Cardiol.*, **45**, 1467–71.
13. Simak J, Gelderman MP, Yu H, Wright V, Baird AE (2006) Circulating endothelial microparticles in acute ischemic stroke: a link to severity, lesion, volume and outcome. *J. Thromb. Haemost.*, **6**, 1296–1302.
14. Ederhy S, et al. (2007) Levels of circulating procoagulant microparticles in nonvalvular atrial fibrillation, *Am. J. Cardiol.*, **100**, 989–994.

15. Boulanger C, Scoazec A, Ebrahimian T, Henry P, Mathieu E, Tedgui A, Mallet Z (2001) Circulating microparticles from patients with myocardial infarction cause endothelial dysfunction, *Circulation*, **104**, 2649–2652.

16. Rozmyslowicz T, Majka M, Kijowski J, Murphy SL, Conover DO, Poncz M, Ratajczak J, Gualton GN, Ratajczak MZ (2003) Platelet and megakaryocyte-derived microparticles transfer CSCR4-null cells and make them susceptible to infection by X4-HIV, *AIDS*, **17**, 33–42.

17. Kim HK, Song KS, Chung JH, Lee KR, Lee SN (2004) Platelet microparticles induce angiogenesis *in vitro*, *BJH*, **124**, 376–384.

18. Biro E, Lok CA, Hack CE, van der Post JA, Schaap MC, Sturk A Nieuwland R (2007) Cell-derived microparticles and complement activation in preeclampsia versus normal pregnancy, *Placenta*, **8–9**, 928–935.

19. Pecora R, Tagami YJ (1969) *Chem. Phys.*, **51**, 3298.

20. Bern BJ, Perora R (1974) Laser light scattering from liquids, *Annu. Rev. Chem.*, **25**, 233–253.

21. Ford NC, Karasz FE, Owen JEM (1970) Rayleigh scattering from polystyrene solutions, *Discuss. Faraday Soc.*, **49**, 228–237.

22. Bloomfield VA (1981) Quasi-elastic light scattering applicationsin biochemistry and biology, *Ann. Rev. Biophys. Biochem.*, **10**, 421–450.

23. Schurr JM (1977) Dynamic light scattering of biopolymer and biocolloids, *Crit. Rev. Biochem.*, **4**, 471–431.

24. Everist M, MacNeil JA, Moulins JR, Leaist DG (2009) Coupled mutual diffusion in solutions of micelles and solubilizates, *Phys. Chem. Chem. Phys.*, **11**, 8173–8182.

25. Hall RS, Johnson CS, Jr. (1980) Experimental evidence that mutual and tracer diffusion coefficients for hemoglobin are not equal, *J. Chem. Phys.*, **72**, 4251–4253.

26. Ware BR (1974) Electrophoretic light scattering, *Adv. Colloid Interface Sci.*, **4**, 1.

27. Smoluchowski M (1908) Molekular-kinetische Theorie der Opaleszenz von Gasen in kritischen Zustande sowie einiger verwandter Erscheinungen, *Ann. Phys.*, **25**, 205.

28. Johnson CS, Gabriel DA (1994) *Laser Light Scattering*, New York, NY: Dover Publications.

29. Chu B (1991) *Laser Light Scattering: Basic Principles and Practice. Second Edition*, Boston, MA: Academic Press.

30. Pecora R (1985) *Dynamic Light Scattering: Applications of Photon Correlation Spectroscopy*, New York, NY: Plenum Press.

31. Schmitz KS (1990) *An Introduction to Dynamic Light Scattering by Macromolecules*, New York, NY: Academic Press.

32. Ware BR, Flygare WH (1971) The simultaneous measurement of the electrophoretic mobility and diffusion coefficient in bovine serum albumin solutions by light scattering, *Chem. Phys. Lett.*, **12**, 81–85.

33. Jung SM, Kinoshita K, Tanoue K, Isohisa I, Yamazaki H (1982) Role of surface negative charge in platelet function related to the hyperreactive state in estrogen-treated prostatic carcinoma, *Thromb. Haemost.*, **47**, 203–209.

34. Pthica BA (1961) The physical chemistry of cell adhesion, *Exp. Cell Res.*, **8**, 123–140.

35. Kerker M (1969) *The Scattering of Light and Other Electromagnetic Radiation*, New York, NY: Academic Press.

36. Maurer-Spurej E, Labrie A, Pittendreigh C, Chipperfield K, Smith C, Heddle N, Liu Y, Yi Q, Barnett M (2009) Platelet quality measured with dynamic light scattering correlates with transfusion outcome in hematologic malignancies, *Transfusion*, **49**(11), 2276–2284.

37. Mauer-Spurej E, Brown K, Labrie A, Marziali A, Glatter O (2006) Portable dynamic light scattering instrument and method for the measurement of blood platelet suspensions, *Phys. Med. Biol.*, **51**, 3747–3758.

38. Mauer-Spurej E, Labrie A, Brown K (2008) Routine quality testing of blood platelet transfusions with dynamic light scattering, *Part. Syst. Char.*, **25**, 99–104.

39. Lawrie AS, Albanyan A, Cardigan RA, Mackie IJ, Harrison P (2009) Microparticle sizing by dynamic light scattering in fresh frozen plasma, *Vox Sang.*, **96**, 206–212.

40. Harrison P, Dragovic R, Albanyan A, Lawrie AS, Murphy M, Sargent I (2009) Application of dynamic light scattering to the measurement of microparticles, *J. Thromb. Haemost.*, **7**(Suppl 2), Abstract No. OC-TU-056.

41. Li X, Gabriel DA (1997) The physical exchange of factor VIII (FVIII) between von Willebrand factor and activated platelets and the effect of the FVIII B-domain on platelet binding, *Biochemistry*, **36**, 10760–10767.

42. Melton L, Li T, Stafford D, Gabriel, DA (2001) Location of the platelet binding site in zymogen coagulation factor IX (FIX), *Blood Coagul. Fibrinolysis*, **12**, 237–243.

Chapter 11

Nanoparticle Tracking Analysis

Chris Gardiner and Rebecca Dragovic
Nuffield Department of Obstetrics and Gynaecology, University of Oxford, John Radcliffe Hospital Oxford OX3 9DU, U.K.
chris.gardiner@obs-gyn.ox.ac.uk, rebecca.dragovic@obs-gyn.ox.ac.uk

11.1 Introduction

There is increasing evidence that only the largest microvesicles may be detected by conventional flow cytometry, and the vast majority of extracellular vesicles (EVs) are smaller than the maximum theoretical resolution of light microscopy (200 nm at a wavelength of 532 nm).[1] However, when light illuminates a particle, some of the light is scattered. This scattered light may be imaged and used to detect smaller EVs. The size of the smallest detectable vesicle is dependent on the image signal-to-noise ratio. The quantity of light scattered by a single small particle is described by the Rayleigh approximation. In this regime the amount of light scattering is given by

$$\sigma_s = \frac{2\pi^5}{3}\frac{d^6}{\lambda^4}\left(\frac{n^2-1}{n^2+2}\right)^2$$

Extracellular Vesicles in Health and Disease
Edited by Paul Harrison, Christopher Gardiner, and Ian L. Sargent

ISBN 978-981-4411-98-1 (Hardcover), 978-981-4411-99-8 (eBook)
www.panstanford.com

where d is the particle diameter, λ is the wavelength of light, and n is the ratio of the particle refractive index to the solvent refractive index.

The amount of light scattered is proportional to the scattering cross section and is also dependent on its refractive index. Using the Rayleigh approximation, the amount of light scattered by a particle one-tenth of the diameter of a larger particle is 10^6 times less than that of the larger particle.[1] Consequently, detecting the light scattered by small particles requires a very bright light source and sensitive optical detection techniques. Furthermore, the amount of light scattered by these small particles is overwhelmed in the presence of much larger particles, so size measurements based on the intensity of scattered light are unsuitable for measuring polydisperse particles such as the EVs found in biological fluids.

Nanoparticle tracking analysis (NTA) is a light-scattering system used for direct, real-time visualization and analysis of particles in solution. It is a relatively new technique that was first marketed in 2006 and overcomes many of the problems associated with other intensity-based light-scattering methods used for EV analysis.[2] NTA identifies each individual particle and tracks its Brownian motion, from which the diffusion coefficient is obtained and then used to calculate the hydrodynamic diameter. The instrument may also be calibrated so that concentration may be estimated. Although there are several different NTA instruments (all manufactured by Nanosight Ltd., Amesbury, U.K.), all use the same principle and have the same basic components (Fig. 11.1a):

- A sample analysis chamber approximately 500 μm deep and with a volume of approximately 0.25 mL
- Conventional optical microscope optics
- A charged-coupled device (CCD) or an electron-multiplying charged-coupled device (complementary metal oxide semiconductor [CMOS]) video camera
- A violet, blue, green, or red laser (405 nm, 488 nm, 532 nm, or 638 nm) producing a finely focused beam

11.2 Principle

The instrument uses a finely focused laser beam, which passes through the suspension of particles via a glass prism (Fig. 11.1b).

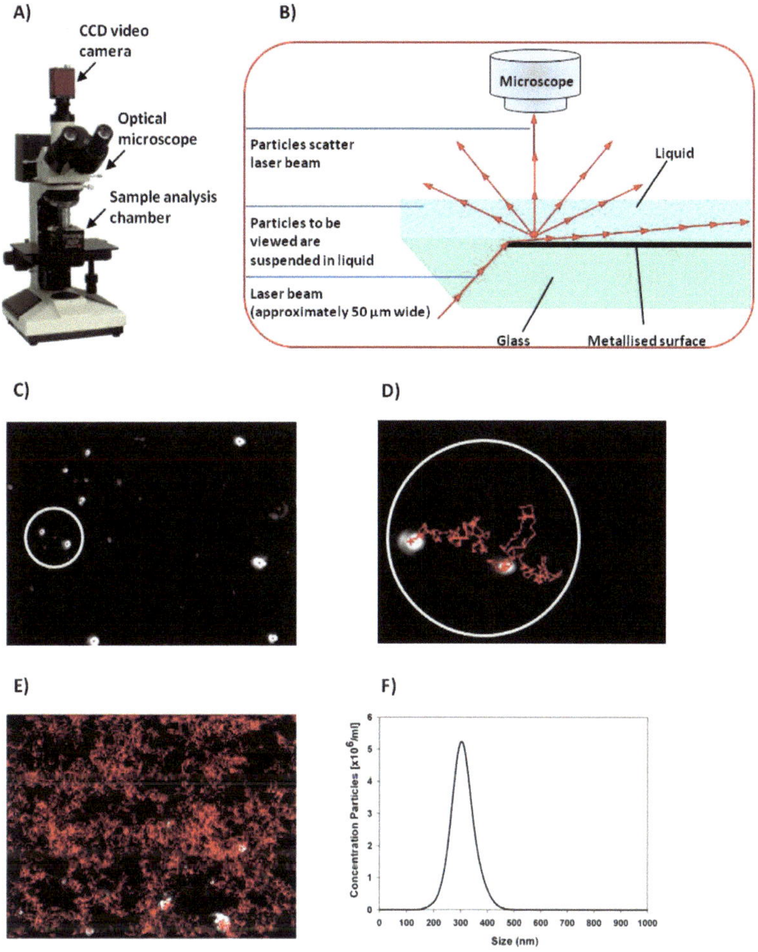

Figure 11.1 NanoSight LM10 instrument and operation of NTA. (A) Image of the NanoSight LM10 NTA instrument. (B) NanoSight instrument configuration. (C) Video capture of light scatter from particles moving under Brownian motion. (D) Tracking of individual particles using NTA; white spot = visualization of particle; red line = track of particle movement under Brownian motion. (E) Thousands of particles tracked on a frame-by-frame basis. (F) NTA software displaying the data of 300 nm beads as a histogram profile of size (nm) vs. particle concentration ($\times 10^6$/mL).

The beam refracts at a low angle as it passes through the sample, illuminating the particles and allowing their visualization by the light they scatter. The particles are viewed as small, bright points of light. A video of the light scattered by the moving particles is recorded at a frame rate of 30 frames per second (fps), and this is then analyzed using the NTA software (Fig. 11.1c). The NTA software identifies each particle and then tracks its Brownian motion (Fig. 11.1d,e). The diffusion coefficient or velocity of each particle is used to calculate particle size by applying the two-dimensional Stokes–Einstein equation:

$$<x, y>^2 = \frac{K_B T t_s}{3\pi\eta d_h}$$

where $<x, y>^2$ is the mean square displacement, K_B is Boltzmann's constant, T is the temperature of the solvent in Kelvin, t_s is the sampling time (i.e., 1/30 fps = 33 ms), η is the viscosity, and d_h is the hydrodynamic diameter. The NTA software displays the data as a histogram of particle diameter (nm) versus particle concentration expressed as particles × 10^6/mL.

A range of parameters can be adjusted both for video capture (i.e., shutter speed, camera gain, and capture duration) and analysis (i.e., blur, detection threshold, and minimum required track length), allowing optimization of particle identification and tracking for any given sample. To accurately track a particle it must be visualized as a single point of light. If the camera image is too bright the particle will be seen as large, overexposed objects with Newton's rings that may produce multiple points of light. These ring patterns may be interpreted by the NTA software as individual particles leading to inaccurate sizing and an overestimation of the particle number. If the camera image is too dim, the light scattered by the particle will fall below the detection threshold and the particle will not be detected.

11.3 Procedure for Measurement of Extracellular Vesicles

Samples are diluted in particle-free/sterile phosphate buffered saline (PBS) immediately prior to analysis. It is important to analyze the sample at the correct dilution. Ideally the particle or vesicle concentration should be between 1 × 10^8/mL and 1 × 10^9/mL.

Below this concentration too few vesicles are tracked, leading to inaccurate measurements with poor reproducibility. If the sample is too concentrated, the vesicle paths cross each other too frequently and the software cannot accurately track them. Moreover, smaller vesicles become obscured by larger ones, leading to an underestimation of the vesicle number.

When using the basic NTA instruments (Nanosight LM10 and LM200), the sample is manually introduced into the sample chamber via a syringe, and it is usual to perform one or more measurements of 30–60 s. With the more complex NS500 instrument, the sample is automatically introduced into the sample chamber, and it may be beneficial to perform several shorter measurements on a fresh volume of sample by automatically advancing the sample between measurements. This approach has several advantages, which will be discussed later.

The focus, camera gain, and shutter speed are adjusted so that all vesicles are visible and are visualized as single points of light. If concentration measurement is to be performed, predefined calibration settings are used. Following the introduction of the sample into the chamber, one or more video recordings are made. If the "script control" function is used on NS500 several short recordings can be captured automatically. An image background file is also created at this stage.

Postacquisition settings are set to optimize visualization of the vesicles. Again, if concentration measurement is required, predefined calibration settings should be used. Postacquisition settings for the most recent NTA software (v2.3) are:

- The **detection threshold**, which determines the minimum gray-scale intensity value of any particle image necessary for it to qualify as a particle to be tracked for analysis.
- The **minimum expected particle size**, which determines the maximum distance (pixels) from the particle's position in a given frame that the software will search for a particle in the next frame (the "search area"). It also establishes an exclusion zone around a particle of the same radius. If another particle enters this exclusion zone then the software excludes the information from both particles. In NTA v2.3 this function may be automatically determined by the software.

- The **Blur** function, which acts as a smoothing function to help eliminate noise, such as diffraction rings surrounding larger particles or from around or within a particle image. An automatic blur setting can be used to apply an appropriate blur width based on the brightness of each pixel, causing brighter particles to be blurred more than dimmer particles. The automatic blur function should always be used by default, although the blur value can be changed manually.
- The **minimum track** length, which defines the minimum number of steps that a particle must take before its size value can be accepted for inclusion in the particle size distribution plot. Entering too large a value will bias the results to larger, slower-moving particles in a mixture of particle sizes or, for a more monodisperse sample, simply reduce the proportion of included tracks.

Because NTA estimates the size of any given particle from measuring the average distance the particle moves, clearly the longer the time the particle is tracked for (the higher the number of frames), the better will be the statistical accuracy of the average and therefore of the particle size estimation. However, the time any given particle remains in the small scattering volume is limited, particularly so for small particles whose Brownian motion is rapid. The average beam residence time for smaller particles is far shorter (perhaps as short as 0.2 s or 6 frames) than for larger particles whose very limited Brownian motion means that they may, on average, remain in the scattering volume for many seconds (>60 frames). Under most circumstances, the automatic minimum track length function should be used, which sets an appropriate value for each individual particle on the basis of the current estimate of its size.

11.4 Calibration

The observed particle size is dependent on using the optimum settings for the size and refractive index of the particles being measured. The importance of correct settings is clearly illustrated in Fig. 11.2. Several groups have verified size measurements made by NTA using National Institute of Standards and Technology–certified polystyrene spherical beads of known size.[3–6]

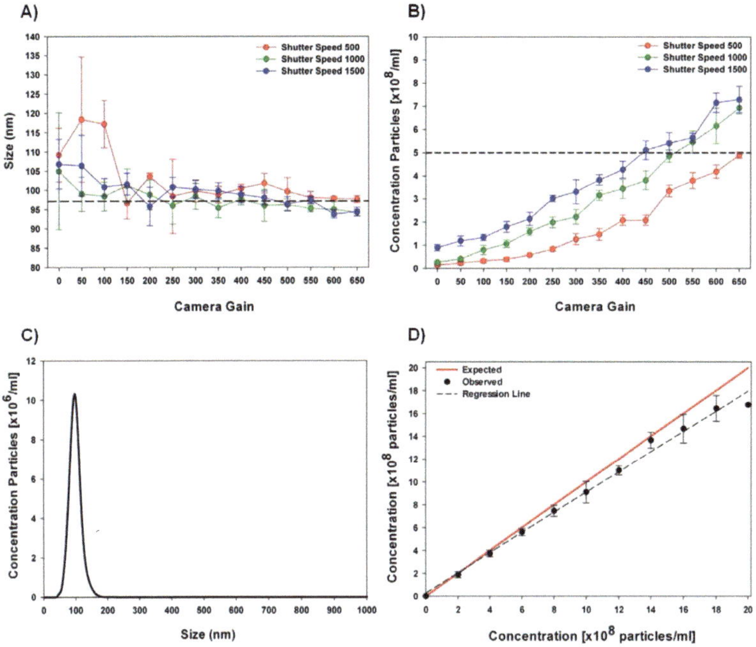

Figure 11.2 NTA of 100 nm polystyrene beads. (A) Modal particle size (mean ± SD) measurement with increasing camera shutter speed and gain settings. The black line denotes the actual size of beads. (B) Measurement of particle concentration (mean ± SD) with increasing camera shutter speed and gain settings. The black line denotes the actual concentration of beads. (C) NTA size vs. concentration profile of 100 nm beads using a shutter speed of 1,000 and camera setting of 500. (D) Expected (red line) vs. observed concentration (black, dotted line) (mean ± SD) of particles using optimal camera shutter and gain settings. Values obtained for each camera shutter speed and camera gain setting are the average of four to five separate video analyses. *Abbreviation*: SD, standard deviation.

To obtain accurate particle concentration and size measurements, different combinations of camera shutter and gain settings can be tested and optimized for each given particle size. Optimization of both the camera shutter speed and gain settings are essential for accurate size (Fig. 11.2a) and concentration measurements (Fig. 11.2b). With optimal settings, precise size and concentration measurements may be obtained over a range of 2–10 × 10^8 particles/mL (Fig. 11.2c,d). The optimum measurement conditions are dependent upon the amount of light scattered and hence the size of the particle being measured. The Rayleigh approximation predicts that a particle with a diameter of 500 nm will scatter 10^6 as much light as a particle with a diameter of only 50 nm. Clearly, this makes it more challenging to accurately size particles of both sizes simultaneously.

Microvesicles and nanovesicles from cell culture supernatants and biological fluids are rarely monodisperse. To examine the effect of polydispersity on NTA measurements a mixture of 100 nm and 300 nm beads was used. Using settings optimal for analyzing 100 nm beads, two distinct particle populations were observed with modal sizes of 94 ± 2.2 nm (average mode ± SD) and 284 ± 5.7 nm (Fig. 11.3a). Using these same camera settings, a constant concentration of 100 nm beads (5×10^8 beads/mL) mixed with increasing concentrations of 300 nm beads ($1–5 \times 10^8$ beads/mL) was analyzed and the particle size and concentration of the two bead populations were measured. Figure 11.3b shows that the 100 nm bead population could be accurately sized at all particle concentrations; however, as the 300 nm bead concentration increased, the sizing precision for this population was compromised and the particles were undersized. An overestimation of the total particle concentration was observed in the presence of the 300 nm beads when used at a concentration in between ($1–4 \times 10^8$ beads/mL) (Fig. 11.3c). Figure 11.3d shows a screenshot of 100 nm beads (5×10^8/mL) and 300 nm beads (1×10^8/mL) analyzed using optimum settings for 100 nm beads. The 100 nm beads are visualized as single points of light; however, the 300 nm beads appear as larger overexposed objects with Newton's rings. At these settings the 300 nm beads may scatter multiple points of light, which are interpreted by the NTA software as individual particles, hence leading to an overestimation of their numbers.

In contrast, using lower image capture settings optimal for 300 nm beads results in only the 300 nm bead population being observed, as the camera was unable to detect the 100 nm beads (Fig. 11.4a). The 300 nm beads were accurately sized at all particle concentrations (Fig. 11.4b); however, due to the 100 nm bead population being undetectable this resulted in an underestimation of the total particle concentration (Fig. 11.4c). A screenshot of the 100 nm beads (5×10^8/mL) and 300 nm beads (5×10^8/mL) shows the 300 nm beads as single points of light, whereas at these camera settings the 100 nm beads are undetectable (Fig. 11.4d). The results indicate that NTA can analyze polydisperse samples, but sizing of the larger 300 nm particles using analysis settings intended for smaller particles is less precise. A high degree of polydispersity also leads to an overestimation of concentration due to secondary scattering events and the inaccurate particle tracking due to the presence of Newton's rings. However, these experiments were performed prior to the introduction of the automated blur function, which greatly reduces interference caused by the overexposure of large particles.

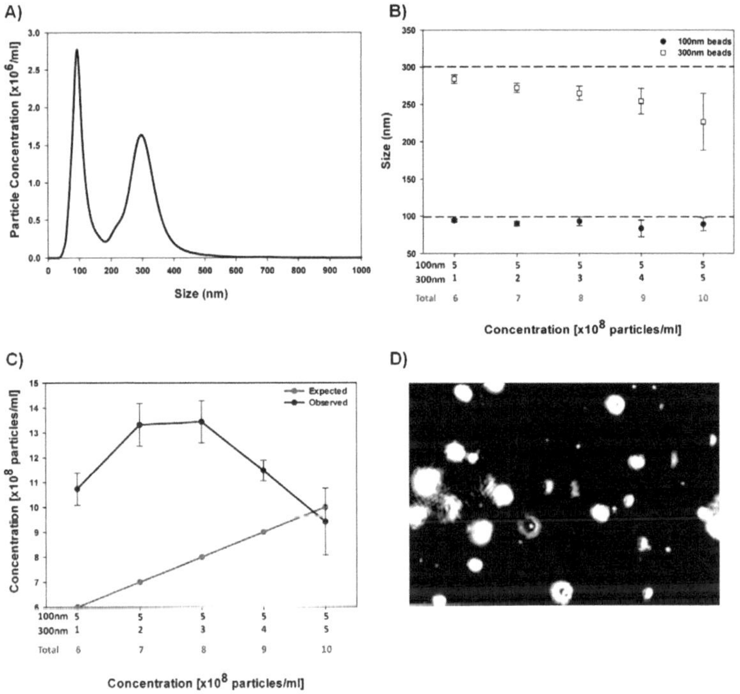

Figure 11.3 NTA using optimized settings for 100 nm particles: 100 nm and 300 nm polystyrene bead mix. (A) NTA size vs. concentration profile. (B) Modal particle size (mean ± SD) measurement with a constant concentration of 100 nm beads (5 × 10^8/mL) and an increasing concentration of 300 nm beads (1–5 × 10^8/mL). Black dotted lines denote the actual size of beads. (C) Expected vs. observed particle concentration (mean ± SD) with a constant concentration of 100 nm beads (5 × 10^8/mL) and an increasing concentration of 300 nm beads (1–5 × 10^8/mL). (D) Screenshot of 100 nm beads (5 × 10^8/mL) and 300 nm beads (1 × 10^8/mL). Values obtained for each bead concentration are the average of four to five separate video analyses.

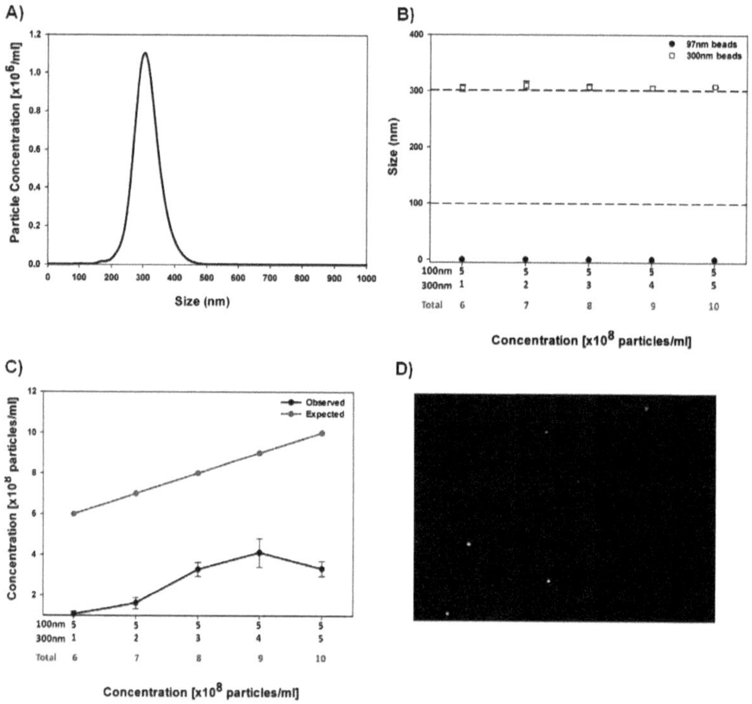

Figure 11.4 NanoSight NTA using optimized settings for 300 nm particles: 100 nm and 300 nm polystyrene bead mix. (A) NTA size vs. concentration profile. (B) Modal particle size (mean ± SD) measurement with a constant concentration of 100 nm beads (5×10^8/mL) and an increasing concentration of 300 nm beads ($1–5 \times 10^8$/mL). Black dotted lines denote the actual size of beads. (C) Expected vs. observed particle concentration (mean ± SD) with a constant concentration of 100 nm beads (5×10^8/mL) and an increasing concentration of 300 nm beads ($1–5 \times 10^8$/mL). (D) Screenshot of 100 nm beads (5×10^8/mL) and 300 nm beads (5×10^8/mL) showing light scatter from the 300 nm beads only, as at these settings the 100 nm beads are undetectable. Values obtained for each bead concentration are the average of four to five separate video analyses.

As previously mentioned, the refractive index of a particle also affects the amount of scattered light. While this does not affect particle size measurements made by NTA (the size measurement being dependent only upon the diffusion coefficient) it would be predicted to affect concentration measurement. Polystyrene beads have a refractive index of 1.59, which is considerably higher than that of cellular microvesicles, which are thought to have a refractive index of approximately 1.39. Silica microspheres with a refractive index of 1.42 have been suggested as a better calibration standard for flow cytometry.[7] A 400 nm polystyrene bead scatters the same amount of light as a 1,000 nm silica bead, so it would be expected that optimizing NTA settings using polystyrene beads would result in an underestimation of particles with a lower refractive index. To show this, silica beads of 100 nm and 540 nm were used to establish ideal acquisition settings for the quantitation of small and large low-refractive-index particles. Known concentrations of 100 nm and 540 nm silica beads and 100 nm and 485 nm polystyrene beads were analyzed using these settings. The procedure was then reversed so that acquisition settings were established using polystyrene beads, and the same beads were the reanalyzed. As can be seen from Table 11.1, calibration using polystyrene beads leads to an underestimation of silica beads and vice versa.

Table 11.1 The effect of low- and high-refractive index beads for establishing optimum calibration settings on concentration measurements

Bead type	Expected concentration	Measured concentration
Polystyrene settings		
Silica 100 nm	5.21	1.84
Polystyrene 100 nm	4.84	5.14
Silica bead settings		
Silica 100 nm	5.21	5.26
Polystyrene 100 nm	4.84	10.94
Polystyrene bead settings		
Silica 540 nm	3.13	0.46
Polystyrene 485 nm	2.99	3.02
Silica bead settings		
Silica 540 nm	3.13	3.36
Polystyrene 485 nm	2.99	7.33

These acquisition settings were then applied to a preparation of blood plasma microvesicles, prepared by ultracentrifugation, diluted

1 in 100 in PBS. Using acquisition settings obtained using 100 nm polystyrene a concentration of 0.66×10^{10}/mL was obtained, while calibration performed with the 100 nm silica beads resulted in a concentration of 1.58×10^{10}/mL. Using low gain and fast shutter speed settings enables the detection of rarer large vesicles, while the smaller (<200 nm) are too dim to be detected. To this end, the same plasma microvesicles diluted 1 in 10 were analyzed using calibration values obtained with 495 nm polystyrene and 540 nm silica beads. This produced concentration measurements of 0.16×10^{8}/mL and 0.40×10^{8}/mL, respectively. It is clear that while polystyrene beads are suitable for verifying size measurements, they should not be used for calibrating vesicle concentration. It is entirely possible that the small difference in refractive index between silica beads and cellular microvesicles may still result in underestimation of microvesicle concentration. However, until a suitable biological standard is available, silica beads appear to be the best current option for calibrating concentration measurements.

11.4.1 Camera Sensitivity

Two different detection devices are available for NTA instruments, CCD and CMOS (older versions had an equivalent electron multiplying charged-coupled device [EMCCD]). The CMOS is 100 times more sensitive than the CCD. The difference between measurements made with these devices is shown in Fig. 11.5, which shows the results for 60 nm, 100 nm, and 200 nm polystyrene beads mixed in a 5:5:2 ratio. While the CMOS clearly resolves all three populations, the size of the 200 nm beads is underestimated. In contrast, the CCD correctly sized the 100 nm and 200 nm beads, but only a fraction of the 100 nm beads and none of the 60 nm beads were detected. Clearly this has important implications for the analysis of smaller microvesicles using NTA.

11.4.2 Sample Heterogeneity

Most biological vesicles are heterogeneous in nature, which, as we have seen, presents problems for NTA due to huge differences in the intensity of scattered light between the smallest and the largest microvesicles. However, in most biological samples tested to date, the number of smaller vesicles (<150 nm) far exceeds the number

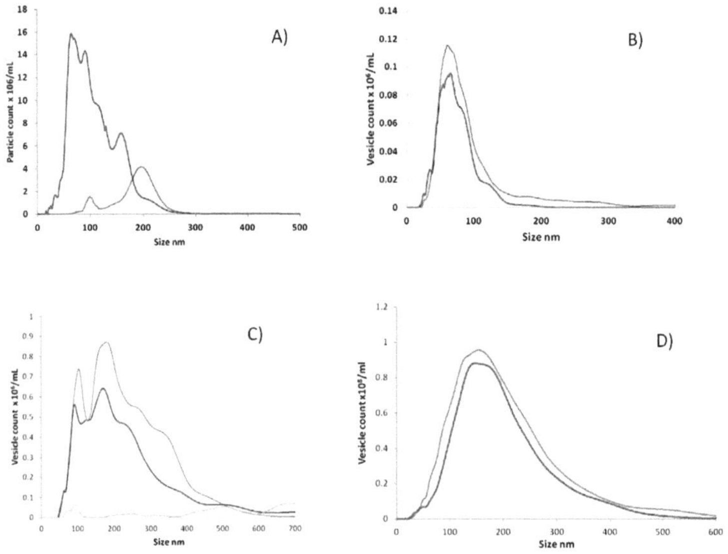

Figure 11.5 (A) A mixture of 60 nm, 100 nm, and 200 nm polystyrene beads analyzed using CCD (blue) and CMOS (red) NTA instruments. (B) NTA analysis of plasma microvesicles by light scatter (blue) and fluorescent labeling with cell tracker QDs (red). (C) NTA size profile of STBMs (blue), fNTA profile of STBMs labeled with antiplacental alkaline phosphatase (NDOG–QD conjugates: red), or isotype control (green). (D) NTA analysis of STBM showing the light scatter profile (blue) and fNTA after labeling of STBM with CellMask™ Orange (red) and overnight incubation to demonstrate stability. *Abbreviations*: QD, quantum dot; fNTA, fluorescence nanoparticle tracking analysis; STBM, syncytiotrophoblast microvesicle.

of larger vesicles. Therefore by analyzing a very dilute suspension at high gain/slow shutter speed, it is usually possible to effectively dilute out the larger vesicles, while analyzing the majority population of smaller microvesicles. The larger vesicles may then be analyzed in a more concentrated suspension using low gain/fast shutter speed settings. Settings established using 100 nm and 540 nm silica beads are ideal for this purpose. The concentration measurements of the different populations may be further optimized by using the on-screen cursors to count vesicles within certain size limits. When using the NS500 instrument, the sampling error can be eliminated by using the script control function in such a way that multiple short videos are recorded on different parts of the same sample by advancing the sample through the sample chamber between measurements.

11.4.3 Fluorescence

Analyzing fluorescently labeled microvesicles using NTA poses two major problems. The first relates to the very small size of microvesicles, which makes it difficult to load sufficient fluorophores onto the vesicle surface to enable detection. It is generally accepted that in flow cytometry an antigen copy number of 200 molecules is required to generate a signal that is bright enough to be detected. Clearly, it is not possible to attach 200 immunoglobulin G (IgG) molecules (each 16 nm long) onto a vesicle only 50 nm in diameter. The second problem is that most fluorophores present on a static sample residing within the sample chamber will suffer severe photobleaching.

There are three strategies to overcome photobleaching for fNTA. The first strategy is the synchronization of the laser power supply to the camera shutter so that the laser is pulsed, thereby reducing the amount of exposure of labeled vesicles to the laser illumination. The second is to use ultrastable fluorophores, for example, fluorescent nanocrystals, which are resistant to photobleaching. The third is to pass the sample through the sample chamber at a constant rate that is slow enough to allow the detection and tracking of microvesicles but fast enough so that the vesicles are not resident in the light path long enough for photobleaching to occur.

A membrane-penetrating peptide conjugated to fluorescent nanocrystals (QTracker cell labeling kit, Invitrogen) was used to fluorescently label cellular-derived vesicles in human blood plasma. This showed that the pelleted vesicles are membrane derived (Fig. 11.5b) and also demonstrated that the majority (>99%) of circulating vesicles in plasma are not cellular in origin, appear to have a lower density, and are not pelleted by ultracentrifugation. These are now known to be lipid vesicles, which are similar in size to EVs and can also be detected by light scatter.[8,9]

It can also be shown that microvesicles can be specifically labelled with antibody-conjugated QDs.[3] In this study, human syncytiotrophoblast vesicles prepared by placental perfusion were labeled with an in-house antibody (NDOG2), which is specific for placental alkaline phosphatase. The antibody was conjugated to functionalized polyethylene glycol amine–coated fluorescent nanocrystals (QDot 605, Invitrogen, Carlsbad, CA) in an antibody:QD molar ratio of 3.5:1. Mouse IgG_1 was labeled in the same way for the isotope control. The syncytiotrophoblast vesicles were diluted 1:100 and incubated with the antibody-conjugated QDs at a final concentration of 10 nM for 15 min. The vesicles were diluted in PBS immediately prior to analysis with NTA, performed in the normal light-scattering mode using a 405 nm violet laser. The analysis was repeated with a long-pass filter in place so that only fluorescently emitted light was detected. Calibration was performed with fluorescent 200 nm beads (Fluoresbrite Plain YG Microspheres, Polysciences Ltd.). Figure 11.5c shows that >90% of vesicles were labeled with NDOG2, while there was minimal binding of the isotype control.

These data demonstrate that antibody–nanoparticle conjugates can be used to successfully phenotype EVs. However, the process is not always straightforward, and several potential obstacles must be overcome:

- The antibody–nanocrystal conjugation can result in loss of antibody–antigen binding, depending upon the conjugation technique used. It is therefore necessary to verify the success of the conjugation process by flow cytometry. The best results are obtained by Fc conjugation, but this can be expensive as large quantities of antibody are required and unconjugated products require separation by fast protein liquid chromatography.
- Some types of nanocrystals aggregate nonspecifically, which can lead to false-positive events. The antibody–nanocrystal conjugates alone should not generate particles of >25 nm. If in doubt, a check for the presence of aggregates should be performed with electron microscopy.
- High concentrations of unbound nanocrystals will result in a high level of background fluorescence when performing fNTA, which makes it difficult to track the labeled vesicles. Performing a titration of the labeled antibody will help to reduce the level of background fluorescence and improve the overall fluorescent signal.

Conventional fluorophores can be employed in fNTA, provided the labelled vesicles do not suffer from excessive photobleaching. This is achieved by adjusting the sample pump to provide a flow rate under which particles travel across the light path in approximately 3 s. The fluorescently labeled objects should be brightly fluorescent for the entire time taken to cross the light path. As can be seen in Fig. 11.5d, 100% bright, stable labeling of vesicles is possible using membrane-specific dyes. In this case CellMask™ Deep Orange (Invitrogen) was used. Antibody labeling of exosomes and small microvesicles has proved to be challenging, as it is difficult to get sufficient labeled antibodies onto the vesicles to distinguish the fluorescent signal from the background fluorescence. The development of high-affinity peptides for antigen detection[10,11] and novel fluorophores[12–14] will undoubtedly lead to improvements in vesicle labeling.

11.4.4 Zeta potential

The zeta potential is the electrokinetic potential of particles in colloidal systems. It is the potential difference between an aqueous medium containing particles and the stationary layer of fluid attached to the dispersed particle. An arbitrary value of 25 mV (positive or negative) is usually cited as that which separates low-charged particles from highly charged particles. Although the zeta potential is not the equivalent of particle surface charge, it is closely related and is much easier to measure. In colloids with low zeta potentials, the particles tend to aggregate, while particles in colloids with high zeta potentials tend to have sufficient repulsion to prevent aggregation. One would therefore expect biological fluids containing microvesicles and exosomes to have comparatively high zeta potentials.

An adapted top plate, supplied as an upgrade for the NanoSight NS500 instrument, allows an electric field to be applied to the sample chamber. The sample is introduced into the chamber between two electrodes. The electric field causes motion of the sample's particles (electrophoresis [EP]), assuming a nonzero effective particle charge, and motion of the aqueous diluent (electro-osmosis [EO]) for the polar liquid. By constraining the behavior of the electro-osmotic flow, through physical isolation with a pinch valve, the closed system assumption that the net EO flow over all chamber depths sums to zero can be made and the two components can be separated. As the total profile (EP and EO both added together) is measured, the measured profile is offset so that it is symmetrical about zero (i.e., no net flow). The average EP particle motion acts to offset the EO profile, so the symmetrical shifted version is a true representation of the EO profile alone (Fig. 11.6a). Once the EO profile is found, the EO velocity at the depth of the first video is known and can be removed from every individual particle velocity recorded at that point. Thus, the velocity that remains is purely electrophoretic in nature.

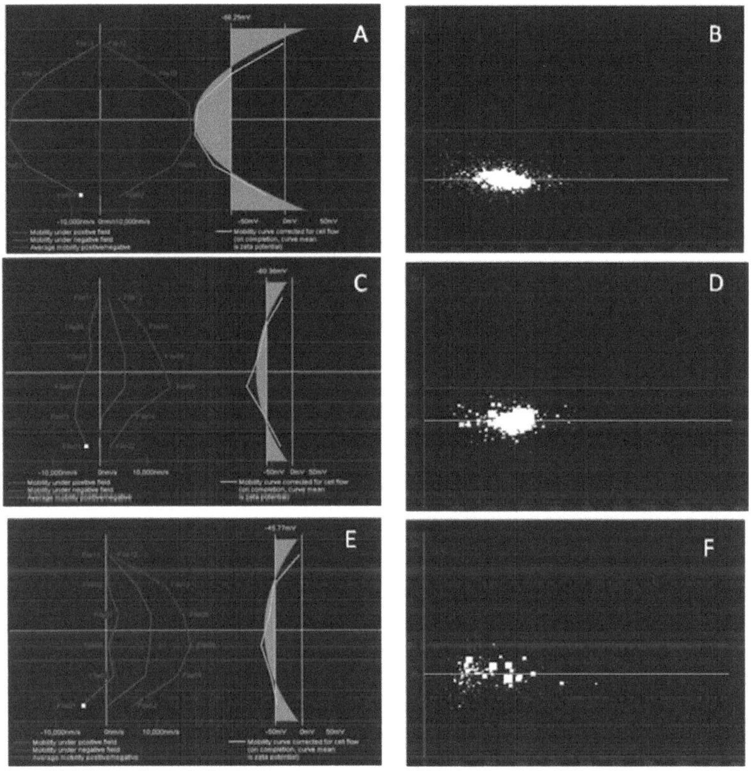

Figure 11.6 (A) In distilled water, Joule heating is minimal, ensuring mirror image mobility curves for positive (red) and negative (blue) electric fields. (B) This results in a narrow distribution of zeta potential measurements for a zeta potential standard. (C) When measuring in PBS the mobility curves are affected by Joule heating, even at a low voltage, and the rate of electrophoretic flow results in a (D) wider spread of the zeta potential for the standard. (E) However, by lowering the voltage it is possible to measure the zeta potential of EVs in PBS, as shown by these purified exosomes (F).

The system was developed using low-ionic-strength aqueous media for nonbiological materials. Clearly, this is not a viable option for EVs where isotonic media are required. However, high-conductivity solutions, such as PBS, are affected by Joule heating, which can produce a large thermal flow. This can be attenuated to some extent by increasing the delay between applying a voltage to the cell and starting the video capture in order to allow the thermal motion to stabilize before capture. Decreasing the voltage reduces Joule heating but also reduces the electrophoretic velocity.

The performance of the system may be assessed using a commercially available quality control material (Zeta Potential Transfer Standard, Malvern Instruments Ltd., Malvern, U.K.). When diluted in distilled water, the expected zeta potential value (for standard particles in their native buffer) of –68 mV was obtained (Fig. 11.6b), but when the charged particles were filtered and resuspended in PBS the measured zeta potential was closer to –60 mV. Furthermore, the mobility curves are not as smooth, and a greater degree of imprecision is observed when size versus zeta is plotted for measurements made in PBS (Fig. 11.6c,d).

11.4.5 Advantages over Other Methods

Electron microscopy, flow cytometry, and Western blotting have been the mainstay of microvesicle and exosome characterization for several years, but each has its drawbacks. Plasma microvesicles were first clearly visualized by transmission electron microscopy (TEM) in 1982[15] while it was not until 1989 that exosomes were characterized by immunogold labeling using TEM.[16] TEM has given us much of our knowledge about microvesicles and exosomes, but it is labor intensive, lengthy, and not quantitative and requires extensive sample preparation. Flow cytometry offers multicolor labeling of microvesicles, but there is increasing evidence that only the largest microvesicles may be detected by conventional flow cytometry.[1] It is also possible that the large number of small EVs measured within biological fluids can also result in significant coincidence events, or "swarming," resulting in a false-positive signal above the threshold limit of the flow cytometer. This could also potentially explain why flow cytometry not only significantly overestimates size but also underestimates the concentration of EVs compared to recent estimates by atomic force microscopy (AFM) and NTA.[17,18] Several studies have used flow cytometry to analyze exosomes that have been isolated using magnetic beads coated with a specific fluorophore-conjugated antibody, targeted against the marker of interest.[19] The major disadvantage associated with using this technique is that it is completely reliant on the expression of a single marker and may therefore only isolate a subpopulation of exosomes. Western blotting and other immunochemical methods can provide valuable information about vesicle composition but do not distinguish between exosomes, larger microvesicles, and soluble antigens.

NTA, unlike TEM, does not require sample preparation other than diluting to an appropriate concentration in PBS. Consequently, NTA is a much faster process, and it also analyzes vesicles in their natural aqueous state. Of course, TEM can give information about the structure of microvesicles, which cannot be obtained by light-scattering techniques. In contrast to flow cytometry, NTA is able to detect the full size range of all EVs, but flow cytometry allows multicolor analysis, which, at the present time, is not possible by NTA. Clearly, flow cytometry and TEM can provide information that NTA cannot, but neither method can provide the fast measurements of size and concentration afforded by NTA.

11.4.6 NTA Applications in the Study of Extracellular Vesicles

Until recently, the only way to image exosomes was TEM, and this showed "cup shaped," membrane-bound vesicles 30–90 nm in diameter. However, it is now generally acknowledged that the dehydration and fixation processes used in TEM cause artifactual changes in both size and shape.[20] Dynamic light scattering (DLS) data are consistent with exosome diameters of 165 nm and 250 nm.[21,22] However, NTA data have shown that exosomes are smaller than suggested by DLS but larger than reported by TEM and that the diameter is dependent on the parent cell type (Table 11.2). This is consistent with measurements made by cryo–electron microscopy (personal communication, A. Brisson).

Table 11.2 Exosome size measurements by NTA

Author	Cell type	Modal diameter (nm)
Sheldon et al., 2010[23]	Glioblastoma U87	120
	HUVEC	114
Sokolova et al., 2011[21]	Human embryonic kidney	120
	Mesenchymal stem cells	110
Soo et al., 2012[24]	Jurkat	76
	Human lyphoblastoid T-cell (CEM)	83
Alvarez et al., 2011[25]	Neuroblastoma (SY5Y)	99
Hoen et al., 2011[26]	Dendritic cell	83
Dragovic et al., 2011[3]	Human plasma	88

Abbreviation: HUVEC, human umbilical vein endothelial cell.

NTA has been used to measure the size, number, and release rate of EVs from a variety of cells.[24,27–30]

As NTA is a rapid method, it is useful for determining EV recovery by different methods.[31] NTA is routinely used to size and quantify exosomes for use in gene therapy research[32] and to assess exosome integrity after gene transfer by electroporation.[33] As the size measurement by NTA is an absolute measurement, it has been used to verify the size of vesicles detected by new sensitive flow cytometry methods.[26]

11.5 Conclusions

NTA is a rapid method for determining the size and concentration of EVs, which requires minimal sample preparation and allows the direct visualization of vesicles as small as 50 nm. Phenotyping of vesicles by fNTA is possible using high-affinity probes with bright, stable fluorophores. The addition of zeta potential measurement to the Nanosight platform provides additional information on vesicle membrane potential. This unique combination of measurements makes multiparametric characterization of EVs possible for the first time and is already transforming the field of EV research.

References

1. van der Pol E, Hoekstra AG, Sturk A, Otto C, van Leeuwen TG, Nieuwland R (2010) Optical and non-optical methods for detection and characterization of microparticles and exosomes, *J. Thromb. Haemost.*, **8**(12), 2596–2607.
2. Carr R, Hole P, Malloy A, Nelson P, Smith J (2009) Applications of nanoparticle tracking analysis (NTA) in nanoparticle research—a mini-review, *Eur. J. Parenter. Pharm. Sci.*, **14**(2), 45–50.
3. Dragovic RA, et al. (2011) Sizing and phenotyping of cellular vesicles using nanoparticle tracking analysis, *Nanomedicine*, **7**, 780–788.
4. Filipe V, Poole R, Kutscher M, Forier K, Braeckmans K, Jiskoot W (2011) Fluorescence single particle tracking for the characterization of submicron protein aggregates in biological fluids and complex formulations, *Pharm. Res.*, **28**(5), 1112–1120.
5. Saveyn H, De BB, Thas O, Hole P, Smith J, Van der Meeren P (2010) Accurate particle size distribution determination by nanoparticle tracking analysis based on 2-D Brownian dynamics simulation, *J. Colloid Interface Sci.*, **352**(2), 593–600.
6. Bendre V, Gautam M, Carr R, Smith J, Malloy A (2011) Characterisation of nanoparticle size and concentration for toxicological studies, *J. Biomed. Nanotechnol.*, **7**(1), 195–196.
7. Chandler WL, Yeung W, Tait JF (2011) A new microparticle size calibration standard for use in measuring smaller microparticles using a new flow cytometer, *J. Thromb. Haemost.*, **9**(6), 1216–1224.
8. Cantero M, Conejo JR, Parra T, Jimenez A, Carballo F, de AG (1998) Interference of chylomicrons in analysis of platelets by flow cytometry, *Thromb. Res.*, **91**(1), 49–52.
9. Ruf H, Gould BJ (1999) Size distributions of chylomicrons from human lymph from dynamic light scattering measurements. *Eur. Biophys. J.*, **28**(1), 1–11.
10. Zheng H, Wang F, Wang Q, Gao J (2011) Cofactor-free detection of phosphatidylserine with cyclic peptides mimicking lactadherin, *J. Am. Chem. Soc.*, **133**(39), 15280–15283.
11. Nord K, Gunneriusson E, Ringdahl J, Stahl S, Uhlen M, Nygren PA (1997) Binding proteins selected from combinatorial libraries of an alpha-helical bacterial receptor domain, *Nat. Biotechnol.*, **15**(8), 772–777.
12. Howarth M, et al. (2008) Monovalent, reduced-size quantum dots for imaging receptors on living cells, *Nat. Methods*, **5**(5), 397–399.

13. Li K, Zhan R, Feng SS, Liu B (2011) Conjugated polymer loaded nanospheres with surface functionalization for simultaneous discrimination of different live cancer cells under single wavelength excitation, *Anal. Chem.*, **83**(6), 2125–2132.
14. Jin Y, Ye F, Wu C, Chan YH, Chiu DT (2012) Generation of functionalized and robust semiconducting polymer dots with polyelectrolytes, *Chem. Commun. (Camb.)*, **48**(26), 3161–3163.
15. George JN, Thoi LL, McManus LM, Reimann TA (1982) Isolation of human platelet membrane microparticles from plasma and serum, *Blood*, **60**(4), 834–840.
16. Johnstone RM, Bianchini A, Teng K (1989) Reticulocyte maturation and exosome release: transferrin receptor containing exosomes shows multiple plasma membrane functions, *Blood*, **74**(5), 1844–1851.
17. van der Pol E, van Gemert MJ, Sturk A, Nieuwland R, van Leeuwen TG (2012) Single vs. swarm detection of microparticles and exosomes by flow cytometry, *J. Thromb. Haemost.*, **10**(5), 919–930.
18. Harrison P, Gardiner C (2012) "Invisible vesicles swarm within the iceberg," *J. Thromb. Haemost.*, **10**, 916–918.
19. Thery C, Amigorena S, Raposo G, Clayton A (2006) Isolation and characterization of exosomes from cell culture supernatants and biological fluids, *Curr. Prot. Cell Biol.*, Chapter 3: Unit 3.22. doi: 10.1002/0471143030.cb0322s30.
20. Gyorgy B, et al. (2011) Membrane vesicles, current state-of-the-art: emerging role of extracellular vesicles, *Cell Mol. Life Sci.*, **68**(16), 2667–2688.
21. Sokolova V, et al. (2011) Characterisation of exosomes derived from human cells by nanoparticle tracking analysis and scanning electron microscopy, *Colloids Surf. B Biointerfaces*, **87**(1), 146–150.
22. Atay S, Gercel-Taylor C, Kesimer M, Taylor DD (2011) Morphologic and proteomic characterization of exosomes released by cultured extravillous trophoblast cells, *Exp. Cell Res.*, **317**(8), 1192–1202.
23. Sheldon H, et al. (2010) New mechanism for Notch signaling to endothelium at a distance by Delta-like 4 incorporation into exosomes, *Blood*, **116**(13), 2385–2394.
24. Soo CY, et al. (2012) Nanoparticle tracking analysis monitors microvesicle and exosome secretion from immune cells, *Immunology*, **136**(2), 192–197.

25. Alvarez-Erviti L, et al. (2011) Lysosomal dysfunction increases exosome-mediated alpha-synuclein release and transmission, *Neurobiol. Dis.*, **42**, 360–367.

26. Nolte-'t Hoen EN, et al. (2011) Quantitative and qualitative flow cytometric analysis of nanosized cell-derived membrane vesicles, *Nanomedicine*, **8**(5), 712–720.

27. Balaj L, et al. (2011) Tumour microvesicles contain retrotransposon elements and amplified oncogene sequences, *Nat. Commun.*, **2**, 180.

28. Aleman MM, Gardiner C, Harrison P, Wolberg AS (2011) Differential contributions of monocyte- and platelet-derived microparticles towards thrombin generation and fibrin formation and stability, *J. Thromb. Haemost.*, **9**(11), 2251–2261.

29. Redman CW, et al. (2012) Review: does size matter? Placental debris and the pathophysiology of pre-eclampsia, *Placenta*, **33**(Suppl), S48–S54.

30. Baietti MF, et al. (2012) Syndecan-syntenin-ALIX regulates the biogenesis of exosomes, *Nat. Cell. Biol.*, **14**(7), 677–685.

31. Momen-Heravi F, et al. (2012) Impact of biofluid viscosity on size and sedimentation efficiency of the isolated microvesicles, *Front Physiol.*, **3**, 162.

32. Maguire CA, et al. (2012) Microvesicle-associated AAV Vector as a novel gene delivery system, *Mol. Ther.*, **20**(5), 960–971.

33. Alvarez-Erviti L, Seow Y, Yin H, Betts C, Lakhal S, Wood MJ (2011) Delivery of siRNA to the mouse brain by systemic injection of targeted exosomes, *Nat. Biotechnol.*, **29**(4), 341–345.

Chapter 12

Extracellular Vesicle Proteomic Analysis

Andrés F. Parguiña[a,b] and Ángel García[a,b]
[a]*Center for Research in Molecular Medicine and Chronic Diseases (CIMUS), Universidade de Santiago de Compostela and Instituto de Investigación Sanitaria (IDIS), Santiago de Compostela, Spain*
[b]*Department of Pharmacology, Faculty of Pharmacy, Universidade de Santiago de Compostela, Santiago de Compostela, Spain*
angel.garcia@usc.es; anxogarcia@hotmail.com

12.1 Introduction

Extracellular vesicles (EVs) are considered miniature versions of their production cell.[1] They are spherical structures that contain hydrophilic soluble components from the cytosol of the donor cell that are limited by a lipid bilayer.[1] Several types of membrane vesicles are released by cells: exosomes are smaller vesicles (40–100 nm in diameter) and originate from the exocytosis of multivesicular bodies; "microvesicles" (MVs) and "apoptotic bodies" are the terms of choice for vesicles larger than 100 nm and 1,500 nm, respectively, which originate from direct budding from the plasma membrane.[2,3] In this chapter we will be focusing on the proteome analysis of EVs released by blood cells.

Extracellular Vesicles in Health and Disease
Edited by Paul Harrison, Christopher Gardiner, and Ian L. Sargent

ISBN 978-981-4411-98-1 (Hardcover), 978-981-4411-99-8 (eBook)
www.panstanford.com

The study of EVs is currently a hot topic. Their discovery dates back to 1967[4] in the case of MVs and 1981[5] for exosomes, and their description as tools of intercellular communication was proposed in 1984.[6] Cells release exosomes constitutively and MVs or apoptotic bodies after activation or apoptosis, respectively. When this happens, the vesicles drift around in the extracellular space, where they could be broken down or migrate to distant places by diffusion, appearing in different biological fluids like saliva, urine, or blood. Their functions are yet to be fully established, but key roles could include antigen presentation and immunostimulatory or inhibitory activities by exosomes, procoagulant activity, contribution to tumor invasiveness, the pathogenesis of rheumatoid arthritis, induction of oncogenic cellular transformation, and fetomaternal communication.[7]

The investigation of EVs is being performed from different perspectives. To understand how and why they perform their functions, one approach is proteomic technology to study the cellular machinery they may contain, that is, proteins. The improvement of mass spectrometry (MS) abilities in recent decades has promoted the use of these techniques in many research fields, and one, where the interest is growing fastest, is the study of MVs. As with other applications of proteomics, the first studies aimed to map the proteome of different subsets of EVs in healthy subjects and/or cell lines (for a recent review see Raimondo et al.[8]). Currently, studies are diversifying, and the interest is shifting toward the comparison of healthy and disease states with the aim of finding biomarkers of these diseases where MVs play important roles.[7]

A prerequisite for successful proteomic analysis is a simple and robust sample preparation method.[9] The field of blood EV proteomics has been hindered by a lack of standardized protocols to obtain EVs from blood. Previous chapters of this book have dealt with various analytical methods applied to EVs. However, there are two issues related to their isolation that are very relevant for proteomic analyses: the EV preparation should not contain any contaminants (e.g., cells or plasma proteins), and the amount of protein obtained should be sufficient for the analysis. In the next section we will highlight some recommendations for blood-derived EV isolation for proteomic analysis.

12.2 Sample Preparation

One of the key steps for a successful proteomic study is to have a reliable, simple, and robust sample preparation protocol.[9] Moreover, when dealing with biological samples, such as MVs, it is important not to distort the results by increasing the technical variation, bearing in mind that the biological one is inevitable. There are two main sources of MVs for research purposes, those obtained from cell cultures and those isolated directly from plasma. The requirements and prerequisites to perform a proteomic study are more demanding in the latter due to the existence of several preanalytical parameters that need to be controlled rigorously, or they could greatly influence the results obtained.[10]

The basic workflow for proteomic analysis of EVs obtained from blood or cell culture is highlighted in Fig. 12.1. When the object of study is EVs obtained from cell cultures, proteomic analyses are usually straightforward. Cells can be stimulated with agonists to produce EVs, or the EVs could be directly recovered from the culture medium. There is no standard method for obtaining EVs, and indeed the diversity of protocols does not help to get consistent results among different laboratories. In the case of cell cultures, before starting it is important to make sure the serum added to the culture medium is free of exosomes. This can be done by ultracentrifuging the serum at 100,000 × g for 60 min prior to its addition to the medium. For EV isolation, an initial centrifugation step at low speed (100–13,000 × g) is usually performed to remove dead cells and any large debris present in the cell culture medium. Then, depending on the population of interest, either exosomes or MVs, different high-speed centrifugations are carried out. If the focus of the study is the proteome of exosomes then ultracentrifugation (usually 100,000–150,000 × g for 60–120 min) is normally performed. If the interest is MVs, there is more variability: some groups use centrifugation at 15,000–45,000 × g, and others use the typical *g* values for exosome isolation (see Table 12.1).[11–13] The problem is that many of the studies using high-speed centrifugation may have exosome contamination in the MV pellets, so if purity is a requisite it is necessary to take measures to purify the samples. One option is sucrose-gradient centrifugation, which separates the EVs in different bands. Another option is to use magnetic beads coated with antibodies directed toward characteristic membrane antigens of the vesicles.[8]

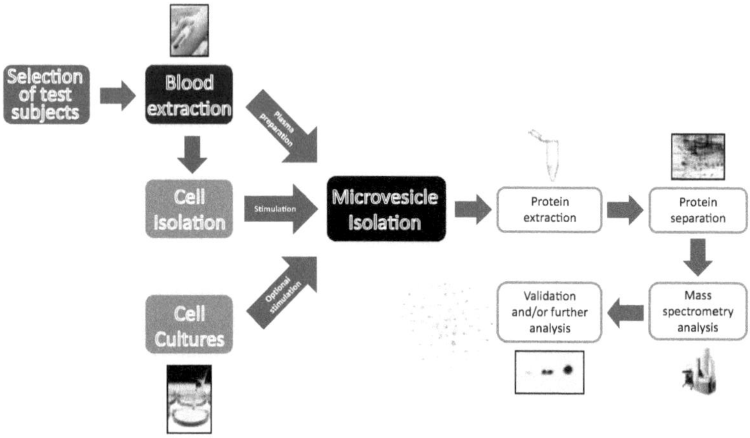

Figure 12.1 Schematic workflow of MV proteomic analysis.

Isolating EVs from blood is a complex process as many factors can influence the proteome of the vesicles. Blood should be withdrawn using a ≥21-gauge needle with a light tourniquet to avoid shear stress and undue pressure that could produce platelet activation, endothelial-derived MV (EMV) generation, or *in vitro* hemolysis. In addition, the first 3 mL of blood should be discarded to eliminate the possible effects of the vascular damage caused by the needle puncture, and blood should be collected in plastic tubes containing citrate or a mixture of citrate, theophylline, adenosine, and dipyridamole.[14] After extracting blood, the samples should be processed as soon as possible, as blood cells can become activated within 2 h and the count of MVs tends to increase with time.[15] Transportation of the samples has also been identified as a preanalytical issue, and it is recommended to use transportation boxes that maintain the tubes in horizontal position.[14] Temperature for the procedure should be kept around 20–25°C to avoid platelet activation. In our experience, light centrifugation for 20 min at 200 × g produces two clear phases in the blood, which allows for collection of the platelet-rich plasma that is later centrifuged 10 min at 1,000 × g to pellet and separate platelets to obtain platelet-poor plasma (PPP). These two centrifugations should be accomplished in the presence of platelet inhibitors such as prostacyclin (final concentration 1 μm) to avoid platelet activation and *ex vivo* MV formation. Then, PPP is centrifuged two more times (10 min at 1,500 × g and 2 min at 15,000 × g) to remove completely the remaining platelets and obtain platelet-free

Table 12.1 List of proteomic studies analyzing blood-related MVs

MVs analyzed	Authors	Species (starting material)	Isolation method	Proteomic approach	Stimulation/ Infection
B-cell EXs	Wubbolts et al. (2003)[48]	Human (lymphocytic cell line: RN (HLA-DR15$^+$)	Ultracentrifugation (70,000 × g for 60 min)	SDS-PAGE/ MALDI-TOF/TOF	–
B-cell MPs	Miguet et al. (2009)[49]	Human (blood with chronic B-cell malignancies)	Centrifugation (15,000 × g for 45 min)	SDS-PAGE/LC-MS/MS (QTOF)	Cells were treated with actinomycin D.
Endothelial progenitor cell MPs	Prokopi et al. (2009)[38]	Human (culture of EPCs)	Centrifugation (20,500 × g for 150 min at 4°C)	SDS-PAGE/HPLC (RP) LTQ-IT	–
Endothelium MPs	Banfi et al. (2005)[12]	Human (culture of HUVECs)	Ultracentrifugation (100,000 × g for 90 min at 10°C)	SDS-PAGE/ MALDI-TOF/TOF	Cells were activated with TNFα.
Endothelium MPs	Sander et al. (2008)[13]	Human (culture of HUVECs)	Ultracentrifugation (100,000 × g for 90 min at 4°C)	2DE/MALDI-TOF/TOF	Cells were activated with PAI-1.

(*Continued*)

Table 12.1 (*Continued*)

MVs analyzed	Authors	Species (starting material)	Isolation method	Proteomic approach	Stimulation/ Infection
Endothelium MPs	Peterson et al. (2008)[37]	Human (culture of HUVECs)	Ultracentrifugation (100,000 × g for 60 min at 4°C)	SDS-PAGE/LC-MS/MS (LTQ)	Cells were activated with both PAI-1 and TNFα.
Erythrocyte MPs	Bosman et al. (2008)[39]	Human (erythrocyte concentrates)	Ultracentrifugation (For MPs = 40,000 × g for 20 min at 4°C; For EXs = 100,000 × g for 60 min at 4°C)	SDS-PAGE/LC-MS/MS (LTQ-FT-ICR)	–
Erythrocyte MPs	Rubin et al. (2008)[41]	Human (erythrocyte concentrates)	Ultracentrifugation (120,000 × g for 90 min at 4°C)	SDS-PAGE/ MALDI-TOF/TOF	–
Erythrocyte MPs	Bosman et al. (2012)[40]	Human (healthy blood)	MVs pelleted at 40,000 × g for 20 min at 4°C, then sorted with flow cytometry	SDS-PAGE/LC-MS/MS (LTQ-FT-ICR)	–
Lymphocyte MPs	Miguet et al. (2006)[11]	Human (lymphocytic cell line: CEM-T)	Centrifugation (15,000 × g for 45 min)	SDS-PAGE/ MALDI-TOF and LC-MS/MS (QTOF)	Cells were treated with phytohemagglutinin or actinomycin D.

MVs analyzed	Authors	Species (starting material)	Isolation method	Proteomic approach	Stimulation/ Infection
Lymphocyte EXs	Li et al. (2012)[50]	Human (leukemia cell line: H9)	Multistep centrifugation (10,000 × g for 30 min followed by 100,000 × g for 70 min)	SILAC/LC-MS/MS (LTQ-Orbitrap)	Cell were infected with HIV-$1_{NL4.3}$.
Macrophage MVs	Kadiu et al. (2012)[47]	Human (macrophages)	Ultracentrifugation (100,000 × g for 60 min) on a 20% sucrose w/v cushion	SDS-PAGE/LC-MS/MS (LTQ-IT)	Cells were infected with HIV-1.
Monocyte MPs	Bernimoulin et al. (2009)[46]	Human (monocytic cell line: THP-1)	Centrifugation (16,000 × g for 45 min)	SDS-PAGE/LC-MS/MS	Cells were activated with P-sel-Ig, LPS, or IgG.
Plaque MPs	Mayr et al. (2009)[45]	Human (atherosclerotic plaques)	Centrifugation (20,500 × g for 150 min at 4°C)	SDS-PAGE/2DE/ LC-MS/MS	–
Plasma EXs	Looze et al. (2009)[27]	Human (healthy blood)	Prefractionation and ultracentrifugation (175,000 × g for 120 min)	SDS-PAGE/LC-MS/MS (LTQ-IT)	–
Plasma MPs	Jin et al. (2005)[22]	Human (healthy blood)	Ultracentrifugation (250,000 × g for 60 min)	2DE/MALDI-TOF/TOF	–

(*Continued*)

Table 12.1 (*Continued*)

MVs analyzed	Authors	Species (starting material)	Isolation method	Proteomic approach	Stimulation/ Infection
Plasma MPs	Abdullah et al. (2009)[18]	Baboon (blood with experimental VT)	Ultracentrifugation (35,000 RPM for 120 min at 4°C)	iTRAQ/2D-LC-MS/MS (SCX, HPLC and MALDI-TOF/TOF)	–
Plasma MPs	Little et al. (2010)[25]	Human (healthy blood)	Ultracentrifugation (100,000 × g for 120 min)	LC-MS/MS (ESI - LTQ-FT IT)	–
Plasma MPs	Ramacciotti et al. (2010)[17]	Human (blood with DVT)	Ultracentrifugation (200,000 × g for 120 min at 4°C)	iTRAQ/2D-LC-MS/MS (SCX HPLC and MALDI-TOF/TOF)	–
Plasma MPs	Watts et al. (2011)[30]	Rat (blood with experimental PE)	Ultracentrifugation (200,000 × g for 900 min at 4°C)	SDS-PAGE/LC-MS/MS (LTQ-IT)	–
Plasma MPs	Ostergaard et al. (2012)[26]	Human (healthy plasma)	Centrifugation (18,890 × g for 30 min at RT)	nano-LC-MS/MS (LTQ-IT)	–
Plasma MPs	Chaichompoo et al. (2012)[32]	Human (blood with β-thal)	Centrifugation (14,000 × g for 45 min)	2DE/QTOF-MS	–

MVs analyzed	Authors	Species (starting material)	Isolation method	Proteomic approach	Stimulation/ Infection
Plasma MVs	Bastos-Amador et al. (2012)[28]	Human (healthy blood)	Ultracentrifugation (110,000 × g for 120 min)	SDS-PAGE/LC-MSE (Q-TOF)	–
Platelet MPs	García et al. (2005)[23]	Human (healthy blood)	Ultracentrifugation (150,000 × g for 90 min at 4°C)	SDS-PAGE/LC-MS/MS (LTQ-IT)	Platelets were activated with ADP.
Platelet MPs	Dean et al. (2009)[35]	Human (healthy blood)	Ultracentrifugation (130,000 × g)	2D-LC-MS/MS (RP-LC and ESI-MS LTQ-IT)	Platelets were activated with thrombin and collagen.
Platelet MPs	Shai et al. (2012)[36]	Human (platelet concentrates)	Ultracentrifugation (100,000 × g for 60 min)	2DE/MALDI-TOF/TOF and LC-MS/MS	Platelets were activated with shear forces or thrombin.
Platelet MPs and plasma MPs	Smalley et al. (2007)[24]	Human (healthy blood)	Ultracentrifugation (150,000 × g for 90 min at 10°C)	ICAT/LC-MS/MS (LTQ-IT)	Platelets were activated with ADP.
Reticulocyte EXs	Carayon et al. (2011)[44]	Rat (anemic blood)	Ultracentrifugation (100,000 × g for 120 min at 4°C)	SDS-PAGE/ LC-MS/MS (QqTOF and LTQ-Orbitrap)	–

Abbreviations: EX, exosome; MP, microparticle.

plasma (PFP). It is important to know that it has been reported that MV counts are lower in PFP than in PPP, probably due to the last centrifugation step.[10] As an alternative, there is a proposed standard protocol that recommends two centrifugations at 2,500 × g for 15 min to obtain PFP directly from the blood.[14] At this point, PFP could be snap-frozen in liquid nitrogen or directly frozen at –80°C until the isolation of EVs is carried out. It is still a debatable subject, but some studies have shown that EV counts are higher in thawed than in fresh PFP, probably due to some remaining platelets breaking down in the process.[16] There seems to be differences also in the thawing process, as it has been reported that it could be better to do so at 37°C rather than on ice due to effects on coagulant activity and formation of ice crystals.[14,16]

For downstream proteomic analysis EVs must be isolated. This step produces some inevitable losses of vesicles but also limits the contamination of the samples with plasma proteins. The most common way to isolate the vesicles is ultracentrifugating the samples. Different research groups have used different speeds and times to achieve the separation. In our lab, after several tests we decided to follow the protocol by Ramacciotti et al. and centrifuge the PFP diluted in 4(2-hydroxyethyl)piperazine-1-ethanesulfonic acid (HEPES) buffer (1.2:3 v/v) at 200,000 × g for 90 min, resuspend the pellet in KBr 0.25 M and keep it for 20 min on ice, and centrifuge again to limit the presence of soluble plasma proteins.[17,18] Finally, the EV pellet is centrifuged again in phosphate buffered saline (PBS) as an additional washing step. This protocol produces around 100 μg of protein, starting with 27.5 mL of blood (Parguiña et al., unpublished). The EV pellet is then dissolved in an appropriate buffer. There is no need for disruption of EVs, so it is usual to solubilize them in buffers such as Laemmli for sodium dodecyl sulfate polyacrylamide gel electrophoresis (SDS-PAGE) or buffers containing urea/thiourea for two-dimensional (2D) gel electrophoresis (2DE).

12.3 Blood Microvesicle and Exosome Proteomics

Proteomics has been applied extensively to the study of EVs from different sources and in different contexts. Exosome proteomics alone accounted for more than 50 publications in 2009.[8] Fields like cancer-related or urinary exosomes have been the subject of many investigations.[19-21] In this chapter we decided to focus our attention on blood-related MVs as blood plasma is the main source of biomarkers for many prevalent diseases, such as cardiovascular disease or cancer. In this section we will revise the main publications in the field, which are summarized in Table 12.1.

12.3.1 Plasma Microvesicles and Exosomes

The simplest way to analyze the proteome of cell-derived MVs is to isolate them from the plasma of healthy donors or patients. As a consequence, the study of plasma MVs is by far the most explored subfield in blood-related MV proteomics. In the last decade several studies have applied proteomic-related techniques to characterize the plasma-membranous vesicle proteome, the major part of them focusing on the proteome of MVs rather than exosomes. It is important to mention, though, that many of the studies regarding MVs most probably have an important representation of exosome proteins as well because of the chosen isolation methodology. The next section will summarize the most important findings in this field.

12.3.1.1 Plasma microvesicle and exosome studies on healthy individuals

Regarding the study of plasma MVs in healthy subjects, Jin et al. performed one of the first proteomic studies in 2005.[22] They isolated plasma MVs from plasma (n = 16) using ultracentrifugation (250,000 × g for 60 min), pooled them, and compared their proteome with that of plasma and isolated platelets. The proteomic techniques of choice were 2DE (pH 3–10, 24 cm) and MS (matrix-assisted laser desorption/ionization—time of flight/time of flight [MALDI-TOF/TOF]). The results showed the identification of 169 protein spots, corresponding to 83 different unique proteins. Thirty of these proteins had not been previously described in plasma from healthy donors. The identified proteins included a wide number of cytoskeletal proteins, platelet-characterizing receptors (e.g., integrin αIIb and β3), and a selection of enriched plasma proteins (e.g., immunoglobulin M [IgM] chains) that may provide a mechanism for clearance of cell membrane vesicles from circulation by the mononuclear phagocyte system and for maintaining MV homeostasis.[22]

Two years later, Smalley et al. complemented their initial study on platelet-derived microvesicles (PMVs)[23] (reviewed later) with a comparison of the proteome of plasma MVs versus PMVs (generated by activation with adenosine diphosphate [ADP], 10 μM).[24] PMVs were isolated either from plasma using ultracentrifugation (150,000 × g for 90 min at 10°C) or from a platelet suspension with a previous step of gel filtration chromatography and also an ultracentrifugation step (150,000 × g for 90 min at 10°C). Samples were obtained from a healthy donor and run in duplicate. Two complementary strategies were performed: the first was MS-MS spectral count (liquid chromatography–tandem mass spectrometry [LC-MS/MS] on a linear trap quadropole–ion trap [LTQ-IT]), which identified peptides from 21 proteins in plasma MVs that were practically absent in PMVs. Those included several proteins associated with apoptosis, iron transport, immune response, and the coagulation process. The authors speculated that some of these proteins might be present during the initial formation of the MVs and bind to the MV complex.[24] Results were confirmed using the second method (isotope-coded affinity tags). Globally the study detected 229 proteins, which included about 75% of the proteins identified in their previous report.[23]

Smalley et al. extended the characterization of plasma MVs, gathering samples from 42 patients, most of them with a history of vascular disease, and analyzing their plasma MV proteome in an attempt to fully characterize it and understand how plasma MV protein composition is affected by demographic variables.[25] The samples were run in triplicate or quadruplicate, depending on sample availability through an LC-MS/MS system with an electrospray ionization (ESI) source coupled to an LTQ-IT mass spectrometer. From a total of 2,669 proteins identified, 458 were identified with at least two unique peptides and 130 were detected in at least 50% of the patients, which constituted the core proteome of plasma MV proteins. Gene enrichment analysis of the core proteome highlighted several gene ontology (GO) terms such as hemostasis and coagulation, cell motility and localization, and the molecular function of binding actin or other cytoskeletal proteins. This complemented and improved upon the previous study and also showed the effectiveness of label-free spectral counting for MS-based protein identifications. It also proved that there was some correlation between proteins identified in the study and parameters such as age and gender. The authors suggest the potential use of plasma MVs as biomarkers if the analyses are focused on patients with a specific range of demographic values.[25]

Recently, Østergaard et al. decided to utilize new techniques to go deeper in the characterization of the plasma MV proteome in healthy individuals.[26] They analyzed the proteome of 12 healthy samples using label-free nano-LC-MS/MS on an LTQ-IT. They obtained MVs directly from the blood, and for that they used centrifugation to obtain PPP and then a final step by centrifugation at 18,890 g for 30 min. With the aim of obtaining a pure plasma MV population the authors performed an evaluation of the plasma MV–washing procedure by using a series of washes with PBS-citrate at the centrifugation speed chosen. They found that even though albumin was still detectable at wash number 10, other abundant plasma proteins disappeared after the fourth centrifugation. They therefore decided five washing steps was the optimal choice to proceed with the analysis of the samples. They analyzed the plasma MV proteome of four pools, each one made by mixing three samples, and detected a number of 334 proteins (from a total of 536) that constituted the core proteome and were present in all the samples. There was an overrepresentation of extracellular, cytoskeletal, cytoplasmic, and organelle proteins. The authors also performed an independent experiment in which every sample was analyzed in triplicate to give an estimate of intra-assay variability (inter- and intraday) of protein identification. Their results indicate that intraday variability is low enough to justify not running repeat samples and that interday variability is dependent on sample preparation and not on analytical variability.[26] For quantification purposes, the authors also suggest a normalization routine by analyzing the PPP samples by flow cytometry and quantifying some cytoskeletal proteins like myosin-9 or integrin αIIb, whose quantities correlate well with MV numbers.

The first proteomic study of plasma exosomes was performed by Looze et al. in 2009.[27] The authors developed a method using a series of fractionation steps, with gel exclusion chromatography columns and rate zonal centrifugation through continuous sucrose gradients, to separate exosomes from plasma lipid fractions of very low-density lipoprotein (VLDL) and intermediate-density lipoprotein (IDL) with which they cosegregate. Then, exosomes were isolated from the fractions with specific gravities of 1.08–1.15 g/mL using ultracentrifugation (175,000 × g for 120 min). Proteins were separated by SDS-PAGE (4–12% TrisNuPAGE) and identified by LC-MS/MS (LTQ-IT). The MS analysis identified 66 proteins, 28 of which were cell associated and 38 extracellular. Several of the identified proteins were related to vesicular trafficking or protein sorting, which reflected with the exosome formation process.[27] The authors also performed an analysis of a member of the nuclear receptor superfamily that regulates normal adipocyte differentiation and proliferation, PPARγ, and demonstrated that it is present in plasma-derived exosomes, which may indicate a new pathway for paracrine signaling by nuclear receptors.[27]

As mentioned above, when the researchers are dealing with MV samples it is difficult to discriminate MVs from exosomes during isolation. Bastos-Amador et al. decided to study the proteome of plasma MVs with a size smaller than 220 nm, which includes exosomes and the smallest MVs.[28] To perform the analysis they isolated MVs using ultracentrifugation (110,000 × g for 120 min) and filtered them using 0.22 μm micropore filters before ultracentrifuging again. They also immunodepleted albumin and immunoglobulins (Igs) from MVs and enriched some of the samples using a 30% sucrose cushion, which then were analyzed using a gel-free LC-MS^E-based proteomics approach. The study allowed the identification of 161 proteins, 52 of which belong to the Ig protein family and may be part of immune complexes reported recently.[29] Among the rest of proteins identified most of them were cytosolic and cytoskeleton-associated ones. In exosome-enriched preparations the level of protein identifications decreased substantially, and the authors argue that this might be due to low levels of plasma exosomes under healthy conditions.[28]

12.3.1.2 Plasma microvesicle and exosome clinical studies

In regard to the analysis of plasma MVs in a clinical context, a preliminary proteomic study of plasma MVs after experimental venous thrombosis (VT) was performed by Abdullah et al. in 2009.[18] Plasma MVs contribute to thrombogenesis, in part due to their surface proteins, so the authors decided to elucidate the nature of these proteins. They induced VT in juvenile baboons (n = 4), and MVs were obtained from their PPP at baseline and at day 2 postinduction and compared using isobaric tags for relative and absolute quantitation (iTRAQ) and 2D-LC/MS-MS (strong cation exchange [SCX], high-pressure liquid chromatography [HPLC], and MALDI-TOF/TOF). After depleting the sample from plasma contaminants and platelets, they used centrifugation (35,000 rpm for 120 min at 4°C) to isolate plasma MVs. A total of eight experiments were performed (two per sample). The analysis provided the identification of 116 proteins, and after the selective criteria seven proteins were considered to be differential between baseline and 2-day samples. These included up-regulated forms of fibrinogen and α1-antichymotrypsin and some down-regulated Igs, which may influence thrombosis through hemostatic plug formation, inhibition of neutrophil adhesion, and immunoregulation, respectively.[18]

Another clinical study was performed on the proteome of plasma MVs after deep venous thrombosis (DVT).[17] Using iTRAQ and 2D-LC/MS (SCX, HPLC, and MALDI-TOF/TOF) the researchers compared plasma MVs obtained from patients with DVT (n = 9) and healthy controls (n = 6). They obtained plasma MVs using ultracentrifugation of PPP (200,000 × g for 120 min at 4°C, twice). Of the 151 proteins identified, 11 displayed enrichment or depletion of more than 20% in all the experiments performed. From these proteins, two were enriched in plasma MVs from DTV patients (galectin-3-binding protein and α2-macroglobulin) and nine were depleted in plasma MVs of DVT patients, including fibrinogen beta and gamma chain precursors. The two proteins enriched in DVT patients might be promoting thrombus formation, either by promoting platelet thrombotic actions (galectin-3-binding protein) or by inhibiting fibrinolysis and coagulation (α2-macroglobulin). Regarding the down-regulation of fibrinogen, the authors speculate that this could be due to its consumption by the thrombus, given that patients might not be in the early stages of thrombus formation, when fibrinogen is known to be enriched, at the time the sample was obtained.[17]

Pulmonary embolism (PE) is the obstruction of the main artery of the lung or one of its branches by embolized clot material that most commonly arises in the legs from DVT. Last year, Watts et al. studied if the MVs produced after induction of PE in rats have particular characteristics.[30] They wanted to test if the differences observed in previous studies in plasma MV composition were a cause or a consequence of the pathologies. With that objective in mind, they designed a proteomics study where rats were induced a moderately severe form of PE using polystyrene microspheres to produce fixed embolization of the pulmonary vasculature. Plasma MVs were isolated using ultracentrifugation (200,000 × g for 900 min at 4°C). Protein samples (controls n = 3, PE n = 3) were run in 10% Bis-Tris gels and sliced to be analyzed by ultraperformance liquid chromatography coupled to MS (LTQ-IT). Their results showed that plasma MVs in PE-induced rats might display a procoagulant phenotype. Several prothrombotic proteins were increased in PE-induced rats (fibronectin, fibrinogen precursors, and von Willebrand factor), and two others were decreased in their plasma MVs, which correlated well with previous results. Another protein was altered (α2-macroglobulin), which may be inhibiting fibrinolysis. The experimental design allowed the authors to be certain that changes were due to experimentally induced PE and not from pre-existing pathology.[30]

Patients with thalassemia, a group of inherited disorders that result in reduced rate or no synthesis of one of the globin chains that make up hemoglobin, have higher counts of plasma MVs than healthy subjects.[31] Chaichompoo et al. hypothesized that MV protein contents may contribute to the "hypercoagulable state" these patients present.[32] They compared the plasma MV proteome of β-thalassemia/hemoglobin E (β-thal/HbE) patients ($n = 15$) with that of healthy subjects ($n = 15$) using 2DE (pH 3–10) and MS (Q-TOF MS/MS).[32] The isolation procedure was based on centrifugation of PFP at 14,000 × g for 45 min. There were 29 differentially regulated proteins between the groups, and most of them were involved in regulation of the redox reaction, phospholipid turnover, and coagulation system. Their results suggest that the already known increased oxidative stress in platelets and red blood cells (RBCs) in β-thal/HbE patients potentially induces plasma MV formation and that plasma MVs may be contributing to increase the risk for thrombotic events these patients suffer.[32]

12.3.2 Cell-Derived Microvesicles and Exosomes

Several studies have investigated the proteome of specific cell-derived MVs. Most of them have produced MVs by stimulation of cell cultures or isolated cells.

12.3.2.1 Platelet-derived microvesicles

PMVs are the most abundant type of MVs present in human blood.[33,34] The first study concerning PMVs analyzed their proteome content after isolating them directly from the blood.[23] Later, we and another research group characterized the size and content of PMVs after generating them with different platelet agonists or stimuli.[35,36]

The study published in 2005 by García et al. studied the proteome of PMVs obtained from a healthy individual.[23] To do so, platelets were activated with the specific agonist ADP to obtain the vesicles and isolated by ultracentrifugation at 150,000 × g for 90 min at 4°C, and their proteome was compared with that of platelets. Proteins were separated using SDS-PAGE (4–20% gradient gel) and then analyzed using LC-MS/MS (LTQ-IT). A total of 578 proteins were identified in the PMV proteome, 380 of which had not been previously described in the platelet proteome. Identifications included transmembrane and cell surface proteins, proteins probably involved in MV formation, and also several signal transduction proteins. This study was expanded later to cover the analysis of plasma MVs (reviewed in section 12.3.1).[24]

Another study concerning PMVs was performed by Dean et al. in 2009. PMVs were isolated from blood obtained from healthy volunteers (n = 32) after platelet activation with two agonists, thrombin and collagen. After removal of platelet debris, PMVs were isolated by ultracentrifugation (130,000 × g). In this study, PMVs were separated into four fractions using gel filtration chromatography and were analyzed using 2D-LC-MS/MS (reverse phase-LC-LTQ-IT). According to their results, the four fractions corresponded to four different size groups of PMVs, which show certain differences among them. These different size classes contain different lipid/protein ratios, different protein components, and different functional effects on platelets and endothelial cells.[35] The authors found that the relative contribution of platelet membrane proteins to PMVs increases as the size of the vesicles decreases and also that α-granule contents were more present in smaller PMVs (around 100 nm), which may qualify them as exosomes. Interestingly, only the two smallest fractions affected platelet function, and only two of the four fractions significantly enhanced cell barrier integrity, and even these populations had a different active component. These results suggest the existence of specialized PMV populations with particular functions, which may be acting independently *in vivo*.[35]

In a collaborative study with the group of David Varon (Jerusalem), we recently performed a comparative proteome analysis of PMVs generated in two different ways.[36] Platelets were isolated from blood (healthy volunteers) and activated to produce PMVs by application of shear (1,800 s^{-1}) or stimulation with thrombin (0.5 U/mL). Isolation of MVs was performed by ultracentrifugation (100,000 × g for 60 min) after platelet removal. Proteins were separated by high-resolution 2DE (pH 4–7, 24 cm) and differentially regulated proteins identified by MALDI-TOF/TOF and LC-MS/MS. We could see that platelets shed different amounts of MVs and with different proteome content, depending on the stimulus. Thirty unique proteins were differentially regulated, which included mainly proteins involved in platelet activation, including membrane receptors and adapters (Fig. 12.2). Twenty-one of the proteins were interconnected in a network related to cell assembly and organization and cell morphology, as detected by ingenuity pathways analysis. Two proteins were chosen for further validation by Western blotting on the basis of their relevance in platelet activation and potential role in PMVs: integrin α6 and Dok-2. Both proteins participate in the molecular mechanisms regulating angiogenesis, so they could be involved in the PMV regulation of endothelial cells function.[36]

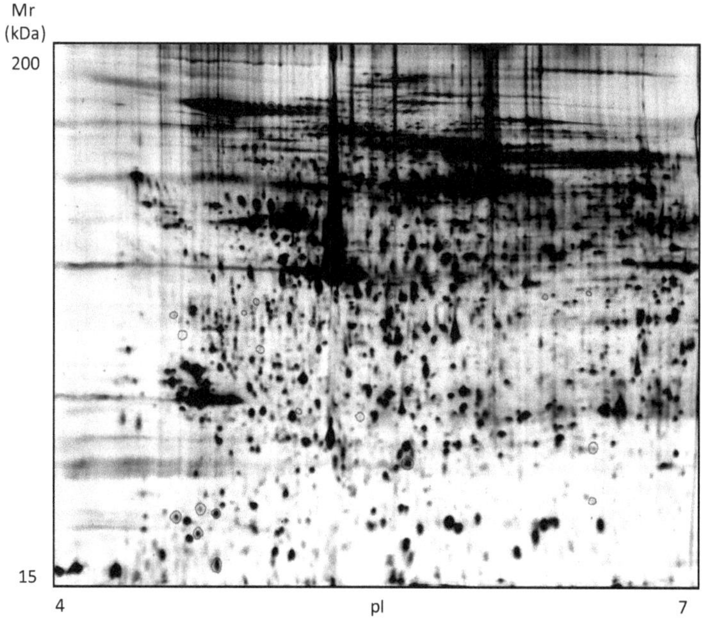

Figure 12.2 High-resolution 2DE-based proteome analysis of platelet-derived MVs. Shown is a representative 2D gel image of proteins from platelet-derived MVs. The first dimension was on IPG strips 4–7, 24 cm. The second dimension was by SDS-PAGE on 11% gels. The figure shows the location on the 2D gels of those spots differentially regulated between MVs originated following platelet activation with shear stress or with thrombin. Reprinted from Ref. 36, Copyright (2012), with permission from Elsevier. *Abbreviation*: IPG, immobilized pH gradient.

12.3.2.2 Endothelium-derived microvesicles

Another subgroup of MVs analyzed by proteomics are those derived from endothelial cells. Endothelial-derived MVs (EMVs) are present in the blood of healthy individuals but also have been detected in higher amounts in some diseases such as hypertension and cardiovascular disorders, pre-eclampsia, and vasculitis. The study of their proteome could provide useful knowledge and lead to new markers for the detection of endothelial dysfunction (ED), a condition where the endothelium loses equilibrium in the production of vasoconstricting and vasodilating substances.

The first study using proteomics on EMVs was by Banfi et al., who isolated them from human umbilical vein endothelial cell (HUVEC) cultures.[12] To do so, they stimulated the cells with tumor necrosis factor-α (TNFα), and after clearing cell fragments, they used ultracentrifugation (100,000 × g for 90 min at 10°C) twice to pellet EMVs from the supernatants. The separation of the proteins was performed using an SDS-PAGE gel (gradient 4–12.5%), and the bands were subjected to MS analysis (MALDI-TOF). The EMV proteome detected was composed of several types of proteins like cytoskeleton-binding proteins (tubulin, actin, myosin), membrane-associated proteins involved in intracellular transport and signaling (caveolin, annexins), and proteins involved in protein folding (calnexin, chaperones) or apoptosis. The authors also checked that EMVs have procoagulant activity and induce it in neighboring cells.

Another study on EMVs was performed by Sander et al.[13] In this case, the EMVs were produced by stimulation of HUVECs with plasminogen activator inhibitor type-1 (PAI-1). The isolation method was also ultracentrifugation (100,000 × g for 90 min at 4°C), and the proteomic analysis was based on 2DE (pH 4–7) and MALDI-TOF/TOF. The authors could identify 58 unique proteins. Four of them were selected for validation by Western blot, namely prohibitin, annexin 5, PDI, and vimentin. Proteins identified were from cytoplasm, nucleus, and membrane origin and were involved in different functions like angiogenesis, cell motility, or cell adhesion. There were several coincidences with the study by Banfi et al. but also distinct features, which could be due to the proteome of EMVs being affected by the upstream stimulus. This could imply that the proteome of EMVs may reflect the original state of the cell in response to a particular environment.[13]

To answer the questions that arose during their previous study, the same research group performed a comparative study between the proteome of EMVs generated from HUVECs with PAI-1 and TNFα.[37] The EMVs were pelleted again by ultracentrifuging the supernatants at 4°C for 60 min at 100,000 × g. Separation of the proteins was in 10% acrylamide gels and identification with an LTQ mass spectrometer. A total of 783 proteins were identified in the control EMV proteome, 679 proteins from PAI-1-generated EMVs and 643 proteins from TNFα-generated EMVs. With regard to the studies previously performed, the coincidence was the identification of 39.7% of Banfi's proteins and 49.4% of Sander's. A total number of 432 proteins were common to the three study groups (PAI-1, TNFα, and unstimulated). Nevertheless, an important number of proteins was characteristic of each study group, highlighting that although EMVs are similar, each population is clearly distinct and these unique aspects of each one may confer them functional differences.[37]

The last study involving endothelial-related MVs involved endothelial progenitor cells (EPCs). There are various aspects related to EPCs, like their nomenclature and phenotype, which are still under discussion. In their study, Prokopi et al. used a proteomic approach (4–12% gradient SDS-PAGE gels and LC-MS/MS with an LTQ-IT) to identify membrane proteins present on MVs in EPC cultures, which might help to understand the open questions about this cell type.[38] Isolation of MVs was achieved using a final centrifugation step (20,500 × g for 150 min at 4°C) of EPC culture supernatants. Six hundred and eighteen proteins were identified in MVs from EPC cultures. The most abundant ones were platelet proteins (integrins αIIb and β3), and the top canonical pathway was integrin signaling. Their results helped to clarify that data from EPCs may be distorted by the presence of platelet proteins and the uptake of those by mononuclear cells. This finding implicates the need for a redefinition of the EPC phenotype and has profound implications for future clinical trials regarding the use of EPCs for stem cell therapy.[38] This study is a good example of how proteomics can be used as a powerful tool for assessing cellular phenotypes and can help in the clarification of issues with clinical relevance.

12.3.2.3 Erythrocyte- and reticulocyte-derived microvesicles and exosomes

In regard to erythrocyte-derived MVs (Ery-MVs), most of the proteomic effort has been oriented toward the understanding of the aging process that erythrocytes undergo when stored for transfusion. Erythrocytes, as other blood cells, produce vesicles during their lifespan, and these vesicles play a role in the aging process and removal of the cells from the circulation. Several proteomic studies have tried to clarify this role. There has also been a study focusing on the first stages of erythrocyte formation, analyzing the secretion of exosomes by reticulocytes during their maturation process.

The first study, conducted by Bosman et al.[39] compared the proteomes of membranes of RBCs and their vesicles at different times during storage. The authors distinguished between MVs (pelleted at 40,000 × g for 20 min at 4°C) and nanovesicles (pelleted at 100,000 × g for 60 min at 4°C). To carry out the analysis they employed SDS-PAGE (10%) and LC-MS/MS (LQT-FT-ICR). The vesicle populations showed differences in size with some overlapping and also in their proteome, characterizing them as distinct entities. The results from Bosman et al. also suggested two possible explanations for the storage-associated vesicle formation process, one triggered by extracellular Igs and the other initiated by intracellular damaged hemoglobin, both of which result in elimination of dangerous cell components by vesiculation. The study also confirmed and expanded previous results on the remodeling that takes places in the RBC membrane during storage and highlights the existence of signaling pathways regulating RBC metabolism and homeostasis.[39] A few years later, the authors extended their study with another investigation regarding EMVs, in this case obtained directly from plasma.[40] The separation of EMVs was carried out by centrifugation of blood (40,000 × g for 20 min at 4°C) to isolate vesicles that were then sorted using fluorescence-activated cell sorting directed to glycophorin A as a marker of erythrocyte origin. The proteome of EMVs was compared to membrane proteomes of erythrocytes that were separated according to cell age. The samples were subjected to the same proteomic strategy as in their previous study (SDS-PAGE and LC-MS/MS). Again, their results proved useful in expanding our knowledge of the role of EVs in RBC aging.

Rubin et al. applied flow cytometry and proteomics (SDS-PAGE and MS) to characterize and analyze the MVs obtained from erythrocyte concentrates stored for transfusion.[41] To isolate the MVs they ultracentrifuged the supernatants three times at 120,000 × g for 90 min at 4°C. They compared the proteome of erythrocyte membranes and that of MVs obtained from the supernatant of

42-day-stored erythrocyte concentrates. An increase in the number of EMVs during storage of erythrocyte concentrates was observed, as it had been previously reported.[42] One of the proteins detected as enriched in MVs was stomatin, which may have a role in membrane microdomain modulation, leading to membrane budding and MV release.[43] Its higher presence was confirmed and detected as constant by Western blotting at day 7 and day 40. There are several hypotheses about why and how the numbers of MVs increase in erythrocyte concentrates with time, so more studies need to be performed to clarify this issue.

An interesting study by Carayon et al. applied proteomic techniques to investigate the alterations in the composition of exosomes secreted by maturing reticulocytes.[44] This maturation process involves the synthesis of large quantities of hemoglobin but also the secretion or disposal of many other cellular components where exosomes play an important role. To investigate this hypothesis, the authors isolated rat reticulocytes and cultured them for a week. Exosomes were then recovered at different time points by ultracentrifugation (100,000 × g for 120 min at 4°C). After the application of a discontinuous sucrose gradient, exosomes' proteins were separated by SDS-PAGE. Identifications were by nano-LC-MS/MS (Qq-TOF) and the LTQ-Orbitrap XL mass spectrometer. The analysis led to the identification of more than 700 proteins, with 522 common to all exosomes. The authors observed changes in the exosome proteome over time, and this was most likely due to the sequential activation of different cellular pathways leading to the production of exosomes,[44] which is in line with previous reports that suggested up to three different mechanisms for the biogenesis of exosomes.

12.3.2.4 Plaque-derived microvesicles

The analysis of MVs derived from *in vivo* human atherosclerotic plaques was comprehensively performed by Mayr et al. using a combination of proteomic, metabolomic, and immunomic techniques.[45] To isolate MVs, plaques obtained from surgical samples were homogenized and centrifuged for 150 min at 20,500 × g at 4°C. SDS-PAGE (4–12%) and LC-MS/MS were the techniques chosen for the proteomic analysis and provided the identification of 151 unique proteins for which a GO analysis associated categories such as membrane, extracellular matrix, protein complexes, and metabolism. The authors also performed a study of MV membrane proteins using CyDye tagging and 2DE, which gave clues about MV origin (most of them were derived from leukocytes but also from smooth muscle cells and erythrocytes) and protein distribution. These results, together with those obtained with other techniques, helped to implicate MVs in modulating tissue inflammation, an important factor in plaque stability and progression.[45]

12.3.2.5 Monocyte- and macrophage-derived microvesicles and exosomes

To the best of our knowledge only two studies have been performed so far investigating monocyte- and macrophage-derived MVs. One of them focused on the comparison of MV proteomes obtained from different stimulations, and the other investigated the role of MVs in human immunodeficiency virus-1 (HIV-1) viral infection.

The cell stimulation study was performed in 2009 by Bernimoulin et al.[46] In this case the focus was on monocytes, which, as many other cell types, shed MVs that promote inflammation, and are highly procoagulant due to the presence of tissue factor and phosphatidylserine. The populations of MVs, compared using a proteomic approach (4–20% SDS-PAGE and LC/MS-MS), were generated from a human monocytic cell line (THP-1), following cell stimulation with different agonists: a physiologic one (soluble P-selectin-Ig chimera, P-sel-Ig), a pathologic one (lipopolysaccharide [LPS]), a control IgG, and no stimulus. Isolation of MVs was achieved using centrifugation (16,000 × g for 45 min) after cell contamination cleaning. The number of MVs was increased three- to fourfold in stimulated cells (P-sel-Ig, LPS) compared to controls. One hundred of the proteins identified were common to all the populations. Many of them were cytoskeletal proteins. Forty-two were only common to stimulated populations, and interestingly, fifty-two unique proteins belonged to the P-sel-Ig MVs and 408 to the LPS MVs. In the LPS-unique protein group there was an important representation of nuclear and mitochondrial proteins and some others associated with metabolism and energy pathways. Meanwhile in the P-sel-Ig group the unique proteins related more to the plasma membrane and involvement in signal transduction and cell communication.[46] This study shows how the protein content of MVs is influenced by the generating stimulus and the existence of distinct MV subpopulations derived from the same cell type.

A group of researchers has recently used proteomics to investigate the role of macrophage-derived MVs (MMVs) in the HIV-1 infection process.[47] In this study the authors cultured monocyte-derived macrophages (MDMs) and infected them with HIV-1. The proteomic procedures were carried out through the use of SDS-PAGE (10%) and LC-MS/MS (LTQ-IT). The isolation of MVs was achieved using ultracentrifugation (100,000 × g for 60 min on a 20% w/v sucrose cushion) of MDM supernatants. The authors could detect differences between the proteome of MMVs and macrophage-derived exosomes but also similarities that qualify them as important tools in long-distance immune response. Also, differences could be established between exosomes and HIV-1 regarding their size, proteomic content, and morphology. HIV-1 was detected in the MV fractions; it was demonstrated that MMVs and macrophage-derived exosomes facilitate transfer of virus and viral constituents from infected macrophages to neighboring uninfected cells.[47]

12.3.2.6 Lymphocyte-derived microvesicles and exosomes

The first study of B-cell-derived EVs was performed by Wubbolts et al. in 2003.[48] The researchers aimed to deepen the understanding of B-cell-derived exosomes and developed a protocol to purify them and study their protein content. Exosomes isolation from RN cell culture supernatants was performed with ultracentrifugation (70,000 × g for 60min). Then, a linear sucrose gradient was applied for 16 h at 270,000 × g, and the positive fractions for major histocompatibility complex (MHC) class II and tetraspanins were further purified with magnetic beads coupled to a monoclonal mouse anti–human MHC class II antibody. Proteins were separated by SDS-PAGE and identified by MALDI-TOF/TOF. Among them, there were human leukocyte antigen (HLA)-encoded proteins and other membrane proteins like CD45 and integrin α4. Other proteins were heat shock proteins, cytoskeletal proteins, and glycolysis-associated proteins, among others. Tetraspanins could not be detected, possibly because of their low abundance and/or the difficulty to recover them from acrylamide gels.[48]

The characterization of MV proteins from lymphoid cells was continued by Miguet et al. in 2006.[11] With the aim of characterizing membrane antigens that could be relevant for the diagnosis and treatment of chronic, mature B-cell lymphocytic malignancies the authors used a proteomic approach, in which they stimulated

MV production from a CEM T-cell line in two different ways, via pharmacological mitogenic activation and via apoptosis induction. To isolate MVs they centrifuged cell supernatants for 45 min at 15,000 × g. Then, they separated the proteins by 8–15% SDS-PAGE gels and identified them by MS (LC-MS/MS and MALDI-MS). An interesting aspect of this study is how the authors combined MS techniques to increase the sequence coverage of the identified proteins. Around 70% of the proteins identified were present in both stimulation approaches, and the differentially expressed ones were basically cytoplasmic proteins. Only three plasma membrane CD antigens were differentially detected (CD81, CD99, and CD107), of which CD99 can be attributed to apoptosis triggering. The method, developed by the authors, for the analysis of membrane proteins by means of studying MVs through proteomics proved useful 3 years later, when they published a subsequent study in which they proposed CD148 as a discriminating biomarker for three types of chronic B-cell lymphocytic malignancies.[49] The researchers obtained MVs from three patients—one with chronic lymphocytic leukemia (CLL), one with small cell lymphoma (SLL), and one with mantle cell lymphomas (MCL)—and applied their proteomic approach. After the appropriate selective criteria, CD148 emerged as a possible discriminating marker for the malignancies, and the authors validated its potential in 158 patients (CLL, n = 93; SLL, n = 24; MCL, n – 41) and 30 healthy controls. It seems CD148 expression might be a good way to discriminate cytologically or phenotypically ambiguous cases of the mature B-cell neoplasms mentioned. This result, as the authors mention, shows the ability of proteomics to propose strong protein candidates to be further confirmed as potential biomarkers.

Li et al. performed the most recent proteomic study about B-cell-derived exosomes.[50] The question asked was if HIV-1 infection of lymphocytes could modify any of the host proteins belonging to the exosomal fraction. The authors utilized stable isotope labeling by amino acids in cell culture (SILAC) to compare the proteomes of the exosomal fractions of HIV-1-infected and HIV-1-uninfected lymphocytic cells. Exosomes were isolated from H9 cell supernatants at 4°C by ultracentrifugation (30 min at 10,000 × g, followed by 70 min at 100,000 × g). Proteins were separated in acrylamide gels (4–12% NuPAGE) and analyzed with a mass spectrometer (LTQ-Orbitrap). Fourteen proteins were differentially expressed, of which

CD38, LDBH, and ANXA5 have been previously associated with proliferation and apoptosis processes. The authors suggest their results support the idea that HIV-1 infection influences the behavior of neighboring cells through the secretion of exosomes.

12.4 Conclusions and Future Perspective

In recent years EVs received increasing interest, given their potential role in several pathologies. Indeed, circulating MVs have been postulated as a source of biomarkers in coagulation, inflammation, and cancer. Moreover, clinical studies have been conducted to assess their contribution to the identification of patients at cardiovascular risk. However, all the studies faced several limitations, with preanalytical steps remaining an important source of variability and artifacts in MV analysis. Great efforts have been recently undertaken by the scientific community to address these limitations. For instance, the International Society for Thrombosis and Haemostasis (ISTH) organized a Scientific and Standardization Committee (SSC) collaborative workshop dedicated to standardize PMV enumeration by flow cytometry with calibrated beads.[51] Moreover, Lacroix et al. recently studied the impact of preanalytical parameters on the measurement of circulating MVs, trying to establish a standardized protocol.[14] Parameters that can influence the analysis include the type of blood collection tube, phlebotomy conditions, transportation practices, centrifugation steps, and freezing.

The high clinical potential of blood MVs as a source of biomarkers is one of the reasons why several research groups have recently applied proteomics to their studies. Most MV analyses done so far are based on flow cytometry. This is a fast way to check the levels of circulating MVs in those pathologies where their number is elevated (e.g., cardiovascular disease). On the other hand, proteomics has the power of identifying MV-related biomarkers. However, the preanalytical parameters mentioned above are even more critical for proteomic analyses. The MV population needs to be pure because any contamination with cell debris or plasma proteins would hamper the analysis. In addition, the amount of protein needed for the analysis requires the isolation of a high number of MVs, which quite often requires a high volume of plasma or cell culture media as starting material. All the above makes MV proteomic analyses quite complicated. Therefore, standardized protocols for MV isolation are definitely needed. As mentioned in this chapter, we can almost say that each research group has its own protocol for MV isolation, with great variations in the centrifugation speed that is used to isolate the same type of MVs. This definitely affects interlab reproducibility of the data. The standardized protocol established by Lacroix et al. is a good starting point to address the above preanalytical issues that can influence the study.[14]

The other main source of interlab variability in MV proteomic studies is the precise proteomic tool that is applied. Obviously, this issue is more complicated to address because different labs will have different proteomic platforms, which will definitely influence the study. The goal would be that the same platforms use the same protocol to achieve maximum reproducibility. Besides that, different proteomic strategies will yield complementary data as in any proteomics study, which is good.

In conclusion, over the last few years several research groups worldwide have applied proteomics to the study of EVs in health and disease. These pioneer studies have faced important analytical difficulties but have also paved the way for the future of the field. Standardized protocols for MV isolation and proteomic analysis will help to improve the studies, making them more reproducible. Several analytical challenges must still be addressed, but the future looks promising.

12.5 Acknowledgments

The authors would like to acknowledge the support given by the Spanish Ministry of Economy and Competitiveness (MINECO) (grant no. SAF2010-22151, cofunded by the European Regional Development Fund [ERDF]).

References

1. Théry C, Ostrowski M, Segura E (2009) Membrane vesicles as conveyors of immune responses, *Nat. Rev. Immunol.*, **9**(8), 581–593.
2. Boulanger CM, Dignat-George F (2011) Microparticles: an introduction, *Arteriorscl. Thromb. Vasc.*, **31**(1), 2–3.

3. Cocucci E, Racchetti G, Meldolesi J (2009) Shedding microvesicles: artefacts no more, *Trends Cell Biol.*, **19**(2), 43–51.
4. Wolf P (1967) The nature and significance of platelet products in human plasma, *BJH*, **13**(3), 269–288.
5. Trams EG, et al. (1981) Exfoliation of membrane ecto-enzymes in the form of micro-vesicles, *Biochim. Biophys. Acta*, **645**(1), 63–70.
6. Bastida E, et al. (1984) Tissue factor in microvesicles shed from U87MG human glioblastoma cells induces coagulation, platelet aggregation, and thrombogenesis, *Blood*, **64**(1), 177–184.
7. György B, et al. (2011) Membrane vesicles, current state-of-the-art: emerging role of extracellular vesicles, *Cell Mol. Life Sci.*, **68**(16), 2667–2688.
8. Raimondo F, et al. (2011) Advances in membranous vesicle and exosome proteomics improving biological understanding and biomarker discovery, *Proteomics*, **11**(4), 709–720.
9. Andrecht S, von Hagen J (2008) *General Aspects of Sample Preparation for Comprehensive Proteome Analysis* in *Proteomics Sample Preparation* (ed. von Hagen J), Weinheim: Wiley-VCH Verlag, 5–20.
10. Yuana Y, Bertina RM, Osanto S (2011) Pre-analytical and analytical issues in the analysis of blood microparticles, *Thromb. Haemost.*, **105**(3), 396–408.
11. Miguet L, et al. (2006) Proteomic analysis of malignant lymphocyte membrane microparticles using double ionization coverage optimization, *Proteomics*, **6**(1), 153–171.
12. Banfi C, et al. (2005) Proteome of endothelial cell-derived procoagulant microparticles, *Proteomics*, **5**(17), 4443–4455.
13. Sander TL, et al. (2008) Protein composition of plasminogen activator inhibitor type 1-derived endothelial microparticles, *Shock*, **29**(4), 504–511.
14. Lacroix R, et al. (2012) Impact of pre-analytical parameters on the measurement of circulating microparticles: towards standardization of protocol, *J. Thromb. Haemost.*, **10**(3), 437–446.
15. Kim HK, et al. (2002) Optimized flow cytometric assay for the measurement of platelet microparticles in plasma: pre-analytic and analytic considerations, *Blood Coagul. Fibrinolysis*, **13**(5), 393–397.
16. Trummer A, et al. (2009) Recovery and composition of microparticles after snap-freezing depends on thawing temperature, *Blood Coagul. Fibrinolysis*, **20**(1), 52–56.

17. Ramacciotti E, et al. (2010) Proteomics of microparticles after deep venous thrombosis, *Thromb. Res.*, **125**(6), e269–274.

18. Abdullah NM, et al. (2009) Microparticle surface protein are associated with experimental venous thrombosis: a preliminary study, *Clin. Appl. Thromb. Hemost.*, **15**(2), 201–208.

19. Moon PG, et al. (2011) Urinary exosomes and proteomics, *Mass Spectrom. Rev.*, **30**(6), 1185–1202.

20. Simpson RJ, Jensen SS, Lim JWE (2008) Proteomic profiling of exosomes: current perspectives, *Proteomics*, **8**(19), 4083–4099.

21. Simpson RJ, et al. (2009) Exosomes: proteomic insights and diagnostic potential, *Expert Rev. Proteomics*, **6**(3), 267–283.

22. Jin M, et al. (2005) Distinct proteome features of plasma microparticles, *Proteomics*, **5**(7), 1940–1952.

23. Garcia BA, et al. (2005) The platelet microparticle proteome, *J. Proteome Res.*, **4**(5), 1516–1521.

24. Smalley DM, et al. (2007) Proteomic discovery of 21 proteins expressed in human plasma-derived but not platelet-derived microparticles, *Thromb. Haemost.*, **97**(1), 67–80.

25. Little KM, et al. (2010) The plasma microparticle proteome, *Semin. Thromb. Haemost.*, **36**(8), 845–856.

26. Østergaard O, et al. (2012) Quantitative proteome profiling of normal human circulating microparticles, *J. Proteome Res.*, **11**(4), 2154–2163.

27. Looze C, et al. (2009) Proteomic profiling of human plasma exosomes identifies PPARgamma as an exosome-associated protein, *Biochem. Biophys. Res. Commun.*, **378**(3), 433–438.

28. Bastos-Amador P, et al. (2012) Proteomic analysis of microvesicles from plasma of healthy donors reveals high individual variability, *J. Proteomics*, **75**(12), 3574–3584.

29. György B, et al. (2011) Detection and isolation of cell-derived microparticles are compromised by protein complexes resulting from shared biophysical parameters, *Blood*, **117**(4), e39–e48.

30. Watts JA, et al. (2012) Proteomics of microparticles after experimental pulmonary embolism, *Thromb. Res.*, **130**(1), 122–128.

31. VanWijk MJ, et al. (2003) Microparticles in cardiovascular diseases, *Cardiovasc. Res.*, **59**(2), 277–287.

32. Chaichompoo P, et al. (2012) Characterizations and proteome analysis of platelet-free plasma-derived microparticles in β-thalassemia/hemoglobin E patients, *J. Proteomics*, **189**(2), 744–754.

33. Horstman LL, Ahn YS (1999) Platelet microparticles: a wide-angle perspective, *Crit. Rev. Oncol. Hematol.*, **30**(2), 111–142.
34. Piccin A, Murphy WG, Smith OP (2007) Circulating microparticles: pathophysiology and clinical implications, *Blood Rev.*, **21**(3), 157–171.
35. Dean WL, et al. (2009) Proteomic and functional characterisation of platelet microparticle size classes, *Thromb. Haemost.*, **102**(4), 711–718.
36. Shai E, et al. (2012) Comparative analysis of platelet-derived microparticles reveals differences in their amount and proteome depending on the platelet stimulus, *J. Proteomics*, doi: 10.1016/j.jprot.2012.02.030 (epub ahead of print).
37. Peterson DB, et al. (2008) Comparative proteomic analysis of PAI-1 and TNF-alpha-derived endothelial microparticles, *Proteomics*, **8**(12), 2430–2446.
38. Prokopi M, et al. (2009) Proteomic analysis reveals presence of platelet microparticles in endothelial progenitor cell cultures, *Blood*, **114**(3), 723–732.
39. Bosman GJCGM, et al. (2008) The proteome of red cell membranes and vesicles during storage in blood bank conditions, *Transfusion*, **48**(5), 827–835.
40. Bosman GJCGM, et al. (2012) The proteome of erythrocyte-derived microparticles from plasma: new clues for erythrocyte aging and vesiculation, *J. Proteomics*, doi: 10.1016/j.jprot.2012.05.031 (epub ahead of print).
41. Rubin O, et al. (2008) Microparticles in stored red blood cells: an approach using flow cytometry and proteomic tools, *Vox Sang.*, **95**(4), 288–297.
42. Keuren JFW, et al. (2006) Effects of storage-induced platelet microparticles on the initiation and propagation phase of blood coagulation, *BJH*, **134**(3), 307–313.
43. Salzer U, et al. (2008) Vesicles generated during storage of red cells are rich in the lipid raft marker stomatin, *Transfusion*, **48**(3), 451–462.
44. Carayon K, et al. (2011) Proteolipidic composition of exosomes changes during reticulocyte maturation, *J. Biol. Chem.*, **286**(39), 34426–34439.
45. Mayr M, et al. (2009) Proteomics, metabolomics, and immunomics on microparticles derived from human atherosclerotic plaques, *Circulation*, **2**(4), 379–388.
46. Bernimoulin M, et al. (2009) Differential stimulation of monocytic cells results in distinct populations of microparticles, *J. Thromb. Haemost.*, **7**(6), 1019–1028.

47. Kadiu I, et al. (2012) Biochemical and biologic characterization of exosomes and microvesicles as facilitators of HIV-1 infection in macrophages, *J. Immunol. (Baltimore, MD: 1950)*, **189**(2), 744–754.

48. Wubbolts R, et al. (2003) Proteomic and biochemical analyses of human B cell-derived exosomes. Potential implications for their function and multivesicular body formation, *J. Biol. Chem.*, **278**(13), 10963–10972.

49. Miguet L, et al. (2009) Proteomic analysis of malignant B-cell derived microparticles reveals CD148 as a potentially useful antigenic biomarker for mantle cell lymphoma diagnosis, *J. Proteome Res.*, **8**(7), 3346–3354.

50. Li M, et al. (2012) Quantitative proteomic analysis of exosomes from HIV-1-infected lymphocytic cells, *Proteomics*, **12**(13), 2203–2211.

51. Lacroix R, et al. (2010) Standardization of platelet-derived microparticle enumeration by flow cytometry with calibrated beads: results of the International Society on Thrombosis and Haemostasis SSC collaborative workshop, *J. Thromb. Haemost.*, **8**(11), 2571–2574.

Chapter 13

Microvesicles and Exosomes in Cancer

Rebecca Karp[a] **and Jeffrey Zwicker**[b]
[a]*Department of Medicine, Beth Israel Deaconess Medical Center, Harvard Medical School, Boston, MA, USA*
[b]*Division of Hemostasis and Thrombosis, Division of Hematology-Oncology, Beth Israel Deaconess Medical Center, Harvard Medical School, Boston, MA, USA*
jzwicker@bidmc.harvard.edu

13.1 Introduction

Cancer microvesicles (MVs) were first described in 1978 when they were detected in the spleen nodules and lymph nodes of a patient with Hodgkin's lymphoma.[1] Over the last several decades the mechanistic activities of extracellular vesicles (EVs) have been investigated, including their role in cell-to-cell communication via direct stimulation by surface-expressed ligands; the transfer of receptors, proteins, and ribonucleic acid (RNA); the delivery of various infections (prions, human immunodeficiency virus [HIV], etc.); and perhaps the transport of whole cell parts such as mitochondria.[2–5] EVs in the malignant state are now implicated as playing a central role in the biology of malignancy, including mechanisms that regulate metastasis, angiogenesis, and immune

Extracellular Vesicles in Health and Disease
Edited by Paul Harrison, Christopher Gardiner, and Ian L. Sargent

ISBN 978-981-4411-98-1 (Hardcover), 978-981-4411-99-8 (eBook)
www.panstanford.com

signaling. Recently, populations of EVs expressing tissue factor (TF) have been identified as a biomarker to predict thrombotic events in patients with malignant disease.

The term "extracellular vesicle" is a general label used to describe vesicular, membrane-bound particles secreted by a whole host of cells ranging from leukocytes aiding in the generation of a fibrin clot to tumor cells implicated in the propagation of a malignant clone.[6–8] EVs were first described in 1967 by Peter Wolf, who referred to "minute particulate material" derived from platelets, which he termed "platelet dust." However, since that time, our knowledge of the diversity, complexity, and significance of EVs has expanded exponentially.[9] In addition to cancer, EVs have been studied in relation to HIV, complement fixation, sepsis, acute coronary syndromes, and even the normal physiologic state.[10–12] There still exists much controversy as to an appropriate nomenclature for these vesicles, with microparticles or MVs being used interchangeably regardless of the cell of origin, mode of production, or cargo contained.[3,13,14] However, there is marked heterogeneity among EVs: an MV formed during apoptotic cell death is distinct, both in structure and in function, from a vesicle carefully packaged and released by a malignant cancer cell. In this chapter, the term "EVs" will be used broadly to refer to all types of membrane-bound vesicles.

13.2 Types of Cancer Vesicles

EVs can be organized into categories based upon their size and the manner in which they are formed: apoptotic blebs (50–5,000 nm diameter), the end products of dying cells; membrane MVs, or ectosomes (50–1,000 nm diameter) that are shed directly from plasma membranes; and exosomes (40–100 nm diameter), membranous vesicles formed through endocytosis that may be recycled via lysosomal degradation, redirected back to the plasma membrane, or further processed and sorted within the cell itself.[3,14–17]

13.2.1 Apoptotic Bodies

Apoptosis effectively ends the life of a cell in its recognizable form. It is characterized by loss of membrane integrity, chromatin condensation, and eventual disintegration of the nucleus.[18] However, oftentimes the fragments released during apoptosis—shards of membrane, organelles, deoxyribonucleic acid (DNA)—find their way into other cells, disseminating the genetic material of the original cell. There is even an evolutionary slant to this ritual—a dying cell's instinct to propagate its own genetic code.

We have mounting evidence that tumor-derived apoptotic EVs play an important role in tumor cell signaling and metastatic spread. Transfer of genetic information between apoptotic malignant cells and neighboring cells has been demonstrated in a variety of cancers, including prostate and cervical.[19,20] Bergmedh et al. described the cellular uptake of tumor apoptotic bodies released by H-ras(V12) and mutated human c-myc-transfected rat fibroblasts. Normal fibroblasts, provided that they are *p53* negative, phagocytose these tumor apoptotic bodies and incorporate tumor DNA into their own genomes.[18] These previously "healthy" cells subsequently develop malignant characteristics, such as loss of contact inhibition and anchorage independence (ability to grow in the absence of adhesion to the extracellular matrix)[21] *in vitro* and a "tumorigenic" phenotype *in vivo*.[18] This was not the case in the presence of *p53*, perhaps because this tumor suppressor gene arrested the cell cycle or induced apoptosis in the mutant fibroblasts.[19,20] Another elegant study by Ehnfors et al. illustrates the uptake of apoptotic tumor cell remnants by adjacent endothelial cells and fibroblasts; again in the setting of *p53* inhibition, these endothelial and fibroblast cells incorporate and replicate tumor DNA, thus proliferating components of the original cancer cell.[22,23]

Thus, there exists a fine balance between tumor cell apoptosis (which in itself would be advantageous to a host) and horizontal spread of apoptotic tumor remnants to "innocent bystanders." The tumor suppressor gene *p53* appears to play a critical role in the regulation of this process.[24] The clinical significance of tumor proliferation via apoptosis continues to be explored, with efforts geared toward prevention of uptake and incorporation of tumor remnants by normal cells and regulation by *p53*.

13.2.2 Membrane Microvesicles

Membrane MVs (also referred to as ectosomes) are appropriately named, being formed by outward blebbing of the plasma membrane.

The largest of the three types of EVs, MVs are formed from membrane lipid rafts—glycoprotein microdomains high in signaling complexes and cholesterol—within the lipid bilayer.[25] In fact, many MVs carry lipid raft–associated proteins such as flotillin-1, TF, and epidermal growth factor receptor (EGFR). This cargo is critical for cell intercommunication and cellular growth and development.

MV formation depends on the unique configuration of the lipid bilayer and its ability to transform in response to certain environmental cues. Cell membrane phospholipid arrangement is quite specialized, with phosphatidylcholine (PC) and sphingomyelin (SM) located on the external layer and phosphatidylserine (PS) and phosphatidyl-ethanolamine (PE) on the inner side.[5] The asymmetric composition of the plasma membrane is maintained through five calcium-dependent enzymes: gelsolin, amino-phospholipid translocase, floppase, scramblase, and calpain. Disruption of membrane asymmetry results in translocation of PS to the outside of the cell and occurs in various physiological and disease states such as cell necrosis, apoptosis, and platelet activation.[26]

Disturbances in phospholipid arrangement also occurs under physiologic conditions during the formation of MVs. An influx in cytosolic calcium activates gelsolin, calpain, and scramblase, leading to the translocation of PS residues to the outside of the cell and MV release.[3,5] This increase in calcium also leads to a gelsolin-mediated transformation of the cell cytoskeleton, specifically the cleavage of actin filaments in platelets and actin-spectrin anchorages in other cell lines. The relevance of phospholipid transport activity and MV generation is exemplified by Scott syndrome, a rare bleeding disorder characterized by deficient scramblase activity, deficient PS exposure, and decreased MV generation.[27,28] The PS located on the surface of these EVs serves as a binding site for the tenase complexes VIIIa–IXa and Va–Xa, which are critical for the activation of factor X and the conversion of prothrombin to thrombin. Patients with Scott syndrome present with moderate to severe bleeding diatheses (almost always provoked), requiring platelet transfusions for even minor surgical procedures.[29] This defect in PS translocation (and membrane vesiculation) has also been described in erythrocyte and lymphocyte cell lines in patients with Scott syndrome.[30–32] And yet, the precise role of phospholipid transport proteins in the generation of platelet M10V remains uncertain as mice deficient in scramblase

(PSLCR1 or PLSCR3) do not exhibit loss of PS asymmetry,[33,34] platelet-derived MV formation occurs in the absence of PS exposure,[35] and a large percentage of circulating platelet EVs lack externalized PS.[28] Recently a mutation in a putative scramblase protein called TMEM16F has been described in Scott syndrome and the phenotype reproduced in knockout mice.[36–38]

13.2.3 Exosomes

Exosomes are the smallest of the three types of EVs, and are formed through a highly coordinated process. These EVs are found in a whole host of cells, including B-cells, T-cells, mast cells, and platelets, and are also present in biologic fluids, including urine, breast milk, epididymal fluid, and pleural fluid.[39] Instead of blebbing outward from the plasma membrane like the larger MVs, exosomes are derived from a complex endomembrane network.[40] These vesicles are processed, recycled, and repackaged via an intricate series of steps, leading ultimately to the fusion of multivesicular bodies with the plasma membrane and the release of cellular cargo. The best-described mechanism of exosome formation is via the endosomal sorting complex required for transport (ESCRT) system.[6,41] The ESCRT is composed of four protein complexes, ESCRT-0, ESCRT-I, ESCRT-II, and ESCRT-III, which are responsible for initial invagination of the plasma membrane (through the formation of clathrin-coated pits) and subsequent sorting of the ubiquitinated cargo proteins. These proteins are eventually incorporated into a membrane vesicle, evolving from early endosomes to multivesicular bodies or late endosomes.[3,6] The vesicular cargo can then follow one of four paths. It can be incorporated back into the plasma membrane; remain within multivesicular bodies in the form of intraluminal vesicles; be directed toward the lysosome for disassembly and reprocessing; or be released from the plasma membrane in the form of an exosome.[6,41] Recent studies have elucidated the pathways regulating exosome biogenesis through the interaction of the cytosolic proteins syntenin and ALIX with ESCRT-III.[42,43]

Exosomes also carry material that is distinct from ectosomes, such as heat shock proteins (HSP-70), tetraspanins (CD81, CD63, CD9, and Tspan8), micro-RNAs (miRNAs) (small noncoding RNAs), messenger

RNAs (mRNAs), ESCRT proteins, and angiogenic proteins.[3,44] They are also implicated in tumor proliferation and metastasis.[45] Yang et al. demonstrated that tumor-associated macrophages from breast cancer cells can modulate the invasiveness of the cancer through the release of exosomes containing oncogenic miRNA (miR-223). After tumor-associated macrophages were activated with interleukin (IL)-4, they secreted vesicular-bound miR-223 that was taken up by the surrounding tumor. The authors suggest that quantification of miR-223 in the blood of patients with breast cancer could help predict disease aggressiveness.[46] Evidence also supports that exosome production and clearance are regulated by *p53*-dependent genes such as *TSAP6* and *CHMP4C* (which promote exosome formation) and *CAV1* and *CHMP4C* (which augmentreceptor endocytosis).[24,47]

13.3 Extracellular Vesicles and Tissue Factor as Mediators of Thrombosis

As many as 5–50% of cancer patients will suffer a thrombotic event during the course of their illness.[48,49] The observed association between malignancy and thrombosis dates back to the 1800s, when the French physician Armand Trousseau reported the case of a 59-year-old gentleman with left-lower-extremity phlebitis who was found to have annular cancer of the pylorus at autopsy. Trousseau stated that "frequent concurrence of phlegmasia alba dolens with an appreciable cancerous tumor, led me to the inquiry whether a relationship of cause and effect did not exist between the two, and whether the phlegmasia was not the consequence of the cancerous cachexia."[50,51] Despite the established link between cancer and thrombosis, the molecular basis for this association has not been definitively established. Recent data suggest a role for tumor-derived EVs in the development of cancer-associated clot formation.[52–54]

13.3.1 Tissue Factor and Clot Formation

TF is a 47 kDa cell surface glycoprotein, which serves as the main catalyst for the extrinsic pathway of blood coagulation and the initiator of thrombin generation and fibrin formation.[55,56] It is present on most nonvascular cells and can be induced on the surface of activated monocytes and neutrophils in the setting of clot development. Coagulation is initiated by the binding of TF to factor VIIa (FVIIa), which anchors FVIIa to the cell membrane and increases its proteolytic activity.[53] And while for decades the notion prevailed that coagulation began with TF release from an injured vascular wall, in 1999 Giesen et al. demonstrated that cell-derived TF (in the form of EVs) also existed within the bloodstream and could be found within a fibrin clot formed in the absence of vascular-derived TF.[57] During thrombus formation, activated platelets accumulate at the site of injury, where they express P-selectin. Subsequently, monocyte-derived TF-bearing EVs latch onto the platelets via P-selectin glycoprotein ligand 1 (PSGL-1), engendering clot formation and propagation.[7,55,58] Plasma membrane–derived EVs may also express increased levels of PS, which serves as a catalytic site for various prothrombotic factors.[59]

13.3.2 Tissue Factor–Bearing Extracellular Vesicles in Cancer

Increased levels of TF-bearing EVs have been found in various types of malignancies and often correlate with survival. A study by Ueno et al. measured TF expression in the plasma of 67 breast cancer patients and found that TF was an independent predictor of distant metastasis and overall survival.[60] Our group previously demonstrated that blood-borne levels of TF-bearing EVs that co-expressed tumor antigens decreased following definitive resection of pancreatic tumors.[7]

Khorana et al. also evaluated TF levels in patients with pancreatic neoplasia and found that individuals with high TF expression had a venous thromboembolism (VTE) rate of 26.3% as compared to a rate of 4.5% in patients with low TF expression. Moreover, pathologic examination of the pancreatic neoplasms themselves demonstrated that tumors with high TF levels were more likely to express vascular endothelial growth factor (VEGF) and have a higher microvessel density (marker of angiogenesis) than tumors without abundant TF.[61]

Researchers have evaluated whether elevated levels of TF-bearing EVs in the plasma of patients with cancer can serve as a biomarker to predict thrombosis in cancer.[59,61] A number of different methodologies have been employed for the detection and measurement of EVs in plasma, including flow cytometry,[62] dynamic light scattering,[63] particle tracking,[64] enzyme-linked immunosorbent assay,[65] atomic force microscopy,[63,66] and MV-based functional assays.[52,67,68] In the absence of an accurate standard for the measurement of TF-bearing MV concentrations, the optimal methodology for MV size measurement, characterization, and enumeration is not known. We previously described the development of an impedance-based flow cytometer to measure EVs in plasma.[7] We performed a case control study on 30 patients with cancer and acute VTE, and 60 cancer patients matched for age, stage, sex, and histologic diagnosis without thrombosis. Individuals with cancer and VTE were four times as likely to have elevated TF-bearing EVs compared to the control population. We recently confirmed these findings in a phase II clinical trial (the MicroTEC study).

The MicroTEC study was a randomized, phase II study whereby patients with advanced pancreatic, colon, or non–small cell lung cancer were followed for the development of venous thromboembolic events.[69] After enrollment, the concentration of circulating TF-bearing EVs was measured for each patient. Individuals with elevated TF-bearing EVs were randomized to enoxaparin primary thromboprophylaxis for 2 months versus observation. Individuals with "low" levels of EVs were followed without primary thromboprophylaxis. The cumulative incidence of VTE at 2 months in the group with high TF-bearing EVs not receiving enoxaparin was 27.3% compared to 5.6% in the enoxaparin-treated arm ($p = 0.06$). The cumulative incidence of VTE in the low TF MV arm was 7.2%. These findings confirm the potential prognostic utility of TF-bearing EVs in predicting thrombotic events in cancer patients. Other groups have published similar results, demonstrating a link between TF-bearing EVs and VTE in breast and pancreatic cancer.[52,70] Our group has now initiated a phase II/III randomized clinical trial to evaluate a novel antithrombotic agent[71] for primary thromboprophylaxis in cancer patients, including those with elevated TF-bearing EVs.

Experimental models also provide evidence that tumors generate TF-bearing EVs. In orthotopic murine models of human pancreatic cancer, tumor-derived TF can similarly be measured in plasma. The concentration of tumor-derived TF appears to correlate with the total tumor burden.[56] Several weeks following subcutaneous injection of green fluorescent protein–labeled tumor cells into mice, tumor-derived EVs can be visualized at the site of thrombus, following a vascular injury.[55]

13.4 Extracellular Vesicles in Cell-to-Cell Communication

Communication is important not only amongst tumor cells themselves but also between malignant cells and their unaffected neighbors. One proposed mechanism by which tumor-derived EVs "talk" with nearby cells is via "horizontal propagation": that is, the uptake of cancerous exosomes by nonmalignant cells (similar to the phagocytosis of apoptotic fragments described earlier).[72] This mechanism is elegantly described by Al-Nedawi et al. in the setting of glioma cells and a mutant form of EGFR found in these tumors (EGFR variant III [EGFRvIII]). When glioma cells were incubated with EGFRvIII containing EVs, downstream cell cycle pathways (i.e., mitogen-activated protein kinase [MAPK] and Akt) were activated, and these cells demonstrated an increased production of VEGF, phosphorylation of the VEGF receptor (VEGFR-2), and increased anchorage independence.[73] Normal glial cells had developed oncogenic characteristics upon exposure to these tumor-derived EVs. Moreover, there was a decrease in downstream EGFR signaling when these cells were exposed to EGFRvIII kinase inhibitors. A corollary study by the same group examined the uptake of squamous cell cancer–derived EGFR-containing EVs by cultured endothelial cells.[74] EGFR cell cycle regulatory pathways were subsequently activated in the transfected endothelial cells. Recent studies have also demonstrated the paracrine activation of endothelial cells, following the generation of glioblastoma-derived MVs.[75]

Human glioma cells can modulate their environment by transmitting mRNA to human brain endothelial cells via EVs.[8] EVs released from tumor cells contained even higher concentrations of certain mRNAs than donor cells; some EVs even contained proteins, cytokines, and transcripts that were not detected in the original cell. Multiple other studies have illustrated the role of EVs in relaying markers for cell development and pluripotentiality. Ratajczak et al. describe embryonic cell MVs as stimulating growth of hematopoetic progenitor cells in culture and enhancing expression of pluripotent and hematopoetic stem cell makers.[76] Martinez et al. found that stimulated T-cells release MVs that contain hedgehog signaling proteins, leading to differentiation of erythroleukemic cells toward the megakaryocytic lineage. Inhibition of the hedgehog pathway negated this transformation.[77]

13.5 Extracellular Vesicles and Angiogenesis

In addition to their role in cancer-mediated thrombosis and cell-to-cell communication, EVs have also been implicated in coagulation-independent cancer-mediated angiogenesis and metastasis.[1,3,78] Platelet-derived EVs are known to promote endothelial cell angiogenesis through a variety of mechanisms, including enhancing endothelial cell chemotaxis and inducing endothelial cell tubule formation.[79] In addition to VEGF, platelet EVs harbor platelet-derived growth factor (PDGF), basic fibroblast growth factor (bFGF), and heparanase, all of which contribute to various stages of new blood vessel formation.[80] In a mouse model of hypoxia-induced retinal angiogenesis, Rhee et al. found that both thrombocytopenia and platelet inhibition via an αIIbß3-integrin antagonist decreased retinal neovascularization by 35–50%. Platelet remnants and platelet EVs were found at sites of "angiogenic sprouts," and *in vitro* platelets promoted capillary development of microvascular EVs.[81] The specific role of tumor-derived exosomes in angiogenesis has been characterized as well. MVs derived from human glioma cells contain a higher concentration of angiogenic proteins (i.e., angiogenin, IL-6, IL-8, tissue inhibitor of metalloproteinase 1 [TIMP-1], VEGF, and TIMP-2) than their parent cells.[82] Moreover, when human brain microvascular endothelial cells were incubated with EVs released from glioma cells, the tubule length of the endothelial cells doubled in 16 h; this result was comparable to the incubation of endothelial cells with angiogenic growth factors alone. The proangiogenic activity of tumor-derived EVs appears to be mediated through the activation of the phophoinositide 3-kinase/AKT pathway.[83]

13.6 Extracellular Vesicles and Tumor Invasiveness and Metastasis

Depending on the type of cargo they ferry, EVs may also participate in cancer invasiveness and metastasis. For instance, EVs harboring proteolytic enzymes, such as the matrix metalloproteinases (MMPs) MMP-2 and MMP-9 and urokinase-type plasminogen activator (uPA), have been shown to promote tumor invasion in cultured malignant ovarian epithelium through degradation of the extracellular matrix.[84] Graves et al. measured the levels of EVs containing these enzymes in the ascitic fluid of women with various stages of ovarian cancer. The authors concluded that women with advanced ovarian cancer had higher concentrations of MVs containing proteolytic enzymes in their ascitic fluid than women without such late-stage disease. Furthermore, inhibition of MMP and uPA prevented MV invasion of the cellular milieu.

13.7 Extracellular Vesicles in Tumor Immune Escape

Tumor-derived exosomes have been implicated in a cancer's ability to escape from routine immune surveillance by manipulating both the innate and the adaptive immune system.[39] Various mechanisms for this phenomenon have been proposed, including MV induction of T-cell apoptosis, attenuation of natural killer (NK) and T-cell function, and proliferation of T-regulatory (T-reg) cells, amongst others.[39,85–87]

Perhaps the best understood way in which tumor-derived exosomes facilitate T-cell apoptosis is through expression of the T-cell-killing molecule Fas ligand (FasL) on their surface. FasL is a transmembrane protein belonging to the tumor necrosis factor (TNF) family; it plays a critical role in apoptosis by regulating T-cell homeostasis, regulating cytoxic T-cell activity, and maintaining certain cells' "immune privilege" status. High concentrations of FasL have been reported in a wide variety of malignancies, including colorectal cancer, prostate cancer, melanoma, and leukemia.[87] Abusamra et al. demonstrated that prostate cancer–derived exosomes carrying membrane-bound FasL inhibited T-cell proliferation and induced apoptosis of CD8+ T-cells. In this manner, the tumor cells were able to escape destruction by the cell-mediated arm of the immune system. Similar results were reported by Huber et al. who established that colorectal cells secrete EVs containing FasL and TNF-related apopotosis-inducing ligand (TRAIL), possibly affording greater immune protection to these tumor cells.[88] EVs from healthy donor serum did not contain these proapoptotic molecules.

Tumor-derived exosomes also exert inhibitory effects on NK and T-cells by diminishing expression of the receptor NKG2D via secretion of the growth factor transforming growth factor β1 (TGFβ1).[89] TGFβ1 is an immunosuppressant molecule, which is known to promote tumor growth, while weakening lymphocyte response. Clayton et al. studied the effect of tumor-derived exosomes produced *in vitro* or in the pleural fluid of patients with mesothelioma on NK and T-cell function. The authors demonstrated that these exosomes secreted high levels of TGFβ1, which led to decreased lymphocyte expression of NKG2D and, hence, inhibited NK/T-cell activation.[89] Lee et al. also describe down-regulated NKG2D expression and suppressed NK cell lytic activity in response to TGFβ1 released by colorectal and lung tumors.[90] Finally, exosomes may be involved in the proliferation and recruitment of T-reg to the site of a tumor. In fact, T-reg production is regulated in part by TGFβ1.[91] Under physiologic conditions, T-reg cells induce immune tolerance to self-antigens and provide a necessary check on other aspects of the immune system. However, in a malignant state, T-reg cells may be tricked into protecting a tumor from immune attack. Expansion of T-regs has been associated with poorer prognosis and shorter survival in patients with ovarian cancer.[92]

The clinical implications of MV-related tumor immune escape are broad. For one, the concentration of EVs in a patient's blood may help predict the aggressiveness of a cancer. Furthermore, perhaps drugs targeted at the interface between EVs and T-cells could be used to curtail tumor escape. Finally, several groups have even proposed a type of dialysis filter by which tumor-derived exosomes would be removed. Using extracorporeal filtration, this device would help diminish the number of circulating EVs, thus affording the host's immune system greater advantage in its struggle against cancer.[86,93]

13.8 Conclusion

Our understanding of the formation and function of EVs has expanded dramatically over the past few decades. The potential role of EVs in health and malignancy continues to be elucidated. Understanding the critical roles EVs play in malignant transformation, progression, and hypercoagulability will hopefully translate into the development of MV-directed therapeutics for the treatment of cancer and its complications.

References

1. Muralidharan-Chari V, Clancy JW, Sedgwick A, D'Souza-Schorey C (2010) Microvesicles: mediators of extracellular communication during cancer progression, *J. Cell Sci.*, **123**, 1603–1611.
2. Ratajczak J, Wysoczynski M, Hayek F, Janowska-Wieczorek A, Ratajczak MZ (2006) Membrane-derived microvesicles: important and underappreciated mediators of cell-to-cell communication, *Leukemia*, **20**, 1487–1495.
3. Rak J (2010) Microparticles in cancer, *Semin. Thromb. Haemost.*, **36**, 888–906.
4. Pawson T (1995) Protein modules and signalling networks, *Nature*, **373**, 573–580.
5. Piccin A, Murphy WG, Smith OP (2007) Circulating microparticles: pathophysiology and clinical implications, *Blood Rev.*, **21**, 157–171.
6. Lee TH, D'Asti E, Magnus N, Al-Nedawi K, Meehan B, Rak J (2011) Microvesicles as mediators of intercellular communication in cancer--the emerging science of cellular "debris," *Semin. Immunopathol.*, **33**, 455–467.
7. Zwicker JI, et al. (2009) Tumor-derived tissue factor-bearing microparticles are associated with venous thromboembolic events in malignancy, *Clin. Cancer Res.*, **15**, 6830–6840.
8. Skog J, et al. (2008) Glioblastoma microvesicles transport RNA and proteins that promote tumour growth and provide diagnostic biomarkers, *Nat. Cell Biol.*, **10**, 1470–1476.

9. Wolf P (1967) The nature and significance of platelet products in human plasma, *BJH*, **13**, 269–288.
10. Mallat Z, et al. (2000) Elevated levels of shed membrane microparticles with procoagulant potential in the peripheral circulating blood of patients with acute coronary syndromes, *Circulation*, **101**, 841–843.
11. Hugel B, et al. (1999) Elevated levels of circulating procoagulant microparticles in patients with paroxysmal nocturnal hemoglobinuria and aplastic anemia, *Blood*, **93**, 3451–3456.
12. Nieuwland R, et al. (2000) Cellular origin and procoagulant properties of microparticles in meningococcal sepsis, *Blood*, **95**, 930–935.
13. Mathivanan S, Ji H, Simpson RJ (2010) Exosomes: extracellular organelles important in intercellular communication, *J. Proteomics*, **73**, 1907–1920.
14. Mathivanan S, Simpson, RJ (2012) Extracellular microvesicles: the need for internationally recognised nomenclature and stringent purification criteria, *J. Proteomics Bioinf.*,.
15. Pilzer D, Gasser O, Moskovich O, Schifferli JA, Fishelson Z (2005) Emission of membrane vesicles: roles in complement resistance, immunity and cancer, *Semin. Immunopathol.*, **27**, 375–387.
16. Denzer K, Kleijmeer MJ, Heijnen HF, Stoorvogel W, Geuze HJ (2000) Exosome: from internal vesicle of the multivesicular body to intercellular signaling device, *J. Cell Sci.*, **113**(Pt 19), 3365–3374.
17. Johnstone RM (2006) Exosomes biological significance: a concise review, *Blood Cells Mol. Dis.*, **36**, 315–321.
18. Bergsmedh A, et al. (2001) Horizontal transfer of oncogenes by uptake of apoptotic bodies, *Proc. Natl. Acad. Sci. U. S. A.*, **98**, 6407–6411.
19. de la Taille A, Chen MW, Burchardt M, Chopin DK, Buttyan R (1999) Apoptotic conversion: evidence for exchange of genetic information between prostate cancer cells mediated by apoptosis, *Cancer Res.*, **59**, 5461–5463.
20. Gaiffe E, et al. (2012) Apoptotic HPV positive cancer cells exhibit transforming properties, *PLOS ONE*, **7**, e36766.
21. Schwartz MA (1997) Integrins, oncogenes, and anchorage independence, *J. Cell Biol.*, **139**, 575–578.
22. Ehnfors J, et al. (2009) Horizontal transfer of tumor DNA to endothelial cells *in vivo*, *Cell Death Differ.*, **16**, 749–757.
23. Holmgren L, et al. (1999) Horizontal transfer of DNA by the uptake of apoptotic bodies, *Blood*, **93**, 3956–3963.

24. Yu X, Riley T, Levine AJ (2009) The regulation of the endosomal compartment by p53 the tumor suppressor gene, *FEBS J.*, **276**, 2201–2212.

25. Del Conde I, Shrimpton CN, Thiagarajan P, Lopez JA (2005) Tissue-factor-bearing microvesicles arise from lipid rafts and fuse with activated platelets to initiate coagulation, *Blood*, **106**, 1604–1611.

26. Zwicker JI, Lacroix R, Dignat-George F, Furie BC, Furie B (2012) Measurement of platelet microparticles, *Methods Mol. Biol.*, **788**, 127–139.

27. Zwaal RF, Comfurius P, Bevers EM (2005) Surface exposure of phosphatidylserine in pathological cells, *Cell Mol. Life Sci.*, **62**, 971–988.

28. Connor DE, Exner T, Ma DD, Joseph JE (2010) The majority of circulating platelet-derived microparticles fail to bind annexin V, lack phospholipid-dependent procoagulant activity and demonstrate greater expression of glycoprotein Ib, *Thromb. Haemost.*, **103**, 1044–1052.

29. Zwaal RF, Comfurius P, Bevers EM (2004) Scott syndrome, a bleeding disorder caused by defective scrambling of membrane phospholipids, *Biochim. Biophys. Acta*, **1636**, 119–128.

30. Morel O, Toti F, Jesel L, Freyssinet JM (2010) Mechanisms of microparticle generation: on the trail of the mitochondrion! *Semin. Thromb. Haemost.*, **36**, 833–844.

31. Bevers EM, et al. (1992) Defective Ca(2+)-induced microvesiculation and deficient expression of procoagulant activity in erythrocytes from a patient with a bleeding disorder: a study of the red blood cells of Scott syndrome, *Blood*, **79**, 380–388.

32. Kojima H, Newton-Nash D, Weiss HJ, Zhao J, Sims PJ, Wiedmer T (1994) Production and characterization of transformed B-lymphocytes expressing the membrane defect of Scott syndrome, *J. Clin. Invest.*, **94**, 2237–2244.

33. Wiedmer T, et al. (2004) Adiposity, dyslipidemia, and insulin resistance in mice with targeted deletion of phospholipid scramblase 3 (PLSCR3), *Proc. Natl. Acad. Sci. U. S. A.*, **101**, 13296–13301.

34. Zhou Q, Zhao J, Wiedmer T, Sims PJ (2002) Normal hemostasis but defective hematopoietic response to growth factors in mice deficient in phospholipid scramblase 1, *Blood*, **99**, 4030–4038.

35. Razmara M, Hu H, Masquelier M, Li N (2007) Glycoprotein IIb/IIIa blockade inhibits platelet aminophospholipid exposure by potentiating

translocase and attenuating scramblase activity, *Cell Mol. Life Sci.*, **64**, 999–1008.

36. Suzuki J, Umeda M, Sims PJ, Nagata S (2010) Calcium-dependent phospholipid scrambling by TMEM16F, *Nature*, **468**, 834–838.

37. Yang H, et al. (2012) TMEM16F forms a Ca2+-activated cation channel required for lipid scrambling in platelets during blood coagulation, *Cell*, **151**, 111–122.

38. Lhermusier T, Chap H, Payrastre B (2011) Platelet membrane phospholipid asymmetry: from the characterization of a scramblase activity to the identification of an essential protein mutated in Scott syndrome, *JTH*, **9**, 1883–1891.

39. Ge R, Tan E, Sharghi-Namini S, Asada HH (2012) Exosomes in Cancer Microenvironment and Beyond: have we Overlooked these Extracellular Messengers? *Cancer Microenv.*, **5**(3), 323–332.

40. Williams RL, Urbe S (2007) The emerging shape of the ESCRT machinery, *Nat. Rev. Mol. Cell Biol.*, **8**, 355–368.

41. Simons M, Raposo G (2009) Exosomes–vesicular carriers for intercellular communication, *Curr. Opin. Cell Biol.*, **21**, 575–581.

42. Hurley JH, Odorizzi G (2012) Get on the exosome bus with ALIX, *Nat. Cell Biol.*, **14**, 654–5.

43. Baietti MF, et al. (2012) Syndecan-syntenin-ALIX regulates the biogenesis of exosomes, *Nat. Cell Biol.*, **14**, 677–685.

44. Vlassov AV, Magdaleno S, Setterquist R, Conrad R (2012) Exosomes: current knowledge of their composition, biological functions, and diagnostic and therapeutic potentials, *Biochim. Biophys. Acta*, **1820**, 940–948.

45. Kosaka N, et al. (2013) Exosomal tumor-suppressive microRNAs as novel cancer therapy: "Exocure" is another choice for cancer treatment, *Adv. Drug Delivery Rev.* **65**(3), 376-382. Epub 2012 Jul 25.

46. Yang M, et al. (2011) Microvesicles secreted by macrophages shuttle invasion-potentiating microRNAs into breast cancer cells, *Mol. Cancer*, **10**, 117.

47. Yu X, Harris SL, Levine AJ (2006) The regulation of exosome secretion: a novel function of the p53 protein, *Cancer Res.*, **66**, 4795–4801.

48. Otten HM, et al. (2004) Symptomatic venous thromboembolism in cancer patients treated with chemotherapy: an underestimated phenomenon, *Arch. Intern. Med.*, **164**, 190–194.

49. Zwicker JI (2010) Predictive value of tissue factor bearing microparticles in cancer associated thrombosis, *Thromb. Res.*, **125**, S89–S91.

50. Trousseau A (1865) Phlegmasia alba dolens, in *Clinique Medicale d e 1'HBtel-Dieu de Paris*, vol. 3, 654.

51. Khorana AA (2003) Malignancy, thrombosis and Trousseau: the case for an eponym, *J. Thromb. Haemost.*, **1**, 2463–2465.

52. Tesselaar ME, Romijn FP, Van Der Linden IK, Prins FA, Bertina RM, Osanto S (2007) Microparticle-associated tissue factor activity: a link between cancer and thrombosis? *J. Thromb. Haemost.*, **5**, 520–527.

53. Zwicker JI, Furie BC, Furie B (2007) Cancer-associated thrombosis, *Crit. Rev. Oncol. Hematol.*, **62**, 126–136.

54. Davila M, et al. (2008) Tissue factor-bearing microparticles derived from tumor cells: impact on coagulation activation, *JTH*, **6**, 1517–1524.

55. Furie B, Furie BC (2008) Mechanisms of thrombus formation, *New Engl. J. Med.*, **359**, 938–949.

56. Zwicker JI, Trenor CC, 3rd, Furie BC, Furie B (2011) Tissue factor-bearing microparticles and thrombus formation, *Arterioscler. Thromb. Vasc. Biol.*, **31**, 728–733.

57. Giesen PL, et al. (1999) Blood-borne tissue factor: another view of thrombosis, *Proc. Natl. Acad. Sci. U. S. A.*, **96**, 2311–2315.

58. Manly DA, Boles J, Mackman N (2011) Role of tissue factor in venous thrombosis, *Annu. Rev. Physiol.*, **73**, 515–525.

59. Aharon A, Brenner B (2009) Microparticles, thrombosis and cancer. Best practice, and research, *Clin. Haematol.*, **22**, 61–69.

60. Ueno T, Toi M, Koike M, Nakamura S, Tominaga T (2000) Tissue factor expression in breast cancer tissues: its correlation with prognosis and plasma concentration, *Br. J. Cancer*, **83**, 164–170.

61. Khorana AA, et al. (2007) Tissue factor expression, angiogenesis, and thrombosis in pancreatic cancer, *Clin. Cancer Res.*, **13**, 2870–2875.

62. Lacroix R, Robert S, Poncelet P, Kasthuri RS, Key NS, Dignat-George F (2010) Standardization of platelet-derived microparticle enumeration by flow cytometry with calibrated beads: results of the International Society on *Thromb. Haemost.* SSC collaborative workshop, *J. Thromb. Haemost.*, **8**, 2571–2574.

63. Gyorgy B, et al. (2011) Detection and isolation of cell-derived microparticles are compromised by protein complexes resulting from shared biophysical parameters, *Blood*, **117**,e39–e48.

64. Aleman MM, Gardiner C, Harrison P, Wolberg AS (2011) Differential contributions of monocyte- and platelet-derived microparticles towards thrombin generation and fibrin formation and stability, *J. Thromb. Haemost.*, **9**, 2251–2261.

65. Lee RD, et al. (2012) Pre-analytical and analytical variables affecting the measurement of plasma-derived microparticle tissue factor activity, *Thromb. Res.*, **129**, 80–85.

66. Yuana Y, et al. (2010) Atomic force microscopy: a novel approach to the detection of nanosized blood microparticles, *J. Thromb. Haemost.*, **8**, 315–323.

67. Manly DA, et al. (2010) Increased microparticle tissue factor activity in cancer patients with venous thromboembolism, *Thromb. Res.*, **125**, 511–512.

68. Thaler J, et al. (2012) Microparticle-associated tissue factor activity, venous thromboembolism and mortality in pancreatic, gastric, colorectal and brain cancer patients, *J. Thromb. Haemost.*, **10**, 1364–1370.

69. Zwicker JI, et al. (2013) Prediction and prevention of thromboembolic events with enoxaparin in cancer patients with elevated tissue factor-bearing microparticles: a randomized-controlled phase II trial (the MicroTEC Study), *Br. J. Haematol.*, **160**(4), 530–537

70. Tesselaar ME, Romijn FP, van der Linden IK, Bertina RM, Osanto S (2009) Microparticle-associated tissue factor activity in cancer patients with and without thrombosis, *J. Thromb. Haemost.*, **7**, 1421–1423.

71. Jasuja R, et al. (2012) Protein disulfide isomerase inhibitors constitute a new class of antithrombotic agents, *J. Clin. Invest.*, **122**, 2104–2113.

72. Antonyak MA, et al. (2011) Cancer cell-derived microvesicles induce transformation by transferring tissue transglutaminase and fibronectin to recipient cells, *Proc. Natl. Acad. Sci. U. S. A.*, **108**, 4852–4857.

73. Al-Nedawi K, et al. (2008) Intercellular transfer of the oncogenic receptor EGFRvIII by microvesicles derived from tumour cells, *Nat. Cell Biol.*, **10**, 619–624.

74. Al-Nedawi K, Meehan B, Kerbel RS, Allison AC, Rak J (2009) Endothelial expression of autocrine VEGF upon the uptake of tumor-derived microvesicles containing oncogenic EGFR, *Proc. Natl. Acad. Sci. U. S. A.*, **106**, 3794–3799.

75. Svensson KJ, et al. (2011) Hypoxia triggers a proangiogenic pathway involving cancer cell microvesicles and PAR-2-mediated heparin-

binding EGF signaling in endothelial cells, *Proc. Natl. Acad. Sci. U. S. A.*, **108**, 13147–13152.

76. Ratajczak J, et al. (2006) Embryonic stem cell-derived microvesicles reprogram hematopoietic progenitors: evidence for horizontal transfer of mRNA and protein delivery, *Leukemia*, **20**, 847–856.

77. Martinez MC, et al. (2006) Transfer of differentiation signal by membrane microvesicles harboring hedgehog morphogens, *Blood*, **108**, 3012–3020.

78. van Doormaal FF, Kleinjan A, Di Nisio M, Buller HR, Nieuwland R (2009) Cell-derived microvesicles and cancer, *Neth. J. Med.*, **67**, 266–273.

79. Burnier L, Fontana P, Kwak BR, Angelillo-Scherrer A (2009) Cell-derived microparticles in haemostasis and vascular medicine, *Thromb. Haemost.*, **101**, 439–451.

80. Brill A, Elinav H, Varon D (2004) Differential role of platelet granular mediators in angiogenesis, *Cardiovasc. Res.*, **63**, 226–235.

81. Rhee JS, et al. (2004) The functional role of blood platelet components in angiogenesis, *Thromb. Haemost.*, **92**, 394–402.

82. Skog J, et al. (2008) Glioblastoma microvesicles transport RNA and proteins that promote tumour growth and provide diagnostic biomarkers, *Nat. Cell Biol.*, **10**, 1470–1476.

83. Kawamoto T, et al. (2012) Tumor-derived microvesicles induce proangiogenic phenotype in endothelial cells via endocytosis, *PLOS ONE*, **7**, e34045.

84. Graves LE, Ariztia EV, Navari JR, Matzel HJ, Stack MS, Fishman DA (2004) Proinvasive properties of ovarian cancer ascites-derived membrane vesicles, *Cancer Res.*, **64**, 7045–7049.

85. Pap E, Pallinger E, Falus A (2011) The role of membrane vesicles in tumorigenesis, *Crit. Rev. Oncol. Hematol.*, **79**, 213–223.

86. Ichim TE, et al. (2008) Exosomes as a tumor immune escape mechanism: possible therapeutic implications, *J. Transl. Med.*, **6**, 37.

87. Abusamra AJ, et al. (2005) Tumor exosomes expressing Fas ligand mediate CD8+ T-cell apoptosis, *Blood Cells Mol. Dis.*, **35**, 169–173.

88. Huber V, et al. (2005) Human colorectal cancer cells induce T-cell death through release of proapoptotic microvesicles: role in immune escape, *Gastroenterology*, **128**, 1796–1804.

89. Clayton A, Mitchell JP, Court J, Linnane S, Mason MD, Tabi Z (2008) Human tumor-derived exosomes down-modulate NKG2D expression, *J. Immunol.*, **180**, 7249–7258.

90. Lee JC, Lee KM, Kim DW, Heo DS (2004) Elevated TGF-beta1 secretion and down-modulation of NKG2D underlies impaired NK cytotoxicity in cancer patients, *J. Immunol.*, **172**, 7335–7340.

91. Wieckowski EU, Visus C, Szajnik M, Szczepanski MJ, Storkus WJ, Whiteside TL (2009) Tumor-derived microvesicles promote regulatory T cell expansion and induce apoptosis in tumor-reactive activated CD8+ T lymphocytes, *J. Immunol.*, **183**, 3720–3730.

92. Curiel TJ, et al. (2004) Specific recruitment of regulatory T cells in ovarian carcinoma fosters immune privilege and predicts reduced survival, *Nat. Med.*, **10**, 942–949.

93. Marleau AM, Chen CS, Joyce JA, Tullis RH (2012) Exosome removal as a therapeutic adjuvant in cancer, *J. Transl. Med.* **10**, 134.

Chapter 14

Extracellular Vesicles and Tissue Factor

Pierre-Emmanuel Rautou and Nigel Mackman
Department of Medicine, Division of Hematology and Oncology, McAllister Heart Institute, University of North Carolina at Chapel Hill, Chapel Hill, NC 27599, USA
nmackman@med.unc.edu

Tissue factor (TF) is a transmembrane glycoprotein that is essential for hemostasis. It binds coagulation serine protease factor VII/VIIa (FVII/VIIa) to form a bimolecular complex that functions as the primary initiator of coagulation *in vivo*. The TF/FVIIa complex activates both FX and FIX and leads to the generation of thrombin and fibrin.[1] TF is constitutively expressed at high levels in adventitial fibroblasts of the vessel wall, which facilitates rapid activation of the coagulation cascade after injury. In a basal state, smooth muscle cells in the media of the vessel wall and monocytes/macrophages also contain small amounts of TF. TF can be induced in these cells and in endothelial cells of the vessel wall under various pathologic conditions, resulting in a prothrombotic state.[2] TF is also expressed in a tissue-specific manner with high levels in a variety of tissues, such as the brain, heart, kidney, and placenta.[1]

Extracellular Vesicles in Health and Disease
Edited by Paul Harrison, Christopher Gardiner, and Ian L. Sargent

ISBN 978-981-4411-98-1 (Hardcover), 978-981-4411-99-8 (eBook)
www.panstanford.com

14.1 Circulating TF

Early studies suggested that blood did not contain significant levels of tissue factor (TF) and that TF was only expressed by extravascular cells in healthy individuals where it formed a "hemostatic envelope" around blood vessels.[3] However, several groups discovered that the blood of healthy individuals contained very low levels of functional TF (so-called blood-borne TF).[4,5] Giesen et al. showed that blood-borne TF contributed to thrombus formation in *ex vivo* models.[4] In the same study, the authors observed TF+ membrane vesicles near the surface of platelets in the thrombus. Three kinds of extracellular vesicles (EVs) can be released by cells in the plasma: apoptotic bodies, microvesicles (MVs), and exosomes. Apoptotic bodies are vesicles released during the terminal steps of programmed cell death and represent condensed and deoxyribonucleic acid (DNA)-rich remnants of the fragmented apoptotic cells. They have a size ranging from 1 μm to 4 μm and may act as passive shuttles delivering DNA and oncogenes to the nuclear compartment of phagocytes. TF has not been observed at the surface of such apoptotic bodies. Conversely, the presence of TF on MVs and to a lesser extent on exosomes has been well described. Blood also contains soluble forms of TF, including alternatively spliced TF. However, the absence of the transmembrane domain greatly reduces the procoagulant activity of this alternatively spliced TF and it will not be discussed.

14.1.1 Tissue Factor–Positive EVs

MVs are 0.1 μm to 1 μm EVs released into the extracellular space following cell activation or apoptosis. They contain proteins and blood from their parental cell and are characterized by surface exposure of negatively charged phospholipids, such as phosphatidylserine (PS).

14.1.1.1 Procoagulant activity of MVs

The presence of PS confers an increased procoagulant activity to MVs because it facilitates the assembly of components of the clotting cascade. This is due to an electrostatic interaction between γ-carboxyglutamic acid domains in many clotting proteins, calcium bridges, and PS on the MV membrane. Clotting proteins that contain a γ-carboxyglutamic acid domain include the vitamin K–dependent factors VII, IX, and X (FVII, FIX, and FX) and prothrombin. PS increases the catalytic efficiency of both tenase and prothrombinase complexes by 200- and 1,000-fold, respectively.[6]

MVs without PS on their outer leaflet have also been detected. Exposure of platelets to a PAR-1 agonist peptide induces the release of both PS-positive and PS-negative MVs.[7] Substantial numbers of circulating MVs express platelet or endothelial markers but appear to lack externalized PS, depending on the absence of annexin V labeling.[8] The reason for the lack of annexin V labeling of some MVs is unknown but could result from different blebbing mechanisms, the presence of membrane enzyme activity capable of restoring membrane asymmetry, or the increased curvature of the membrane on small EVs preventing their labeling. Alternatively, PS might indeed be expressed at the surface of annexin V–negative MVs but already engaged in other molecular interactions, such as with antibodies to oxidized lipids, phospholipase activities, lactadherin, or proteins of the coagulation cascade, and therefore be "unavailable" for annexin V binding. Another possibility could be that these "PS-negative MVs" are in fact immune complexes. Indeed, Gyorgy et al. have shown that measuring particles in synovial fluid from patients with rheumatoid arthritis by flow cytometry detected both MVs and immune complexes, causing false-positive signals.[9] However, this potential artifact was not observed in healthy humans and coronary artery disease patients plasma.[10] Importantly, PS-negative MVs would be expected to have a low procoagulant activity (Fig. 14.1).

Conversely, the presence of TF on MVs dramatically increases their procoagulant activity (Fig. 14.1). TF has a high affinity for FVII/FVIIa, and therefore TF+ MVs in blood will readily bind FVII/FVIIa. Whether or not the blood of healthy subjects contains significant levels of functional TF remains highly controversial. Some investigators believe that there is no functional TF in unstimulated blood of healthy individuals.[11] Similarly, others have failed to detect measureable TF activity in plasma or TF associated with MVs isolated from unstimulated whole blood.[12] One study found that isolated MVs from healthy controls generated thrombin in a TF-independent manner.[13] Some investigators reported that if there is TF-related activity either in blood or in plasma from healthy humans, the concentration of active TF does not exceed 20 fmol/L.[14] In support of this statement, blood or plasma activated with (sub)picomolar concentrations of functional TF clots within several minutes.[14] However, other groups have reported very low levels of TF activity in MVs in the blood of healthy individuals, although these levels are close to the detection limit of the assays.[15–17] It was reported that 95% of TF activity in the blood of healthy humans was present in an inactive, encrypted form on peripheral blood mononuclear cells and only 5% was present on MVs.[18]

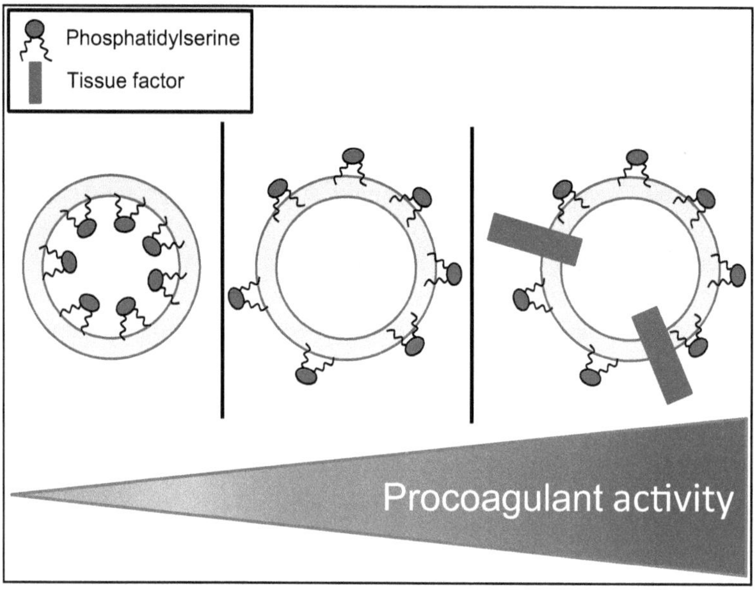

Figure 14.1 MVs have a different procoagulant activity, depending on the exposure of PS and TF at their surface.

The detection of circulating TF+ MVs in normal subjects implies some regulation of TF activity to prevent inadvertent activation of the coagulation cascade. TF has a high affinity for FVII/FVIIa, and therefore TF+ MVs in blood readily bind FVII/FVIIa to initiate blood coagulation. The TF–FVIIa complex is regulated by the tissue factor pathway inhibitor (TFPI). This Kunitz-type inhibitor is primarily synthesized by endothelial cells and circulates in blood to prevent inappropriate activation of the coagulation cascade. It inhibits the TF–FVIIa complex in an FXa-dependent manner.[19] Therefore, it is likely that some of the TF–FVIIa complexes present on MVs in blood will be inhibited by TFPI. TF in blood may also be found in low- and high-activity states, which may be due to differences in the conformation of TF so as not to cause thrombosis in the absence of specific stimuli, as reviewed previously.[20] These different states were hypothesized because disruption of TF+ cells increased TF activity without a change in the TF antigen.[20] Several mechanisms of TF activation have been postulated: PS exposure, decreased levels of TFPI, and post-translational modifications, including glycosylation, removal of glutathione groups, and disulfide bond oxidation.[21] Alternatively, the concentration of TF+ MVs may be below the threshold required to activate coagulation until they are concentrated at an injury site.

14.1.1.2 Cellular Sources of Tissue Factor EVs in Blood

14.1.1.2.1 *Monocytes*

A subset of unstimulated monocytes from healthy individuals has been shown to express TF.[22] LPS stimulation of monocytes induces TF expression and is associated with the release of TF+ MVs.[3,11] MVs derived from LPS-stimulated monocytes also express low levels of TFPI.[23] Importantly, plasma from patients with meningococcal sepsis and sickle cell disease and from humans with endotoxemia contain MVs that express both TF and monocyte markers.[15,24] These studies indicate that monocytes are likely to be the major source of TF+ MVs in health and disease.

14.1.1.2.2 *Neutrophils*

Neutrophils have also been reported to express TF in response to complement C5a.[25,26] However, monocyte-derived TF+ MVs can readily bind to neutrophils, which may explain some of the reports of neutrophil TF expression.[22] We found that deletion of the TF gene in myeloid cells reduced fetal loss in a mouse model of antiphospholipid antibody syndrome, but it was not possible to distinguish between TF expression by neutrophils versus monocytes in these studies.[26]

14.1.1.2.3 *Endothelial cells*

Cultured endothelial cells express TF in response to a variety of agonists, including cytokines and LPS.[27] However, there is limited evidence that endothelial cells express TF *in vivo*. Studies with animal models of endotoxemia and sepsis have reported TF expression in endothelial cells in the splenic vasculature and at branch points in the aorta.[28] It is possible that part or all of this TF staining is due to the binding of monocyte-derived TF+ MVs that are known to be present in septic animals. Indeed, THP-1 cell-derived MVs are enriched in TF and P-selectin glycoprotein 1 (PSGL-1), which would facilitate docking onto activated endothelial cells by binding to P-selectin, as reported with activated platelets.[29] The fact that TF staining of endothelial cells was restricted to granular structures that also contained the leukocyte marker PSGL-1 supports this hypothesis.[28] Furthermore, we found that a selective deletion of the TF gene in endothelial cells did not reduce the activation of coagulation in a mouse endotoxemia model.[30] However, in sickle cell mice, TF expression was observed on endothelial cells of the pulmonary veins.[31] In addition, endothelial cell–derived MVs (CD144+) were observed in sickle cell patients in crisis and expressed TF.[15] Interestingly, the TF activity of MVs derived from activated endothelial cells was markedly increased by inhibition of TFPI, whereas there was only a modest change using MVs from stimulated monocytes, suggesting that TF+ MVs from different cellular sources have different TF activity.[23] These studies indicate that endothelial cells may release TF+ MVs in certain diseases.

14.1.1.2.4 *Platelets*

While platelets have been reported to express TF,[32,33] other investigators failed to detect any TF in resting or activated platelets.[11,34] There are several explanations for these conflicting results. Some of the studies did not use inhibitory anti-TF antibodies to demonstrate that the procoagulant activity of the platelets is indeed due to TF. This is important because high concentrations of FVIIa can activate FX in a TF-independent manner.[35] In addition, the presence of TF on platelets may be due, in part, to the binding of monocyte-derived TF+ MVs to activated platelets. Indeed, TF-bearing MVs from LPS-treated THP-1 cells fuse with activated platelets.[29] It is more difficult to explain the reports of TF pre–messenger ribonucleic acid (pre-mRNA) and mRNA expression in platelets and de novo synthesis of TF protein by platelets, although monocyte contamination of the platelet preparations is always a concern.[33] In conclusion, platelets may express very low levels of TF, but it seems unlikely that they provide a major contribution to the pool of TF+ MVs present in healthy individuals and patients.

14.1.2 TF+ Exosomes

Exosomes originally form as intracellular luminal vesicles (50–90 nm) within the endosome. They fuse with the cell plasma membrane and release their contents outside the cell. Exosomes contain a distinct set of proteins, such as heat shock proteins, major histocompatibility complex class I and II molecules, and tetraspanins.

Some recent studies reported the presence of TF on exosomes. Davila et al. found that 10–20% of the TF activity released by breast carcinoma cell lines was associated with sedimentable particles smaller than 100 nm, which may be exosomes.[36] Similarly, exposure of human glioblastoma cells to hypoxia was associated with a dramatic increase in TF levels in the medium, the vast majority (approximately 80%) of which was associated with the vesicular fraction. Electron microscopy showed that these vesicles were 60–100 nm in size, expressed TF, and stained for markers of exosomes, that is, the tetraspanins CD63 and CD81. However, the vesicle preparation in this study did not remove MVs, and the costaining of TF with markers of exosomes does not allow one to conclude that the TF+ vesicles are really exosomes and not small MVs.[37] Another group detected exosomes with TF at their surface in saliva from healthy humans. These exosomes shortened the clotting time in a factor VII (FVII)-dependent manner. In contrast, exosomes derived from the plasma of the same individuals did not change clotting time.[38] In conclusion, exosomes may contain TF in specific body fluids or in pathological conditions.

14.2 Methods of Tissue Factor EV Detection

EV-associated TF is most commonly measured on circulating MVs in plasma, although other body fluids have also been shown to contain TF-bearing MVs that may have diagnostic and/or prognostic relevance. Not all methods of MV TF detection require an MV isolation step, but at a minimum, there is a need to prepare cell-free samples from blood or the body fluid under study. The centrifugation speed, which must be sufficient to remove contaminating cells, but not the MVs of interest, is a critical variable.[39] Other important preanalytical variables include (1) the venipuncture method by which the sample was obtained, since contamination by TF originating from skin and vessel wall cells can be significant, given the minute quantities of TF being measured; (2) the anticoagulant into which the samples are drawn, since calcium chelators may dissociate bound TFPI, among other effects; and (3) whether the samples are analyzed fresh or after freezing, since freeze-thawing may increase PS exposure on the MVs.[40] When employed, the specific conditions used to isolate MVs from plasma also vary considerably among research laboratories. MV TF measurement thus remains an important area for future standardization.[40]

Broadly speaking, MV TF assays can be classified as either immunological or functional.

14.2.1 Immunological

Immunological methods detect the TF antigen on the surface of MVs using specific antibodies or other probes that have been labeled with a detector, such as a fluorescent tag.

In flow cytometry, an operational definition is used to detect MVs. The usual criteria include events that are 0.1–1 μm in size and the expression of PS, which is usually detected by binding of labeled annexin V or lactadherin. Some investigators consider that PS exposure is not required and can be replaced by expression of antigens that characterize the cell of origin and/or the ability to be removed by high-speed centrifugation (at least 18,000 g for 30 min).[41] TF may be detected on MVs using specific labeled antibodies, but this approach clearly imparts no information about whether the TF is functionally active, and it is at best semiquantitative. Flow cytometry also suffers from the fact that particles having a diameter below 0.5 μm are smaller than the wavelength of the laser light used to detect them and therefore may not be detectable.

Therefore, there has been recent interest in other physical methods to detect MVs. These approaches have included dynamic light scattering (see Chapter 4), nanoparticle tracking analysis (see Chapter 11), capillary electrophoresis with laser-induced fluorescence detection, impedance-based flow cytometry (see Chapter 8), and atomic force microscopy (see Chapter 9). In theory, all of these methods can be adapted to detect TF and other antigens on the MV surface, although these studies are largely still in their infancy.[21] However, the low abundance of TF+ MVs makes them difficult to detect in plasma.

Capture assays (see Chapter 7) such as enzyme-linked immunosorbent assay (ELISA) may also be an approach. However, the commercial kit has low sensitivity and/or a high background.[40]

14.2.2 Functional

Two distinct types of assays have been developed for measurement of MV TF activity in human plasma using either antibody-mediated MV capture or isolation by centrifugation.

In the first of these, a monoclonal antibody (1B10) that recognizes an antigen expressed on multiple cell types, but not platelets or erythrocytes, is used to capture MVs from platelet-poor plasma.[15] After washing to remove excess plasma, FVIIa and FX are added to the well and the FXa generated is then measured using a specific chromogenic substrate. Because FVIIa at high concentrations will activate FX in the absence of TF, it is important to ensure assay specificity. This is done by the addition of an inhibitory antibody to TF in parallel wells. The difference in FXa generation between the wells without and with the inhibitory antibody is then taken as a measure of TF-dependent procoagulant activity. This procoagulant activity can be converted to TF activity by comparison with a standard curve constructed using a TF standard, often recombinant TF relipidated in artificial phospholipid vesicles. Alternatively, TF activity may be expressed simply as the TF-dependent FXa generation rate, without reference to a standard. This assay is quite time consuming and has not been widely adopted. Another strategy based on the capture of TF+ MVs with an anti-TF antibody followed by the measure of TF activity has also been described.[40]

A second type of approach is to use high-speed centrifugation to isolate MVs with or without platelets from platelet-rich plasma, platelet-poor plasma, or platelet-free plasma.[16,17] We add 1 mL of Hanks balanced saline solution with bovine serum albumin to 200 μL of platelet-poor or platelet-free plasma and centrifuge the sample at 20,000 g for 15 min to pellet the MVs.[16] MVs are washed two times and then resuspended in a buffer before measuring FXa generation in the presence and absence of an inhibitory anti-TF antibody. The principle of this assay is shown in Fig. 14.2. We have adapted this assay to measure circulating MV TF activity in mice.[40]

Finally, global assays of thrombin generation, such as calibrated automated thrombography, have been adapted to measure endogenous MV TF activity in platelet-rich or platelet-poor plasma.[42] However, the sensitivity of this approach is low and unlikely to be useful for measuring levels of MV TF activity in clinical samples.

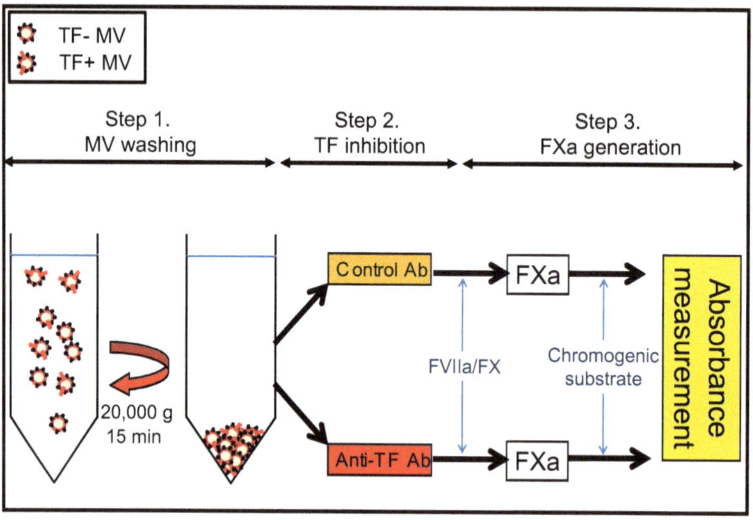

Figure 14.2 MV TF activity assay on MVs isolated from plasma by centrifugation. MVs in plasma are pelleted by centrifugation and then washed two times before being incubated with FVIIa and FX in the presence of either a control antibody (Ab) or an inhibitory anti-TF antibody. The amount of FXa generated is measured using the FXa substrate S-2765.

14.3 MV TF as Biomarkers of Thrombotic Risk

Increased circulating levels of TF+ MVs have been observed in numerous diseases associated with elevated thromboembolic risk and/or activation of coagulation (Table 14.1), in particular in cancer and cardiovascular diseases (see below) but also in sepsis, endotoxemia, and sickle cell disease (Table 14.1).

We will focus here on studies that measured levels of TF+ MVs using flow cytometry or impedance or using activity assays.

Table 14.1 Selected clinical studies assessing circulating TF+ MVs in patients

Disease state	Specific condition	TF+ flow cytometry	MV TF activity	Major finding of study	Reference
Cancer	Colorectal	Yes	No	Increased TF+ MVs in cancer vs. healthy controls, correlated with D-dimers	51
Cancer	Pancreatic and	Yes	Yes	Increased TF+ MVs and increase in MV TF activity in cancer vs. controls	17
	breast			Increased MV TF activity associated with decreased survival	
Cancer	Multiple forms	No	Yes	Increased MV TF activity in metastatic cancer patients vs. healthy controls	52
Cancer	Pancreatic	No	Yes	Increasing MV TF activity and TF antigen predictive of VTE in cancer patients	16
Cancer	Prostate	Yes	Yes[a]	MV TF activity increased and correlated with D-dimers	53
Cancer	Multiple myeloma	No	Yes	MV TF activity increased in myeloma vs. control and reduced with chemotherapy	54
Cancer	Multiple forms	No	Yes	Cancer patient with VTE with higher MV TF activity vs. controls Increased MV TF activity resulting in decreased survival vs. patients with low MV TF activity	55

(*Continued*)

Table 14.1 (*Continued*)

Disease state	Specific condition	TF+ flow cytometry	MV TF activity	Major finding of study	Reference
Cancer	Multiple forms	Yes[b]	No	Increased TF+ MVs in cancer patients vs. controls and predictive of VTE	50
Cancer	Multiple forms	No	Yes	Increased MV TF activity in patients with cancer and VTE vs. cancer without VTE	57
Cancer	Multiple forms	Yes	No	Increased TF+ MVs in cancer patients with and without VTE vs. healthy controls	56
Acute coronary syndrome	Myocardial infarction	Yes	Yes[c]	Decreased TFPI+ MVs and increased MV TF activity after thrombolysis	45
Acute coronary syndrome	Myocardial infarction	No	Yes	MV TF activity increased in patients with persistent occlusion vs. controls	46
Acute coronary syndrome	Myocardial infarction	No	Yes	MV TF activity increased in failed vs. successful thrombolysis in patients	47
Acute coronary syndrome	Myocardial infarction	No	Yes[c]	MV TF activity increased in lesion blood vs. postangioplasty	48
Diabetes	Type II diabetes	Yes	No	2 times increased TF+ MVs in patients with type II diabetes vs. healthy controls	44
Diabetes	Type II diabetes	Yes	No	3 times increased TF+ MVs in patients with type II diabetes vs. healthy controls	60

Disease state	Specific condition	TF+ flow cytometry	MV TF activity	Major finding of study	Reference
Hypercholesterolemia	Familial hypercholesterolemia	No	Yes	Increased MV TF activity in familial hypercholesterolemia patients vs. healthy controls	43
Sepsis	Meningococcal	Yes	No	Increased TF+ MVs in patients with meningococcal sepsis	24
Endotoxemia	LPS administration	Yes	Yes	Increased TF+ MVs and MV TF activity in healthy volunteers given endotoxin	15
Sickle cell	Sickle cell disease	Yes	Yes[a]	Increased TF+ MVs and MV TF activity in sickle cell patients vs. controls	8

[a] MV TF activity assessed by a one-stage clotting reaction with inclusion of an anti-TF antibody.

[b] MVs assessed by flow impedance.

[c] Assay considered controversial and may not be indicative of specific MV TF activity.

Abbreviation: VTE, venous thromboembolism.

14.3.1 Cardiovascular Risk Factors and Diseases

Some cardiovascular risk factors are associated with increased circulating TF+ MVs. Patients with familial hypercholesterolemia have elevated levels of plasma MV TF activity.[43] Diabetes patients have 2 fold more TF+ MVs than healthy controls. The number of TF+ MVs also correlates with the body mass index, fasting plasma glucose, and insulin and with tumor necrosis factor alpha and serum high-density lipoprotein (HDL) cholesterol in diabetes patients.[44]

Several studies have shown an increase in circulating TF+ MVs in patients with acute coronary syndrome (Table 14.1).[45–48] Morel et al. observed higher MV TF activity in blood collected within the occluded coronary artery than in peripheral blood from the same patients.[48] These TF+ MVs are likely released from atherosclerotic plaques at the time of rupture. Indeed, human atherosclerotic plaques contain at least 200-fold more MVs than the blood from the same patients, and more than 50% of these MVs are TF+.[49] It is unclear whether levels of TF+ MVs will have a value in predicting future thrombotic events because the major source of TF in these cases is likely the plaque itself.

14.3.2 Cancer

Using an impedance-based flow cytometer modified specifically for MV enumeration and characterization, Zwicker et al. observed that two-thirds of patients with pancreatic cancer, one of the most prothrombotic malignancies, have very high levels of circulating TF+ MVs. TF+ MVs were associated with a fourfold increased risk of thrombosis in these patients.[50] The MVs in pancreatic cancer patients are derived, in part, from the tumor. Indeed, circulating MV levels normalized following surgical resection of the pancreatic cancer and MVs coexpressed the epithelial tumor antigen MUC1. Other groups have similarly observed high levels of MV TF activity in plasma samples from patients with cancer and its association with thrombotic events (Table 14.1).[16,17,50–57] Prospective trials are underway to determine whether the presence of TF+ MVs predicts venous thromboembolic events and whether primary thromboprophylaxic strategies are beneficial in such patients.[58]

In conclusion, numerous reports have suggested that TF+ MVs are a potential biomarker in diseases associated with thrombosis. However, methods that measure TF + MVs analysis currently suffer from a high degree of variability.[40] Development of instruments that can more accurately quantify TF+ MVs levels in plasma and standardization of functional assays will significantly help to advance the field.

TF+ EVs are a promising biomarker for thrombosis risk, but further studies are needed. Data obtained from mouse studies are often problematic because in many cases TF+ MV concentrations injected are much higher than the concentrations present *in vivo,* even in pathologic conditions. Moreover, the rapid MV clearance makes interpretation of the results difficult.[59]

Acknowledgments

This work was supported by the Philippe Foundation (P.-E.R.) and a grant (HL095096) (N.M.) from the National Institutes of Health.

References

1. Mackman N (2009) The role of tissue factor and factor VIIa in hemostasis, *Anesth. Analg.*, **108**, 1447–1452.
2. Osterud B, Bjorklid E (2006) Sources of tissue factor, *Semin. Thromb. Haemost.*, **32**, 11–23.
3. Drake TA, Morrissey JH, Edgington TS (1989) Selective cellular expression of tissue factor in human tissues. Implications for disorders of hemostasis and thrombosis, *Am. J. Pathol.*, **134**, 1087–1097.
4. Giesen PL, Rauch U, Bohrmann B, Kling D, Roque M, Fallon JT, Badimon JJ, Himber J, Riederer MA, Nemerson Y (1999) Blood-borne tissue factor: another view of thrombosis, *Proc. Natl. Acad. Sci. U. S. A.*, **96**, 2311–2315.
5. Koyama T, Nishida K, Ohdama S, Sawada M, Murakami N, Hirosawa S, Kuriyama R, Matsuzawa K, Hasegawa R, Aoki N (1994) Determination of plasma tissue factor antigen and its clinical significance, *BJH,* **87**, 343–347.
6. Morel O, Toti F, Hugel B, Bakouboula B, Camoin-Jau L, Dignat-George F, Freyssinet JM (2006) Procoagulant microparticles: disrupting the vascular homeostasis equation? *Arterioscler. Thromb. Vasc. Biol.,* **26**, 2594–2604.

7. Perez-Pujol S, Marker PH, Key NS (2007) Platelet microparticles are heterogeneous and highly dependent on the activation mechanism: studies using a new digital flow cytometer, *Cytometry A,* **71**, 38–45.
8. Shet AS, Aras O, Gupta K, Hass MJ, Rausch DJ, Saba N, Koopmeiners L, Key NS, Hebbel RP (2003) Sickle blood contains tissue factor-positive microparticles derived from endothelial cells and monocytes, *Blood,* **102**, 2678–2683.
9. Gyorgy B, Modos K, Pallinger E, Paloczi K, Pasztoi M, Misjak P, Deli MA, Sipos A, Szalai A, Voszka I, Polgar A, Toth K, Csete M, Nagy G, Gay S, Falus A, Kittel A, Buzas EI (2011) Detection and isolation of cell-derived microparticles are compromised by protein complexes resulting from shared biophysical parameters, *Blood,* **117**, e39–e48.
10. Amabile N, Renard JM, Caussin C, Boulanger CM (2012) Circulating immune complexes do not affect microparticle flow cytometry analysis in acute coronary syndrome, *Blood,* **119**, 2174–2175.
11. Butenas S, Bouchard BA, Brummel-Ziedins KE, Parhami-Seren B, Mann KG (2005) Tissue factor activity in whole blood, *Blood,* **105**, 2764–2770.
12. Osterud B, Breimo ES, Olsen JO (2008) Blood borne tissue factor revisited, *Thromb. Res.,* **122**, 432–434.
13. Berckmans RJ, Nieuwland R, Boing AN, Romijn FP, Hack CE, Sturk A (2001) Cell-derived microparticles circulate in healthy humans and support low grade thrombin generation, *Thromb. Haemost.,* **85**, 639–646.
14. Butenas S, Orfeo T, Mann KG (2009) Tissue factor in coagulation: which? Where? When? *Arterioscler. Thromb. Vasc. Biol.,* **29**, 1989–1996.
15. Aras O, Shet A, Bach RR, Hysjulien JL, Slungaard A, Hebbel RP, Escolar G, Jilma B, Key NS (2004) Induction of microparticle- and cell-associated intravascular tissue factor in human endotoxemia, *Blood,* **103**, 4545–4553.
16. Khorana AA, Francis CW, Menzies KE, Wang JG, Hyrien O, Hathcock J, Mackman N, Taubman MB (2008) Plasma tissue factor may be predictive of venous thromboembolism in pancreatic cancer, *J. Thromb. Haemost.,* **6**, 1983–1985.
17. Tesselaar ME, Romijn FP, Van Der Linden IK, Prins FA, Bertina RM, Osanto S (2007) Microparticle-associated tissue factor activity: a link between cancer and thrombosis? *J. Thromb. Haemost.,* **5**, 520–527.

18. Johnson GJ, Leis LA, Bach RR (2009) Tissue factor activity of blood mononuclear cells is increased after total knee arthroplasty, *Thromb. Haemost.*, **102**, 728–734.

19. Broze GJ, Jr. (1995) Tissue factor pathway inhibitor, *Thromb. Haemost.*, **74**, 90–93.

20. Bach RR (2006) Tissue factor encryption, *Arterioscler. Thromb. Vasc. Biol.*, **26**, 456–461.

21. Zwicker JI, Trenor CC, 3rd, Furie BC, Furie B (2011) Tissue factor-bearing microparticles and thrombus formation, *Arterioscler. Thromb. Vasc. Biol.*, **31**, 728–733.

22. Egorina EM, Sovershaev MA, Olsen JO, Osterud B (2008) Granulocytes do not express but acquire monocyte-derived tissue factor in whole blood: evidence for a direct transfer, *Blood*, **111**, 1208–1216.

23. Bajaj MS, Ghosh M, Bajaj SP (2007) Fibronectin-adherent monocytes express tissue factor and tissue factor pathway inhibitor whereas endotoxin-stimulated monocytes primarily express tissue factor: physiologic and pathologic implications, *J. Thromb. Haemost.*, **5**, 1493–1499.

24. Nieuwland R, Berckmans RJ, McGregor S, Boing AN, Romijn FP, Westendorp RG, Hack CE, Sturk A (2000) Cellular origin and procoagulant properties of microparticles in meningococcal sepsis, *Blood*, **95**, 930–935.

25. Kambas K, Markiewski MM, Pneumatikos IA, Rafail SS, Theodorou V, Konstantonis D, Kourtzelis I, Doumas MN, Magotti P, Deangelis RA, Lambris JD, Ritis KD (2008) C5a and TNF-alpha up-regulate the expression of tissue factor in intra-alveolar neutrophils of patients with the acute respiratory distress syndrome, *J. Immunol.*, **180**, 7368–7375.

26. Redecha P, Tilley R, Tencati M, Salmon JE, Kirchhofer D, Mackman N, Girardi G (2007) Tissue factor: a link between C5a and neutrophil activation in antiphospholipid antibody induced fetal injury, *Blood*, **110**, 2423–2431.

27. Parry GC, Mackman N (1995) Transcriptional regulation of tissue factor expression in human endothelial cells, *Arterioscler. Thromb. Vasc. Biol.*, **15**, 612–621.

28. Lupu C, Westmuckett AD, Peer G, Ivanciu L, Zhu H, Taylor FB, Jr., Lupu F (2005) Tissue factor-dependent coagulation is preferentially up-regulated within arterial branching areas in a baboon model of Escherichia coli sepsis, *Am. J. Pathol.*, **167**, 1161–1172.

29. Del Conde I, Shrimpton CN, Thiagarajan P, Lopez JA (2005) Tissue-factor-bearing microvesicles arise from lipid rafts and fuse with activated platelets to initiate coagulation, *Blood,* **106**, 1604–1611.
30. Pawlinski R, Wang JG, Owens AP, 3rd, Williams J, Antoniak S, Tencati M, Luther T, Rowley JW, Low EN, Weyrich AS, Mackman N (2010) Hematopoietic and nonhematopoietic cell tissue factor activates the coagulation cascade in endotoxemic mice, *Blood,* **116**, 806–814.
31. Solovey A, Kollander R, Shet A, Milbauer LC, Choong S, Panoskaltsis-Mortari A, Blazar BR, Kelm RJ, Jr., Hebbel RP (2004) Endothelial cell expression of tissue factor in sickle mice is augmented by hypoxia/reoxygenation and inhibited by lovastatin, *Blood,* **104**, 840–846.
32. Camera M, Brambilla M, Toschi V, Tremoli E (2010) Tissue factor expression on platelets is a dynamic event, *Blood,* **116**, 5076–5077.
33. Panes O, Matus V, Saez CG, Quiroga T, Pereira J, Mezzano D (2007) Human platelets synthesize and express functional tissue factor, *Blood,* 109, 5242–5250.
34. Bouchard BA, Mann KG, Butenas S (2010) No evidence for tissue factor on platelets, *Blood,* **116**, 854–855.
35. Bom VJ, Bertina RM (1990) The contributions of Ca2+, phospholipids and tissue-factor apoprotein to the activation of human blood-coagulation factor X by activated factor VII, *Biochem J.,* **265**, 327–336.
36. Davila M, Amirkhosravi A, Coll E, Desai H, Robles L, Colon J, Baker CH, Francis JL (2008) Tissue factor-bearing microparticles derived from tumor cells: impact on coagulation activation, *J. Thromb. Haemost.,* **6**, 1517–1524.
37. Svensson KJ, Kucharzewska P, Christianson HC, Skold S, Lofstedt T, Johansson MC, Morgelin M, Bengzon J, Ruf W, Belting M (2011) Hypoxia triggers a proangiogenic pathway involving cancer cell microvesicles and PAR-2-mediated heparin-binding EGF signaling in endothelial cells, *Proc. Natl. Acad. Sci. U. S. A.,* **108**, 13147–13152.
38. Berckmans RJ, Sturk A, van Tienen LM, Schaap MC, Nieuwland R (2011) Cell-derived vesicles exposing coagulant tissue factor in saliva, *Blood,* **117**, 3172–3180.
39. Lacroix R, Judicone C, Poncelet P, Robert S, Arnaud L, Sampol J, Dignat-George F (2012) Impact of pre-analytical parameters on the measurement of circulating microparticles: towards standardization of protocol, *J. Thromb. Haemost.,* **10**, 437–446.
40. Key NS, Mackman N (2010) Tissue factor and its measurement in whole blood, plasma, and microparticles, *Semin. Thromb. Haemost.,* **36**, 865–875.

41. Key NS (2010) Analysis of tissue factor positive microparticles, *Thromb. Res.,* **125**(Suppl 1, S42–S45.

42. Ollivier V, Wang J, Manly D, Machlus KR, Wolberg AS, Jandrot-Perrus M, Mackman N (2010) Detection of endogenous tissue factor levels in plasma using the calibrated automated thrombogram assay, *Thromb. Res.,* **125**, 90–96.

43. Owens AP, 3rd, Passam FH, Antoniak S, Marshall SM, McDaniel AL, Rudel L, Williams JC, Hubbard BK, Dutton JA, Wang J, Tobias PS, Curtiss LK, Daugherty A, Kirchhofer D, Luyendyk JP, Moriarty PM, Nagarajan S, Furie BC, Furie B, Johns DG, Temel RE, Mackman N (2012) Monocyte tissue factor-dependent activation of coagulation in hypercholesterolemic mice and monkeys is inhibited by simvastatin, *J. Clin. Invest.,* **122**, 558–568.

44. Diamant M, Nieuwland R, Pablo RF, Sturk A, Smit JW, Radder JK (2002) Elevated numbers of tissue-factor exposing microparticles correlate with components of the metabolic syndrome in uncomplicated type 2 diabetes mellitus, *Circulation,* **106**, 2442–2447.

45. Steppich B, Mattisek C, Sobczyk D, Kastrati A, Schomig A, Ott I (2005) Tissue factor pathway inhibitor on circulating microparticles in acute myocardial infarction, *Thromb. Haemost.,* **93**, 35–39.

46. Huisse MG, Lanoy E, Tcheche D, Feldman LJ, Bezeaud A, Angles-Cano E, Mary-Krause M, de Prost D, Guillin MC, Steg PG (2007) Prothrombotic markers and early spontaneous recanalization in ST-segment elevation myocardial infarction, *Thromb. Haemost.,* **98**, 420–426.

47. Huisse MG, Ajzenberg N, Feldman L, Guillin MC, Steg PG (2009) Microparticle-linked tissue factor activity and increased thrombin activity play a potential role in fibrinolysis failure in ST-segment elevation myocardial infarction, *Thromb. Haemost.,* **101**, 734–740.

48. Morel O, Pereira B, Averous G, Faure A, Jesel L, Germain P, Grunebaum L, Ohlmann P, Freyssinet JM, Bareiss P, Toti F (2009) Increased levels of procoagulant tissue factor-bearing microparticles within the occluded coronary artery of patients with ST-segment elevation myocardial infarction: role of endothelial damage and leukocyte activation, *Atherosclerosis,* **204**, 636–641.

49. Leroyer AS, Isobe H, Leseche G, Castier Y, Wassef M, Mallat Z, Binder BR, Tedgui A, Boulanger CM (2007) Cellular origins and thrombogenic activity of microparticles isolated from human atherosclerotic plaques, *J. Am. Coll. Cardiol.,* **49**, 772–777.

50. Zwicker JI, Liebman HA, Neuberg D, Lacroix R, Bauer KA, Furie BC, Furie B (2009) Tumor-derived tissue factor-bearing microparticles

are associated with venous thromboembolic events in malignancy, *Clin. Cancer Res.*, **15**, 6830–6840.

51. Hron G, Kollars M, Weber H, Sagaster V, Quehenberger P, Eichinger S, Kyrle PA, Weltermann A (2007) Tissue factor-positive microparticles: cellular origin and association with coagulation activation in patients with colorectal cancer, *Thromb. Haemost.*, **97**, 119–123.
52. Tilley RE, Holscher T, Belani R, Nieva J, Mackman N (2008) Tissue factor activity is increased in a combined platelet and microparticle sample from cancer patients, *Thromb. Res.*, **122**, 604–609.
53. Haubold K, Rink M, Spath B, Friedrich M, Chun FK, Marx G, Amirkhosravi A, Francis JL, Bokemeyer C, Eifrig B, Langer F (2009) Tissue factor procoagulant activity of plasma microparticles is increased in patients with early-stage prostate cancer, *Thromb. Haemost.*, **101**, 1147–1155.
54. Auwerda JJ, Yuana Y, Osanto S, de Maat MP, Sonneveld P, Bertina RM, Leebeek FW (2011) Microparticle-associated tissue factor activity and venous thrombosis in multiple myeloma, *Thromb. Haemost.*, **105**, 14–20.
55. Tesselaar ME, Romijn FP, van der Linden IK, Bertina RM, Osanto S (2009) Microparticle-associated tissue factor activity in cancer patients with and without thrombosis, *J. Thromb. Haemost.*, **7**, 1421–1423.
56. Campello E, Spiezia L, Radu CM, Bulato C, Castelli M, Gavasso S, Simioni P (2011) Endothelial, platelet, and tissue factor-bearing microparticles in cancer patients with and without venous thromboembolism, *Thromb. Res.*,.
57. Manly DA, Wang J, Glover SL, Kasthuri R, Liebman HA, Key NS, Mackman N (2010) Increased microparticle tissue factor activity in cancer patients with venous thromboembolism, *Thromb. Res.*, **125**, 511–512.
58. Zwicker JI (2010) Predictive value of tissue factor bearing microparticles in cancer associated thrombosis, *Thromb. Res.*, **125**(Suppl 2), S89–S91.
59. Rautou PE, Mackman N (2012) Del-etion of microvesicles from the circulation, *Circulation*, **125**, 1601–1604.
60. Sommeijer DW, Hansen HR, van Oerle R, Hamulyak K, van Zanten AP, Meesters E, Spronk HM, ten Cate H (2006) Soluble tissue factor is a candidate marker for progression of microvascular disease in patients with Type 2 diabetes, *J. Thromb. Haemost.*, **4**, 574–580.

Chapter 15

Extracellular Vesicles in Normal Pregnancy and Pre-Eclampsia

Ian L. Sargent, Rebecca A. Dragovic, Dionne S. Tannetta, and Christopher W. Redman
Nuffield Department of Obstetrics and Gynaecology, University of Oxford, John Radcliffe Hospital, Oxford, OX3 9DU, U.K
ian.sargent@obs-gyn.ox.ac.uk

15.1 Introduction

Extracellular vesicles (EVs) are known to be shed by a large number of cell types, both as part of normal physiology and as part of disease processes. Pregnancy adds a new dimension to this subject, as uniquely there is a foreign body, the fetus, added to the equation, which contributes to the populations of EVs circulating within the mother. More correctly, it is not the fetus that is the primary source of these vesicles but the placenta, which is comprised of fetal tissue and forms the interface between mother and baby. In human pregnancy the placenta is of the hemochorial type, where the surface epithelium (the syncytiotrophoblast) is in direct contact with the maternal blood. It is at this interface that the exchange of nutrients,

Extracellular Vesicles in Health and Disease
Edited by Paul Harrison, Christopher Gardiner, and Ian L. Sargent

ISBN 978-981-4411-98-1 (Hardcover), 978-981-4411-99-8 (eBook)
www.panstanford.com

gases, and waste products between mother and baby occurs. By its nature, therefore, any syncytiotrophoblast vesicles released from the placenta will pass directly into the maternal circulation, where they will engage with the maternal immune and cardiovascular systems. The consequences of this vesicle shedding are believed to be important for the maintenance of normal pregnancy, constituting a major signaling mechanism between mother and fetus. Conversely, abnormal shedding of vesicles has been implicated in disorders of pregnancy, in particular pre-eclampsia. Although often ignored by mainstream science, pregnancy affords a unique opportunity to study EVs in normal physiology and disease. Firstly, the placenta carries specific markers that allow the vesicles to be distinguished from those produced by other cell types in the blood. Secondly, unlike any other condition, it is known exactly when pregnancy begins and when it finishes, allowing changes in vesicle shedding to be followed throughout the normal and disease states, and thirdly, the source of the vesicles, the placenta, is uniquely available at the end of the pregnancy for study. Importantly, circulating syncytiotrophoblast vesicles allow us to take, in effect, noninvasive placental biopsies at different stages of pregnancy, providing information about its function.

This chapter reviews current progress in identifying placental vesicles in maternal blood, their functional effects on immune and endothelial cells and the coagulation system *in vitro*, characterization of the molecular cargo they carry and how they may differ between normal pregnancy and pre-eclampsia. Changes in circulating cellular vesicles of maternal origin in relation to pre-eclampsia are also discussed.

15.2 Identification of Placental Vesicles in Maternal Blood

The deportation of syncytiotrophoblast material into the maternal circulation has been recognized for many years.[1] The syncytiotrophoblast is a unicellular sheet (syncytium) formed from fusion of monocellular cytotrophoblasts. This presents particular challenges in terms of its *ex vivo* isolation and culture, as mentioned in subsequent sections. Multinucleate syncytial trophoblast sprouts (approx. 80–200 μm in size) have been identified in uterine vein and inferior vena cava blood and in lung capillaries. By virtue of its size, the majority of this material is trapped in the capillary bed of the mother's lungs, with very little entering the peripheral circulation.[2] This led us to explore the possibility that smaller syncytiotrophoblast vesicles are also released, which are small enough to pass through the lung capillaries and enter the peripheral circulation. This was achieved using an in house enzyme-linked immunosorbent assay (ELISA) in which maternal plasma was ultracentrifuged to pellet the vesicles, which were then detected using a monoclonal antibody to placental alkaline phosphatase (PLAP) (NDOG2), specific for the syncytiotrophoblast.[3] Increased levels of placental vesicles were found in the uterine vein blood compared to the peripheral blood, consistent with their placental origin. Levels increased with advancing pregnancy[4] and labor,[5] returning to zero in most cases by 48 h postdelivery.

Flow cytometry has been used to confirm the particulate nature of the syncytiotrophoblast material pelleted from maternal plasma using two syncytiotrophoblast-specific monoclonal antibodies, anti-PLAP, as used in our ELISA, and ED822, which recognizes an as yet unknown antigen on the apical surface of the syncytiotrophoblast.[3,6] Human leukocyte antigen-G (HLA-G)-positive vesicles have also been reported to be present in the maternal circulation.[7,8] HLA-G is a specialized form of a class I major histocompatibility complex (MHC) antigen, which is almost uniquely expressed on invasive extravillous cytotrophoblasts of the placenta but not the syncytiotrophoblast. Orozco et al. in 2009 investigated the presence of PLAP- and HLA-G-positive vesicles in parallel.[7] They found comparable levels of HLA-G- and PLAP-positive vesicles, with more of the former in the first and second trimesters but more of the latter in the third trimester, consistent with the increased shedding of syncytiotrophoblast vesicles later in pregnancy reported by others.[4] Pap et al. also reported the presence of HLA-G-positive vesicles in the third trimester.[8] However, it is difficult to understand how so many HLA-G-positive vesicles could be present in the maternal circulation, given that the only extravillous cytotrophoblast exposed to the maternal blood is that which invades the maternal spiral arteries at the end of the first trimester. This is present in very low amounts, with a surface area in the order of 10 cm^2 compared to 12 m^2 for the syncytiotrophoblast at term (Professor Graham Burton, University of Cambridge, personal communication). PLAP-positive vesicles have also been isolated directly from pregnancy plasma, taken between 26 and 28 weeks' gestation, using agarose beads coated with an anti-PLAP antibody.[9]

15.2.1 Circulating Placental Vesicles in Pre-Eclampsia

Pre-eclampsia is a disorder of pregnancy that affects 2.5–3.0% of women. No other complication of pregnancy is both so common and so dangerous for mother and baby. For both, it is potentially lethal or detrimental to long-term health and may require preterm delivery of the baby with all its associated consequences. Its first (preclinical) stage comprises deficient remodeling of the uteroplacental circulation (8–18 weeks), dysfunctional perfusion and placental oxidative stress.[10] This stimulates the release of placental factors into maternal blood that cause the second, clinical, stage (after 20 weeks). The latter results from maternal systemic vascular inflammation, which leads to the maternal syndrome of hypertension, proteinuria, edema, activation of the coagulation system, and, in worse cases, eclampsia, characterized by fits. We have shown that a maternal systemic inflammatory response is intrinsic to normal pregnancies but more severe in pre-eclampsia, including metabolic, clotting, and complement disturbances.[11] Proinflammatory factors released by the oxidatively stressed syncytiotrophoblast into the maternal circulation link the two stages. While a number of soluble factors have been implicated in this process (e.g., sFlt-1 and sEndoglin, the soluble receptors for vascular endothelial growth factor [VEGF] and transforming growth factor β [TGFβ]), we believe that syncytiotrophoblast vesicles play an important role.[12]

Evidence for increased levels of placental vesicles in the maternal circulation in pre-eclampsia comes from a number of studies of pregnancy plasma. Using our anti-PLAP ELISA, we showed that the levels of placental vesicles were significantly increased in both uterine vein and peripheral vein blood of pre-eclamptic women at cesarean section compared to normal pregnant controls.[3] We and others have shown that the increase in placental vesicle shedding was more pronounced in the more severe early-onset (before 34 weeks) compared to late-onset pre-eclampsia.[13,14] Interestingly, no increase in placental vesicle levels, compared to normal pregnant controls, was seen in women with fetal growth restriction (FGR) in the absence of any other features of the maternal syndrome (i.e., normotensive FGR). Some types of FGR have the same placental pathology as pre-eclampsia but without the maternal features. This finding suggests that vesicle shedding is key to the development of the maternal syndrome of pre-eclampsia. Using the same ELISA, Reddy et al. showed a significant increase in placental vesicle shedding in pre-eclamptic women at the end of labor, which could account for the postpartum worsening of the disease, which is sometimes seen.[5]

Increased levels of placental vesicles in the plasma of pre-eclamptic women were also found in flow cytometry studies by Lok et al. in 2008[6] using the anti-syncytiotrophoblast monoclonal antibody ED822. In contrast, others who used an anti-PLAP antibody to identify syncytiotrophoblast vesicles and an anti-HLA-G antibody to identify an extravillous cytotrophoblast found no difference in the levels between women with pre-eclampsia and normal pregnant controls.[7] However, these samples were taken from women with late-onset pre-eclampsia, and therefore their findings reflect those of Goswami et al., who found significant differences in placental vesicles only in early- but not late-onset disease.[13] More recently a study of women with severe pre-eclampsia, using an anti-PLAP antibody, showed no significant difference in placental vesicles compared to the normal pregnant controls.[15]

15.2.2 Functional Characterization of Placental Vesicles In Normal Pregnancy and Pre-Eclampsia

Changes in the numbers of circulating placental vesicles throughout normal pregnancy and pre-eclampsia suggest that they may play important functional roles in both. To study this requires the isolation of large amounts of pure vesicles. Ideally these would be obtained directly from maternal blood. However, maternal plasma contains not only placental vesicles but others derived from many cell types within the vascular compartment, including erythrocytes, leukocytes, and endothelial cells, with the most prominent population (>90%) derived from platelets.[6] It is therefore necessary to isolate placental vesicles from this large background contamination. While this has been reported using anti-PLAP antibody-conjugated magnetic beads,[9] the yields are low, and detailed analysis would require a substantial volume of blood to be taken from the pregnant woman. However, as mentioned above, pregnancy has a distinct advantage in terms of studying EVs in that their source, the placenta, can be easily obtained at delivery and vesicles prepared from it *ex vivo*.

There are three main methods for preparing EVs from the placenta: *mechanical separation*, *explant culture*, and *placental perfusion*.

15.2.2.1 Mechanical separation

In this technique the maternal decidua overlying the chorionic villi of the placenta is scraped off and the chorionic villi dissected away from the chorionic plate. The villi are then stirred in a buffer for 1 h, during which time the vesicles are shed.[16] While this method gives high yields of vesicles, the need to cut through the tissues may lead to the shedding of vesicles from placental endothelial and stromal cells, which may contaminate the preparations.

15.2.2.2 Explant cultures

Chorionic villi are dissected from the chorionic plate of the placenta and cultured for varying periods of time. The shed placental vesicles are harvested from the culture medium.[17] While this preparation appears to be more biologically relevant than the mechanical method, there are concerns about the health of the tissue in these cultures, as the syncytiotrophoblast is initially shed and then regenerates over a period of time.[18] Furthermore there are a number of nontrophoblast cell types present in the explants, which may also shed contaminating vesicles into the culture supernatant.

15.2.2.3 Placental perfusion

Placental vesicles may also be produced by dual placental perfusion.[19] In this system the fetal artery and vein of a placental lobule are cannulated and perfused with a buffer to maintain the fetal circulation. Catheters are then introduced into the intervillous space on the maternal side, which is perfused with a buffer to mimic the maternal blood flow. Placental vesicles are shed into the perfusate and pelleted by ultracentrifugation for subsequent analysis.[4,20-22]

15.2.2.4 Trophoblast cell lines and primary trophoblast preparations

In addition, vesicles can be prepared from trophoblast cell lines.[23,24] Of particular relevance is the choriocarcinoma cell line BeWo, as it can be induced to fuse by treatment with forskolin or dbCAMP, thereby mimicking the formation of the syncytiotrophoblast by fusion of villous cytotrophoblast *in vivo*.[25,26] Primary villous cytotrophoblasts isolated from placentas also syncytialize *in vitro*, to a far greater extent than BeWo cells, and have been shown to release syncytiotrophoblast-derived vesicles into the culture supernatant.[27]

It is important to note that vesicles prepared by these different methods are not comparable. They have been shown to have different functional properties.[20] The reader should be aware of the type of preparation used in a study when interpreting the results.

15.3 Functional Effects of Placental Vesicles In Vitro

15.3.1 Immunosuppresive Effects of Placental Vesicles

It has long been known that maternal cell–mediated immune responses are suppressed during pregnancy as part of a wider mechanism to prevent fetal rejection. A range of immunosuppresive factors are released by the placenta into the maternal circulation, and there is growing evidence that placental vesicles are involved in this process. T-cell responses are significantly inhibited by placental vesicles in terms of phytohemagglutinin (PHA)- and mixed lymphocyte response–induced proliferation,[21,28,29] Fas ligand–mediated lymphocyte apoptosis, and CD3-zeta loss[9,29–31] and interleukin (IL)-2-induced signal transducer and activator of transcription 3 (STAT3) phosphorylation.[8] Placental vesicles have also been shown to suppress natural killer (NK) cell killing of K562 tumor target cells by blocking their activating NKG2D receptors via MHC class I chain-related (MIC A/B) proteins and UL-16 binding proteins (ULBPs) expressed on their surface.[32,33]

15.3.2 Proinflammatory Effects of Placental Vesicles

While the immunosuppressive effects of placental vesicles on T-cell and NK cell responses go some way to explaining how the placenta avoids immune rejection, the situation is more complicated than this. Although T- and NK cell responses are suppressed in pregnancy, maternal innate immune responses are activated to bring about an inflammatory state.[34] While at first this may appear counterintuitive, inflammatory cytokines in the right amounts are known to be beneficial to the processes of implantation and placentation and may also help the mother to fight infection as her T- and NK-cell-mediated immune responses are suppressed. However, in pre-eclampsia the mild inflammatory state of normal pregnancy decompensates, leading to a systemic inflammatory response in the mother, with associated endothelial dysfunction and activation of the clotting system. This is believed to be the cause of the maternal syndrome.[11]

There is growing evidence that placental vesicles play a role in the maternal inflammatory response. We and others have shown that they are rapidly taken up and internalized by monocytes *in vitro*[4,35–37] with the subsequent release of a range of proinflammatory cytokines (including tumor necrosis factor-α [TNFα], MIP-1α, IL-1α, IL-1β, IL-6, IL-8, IL-12, IL-18, and interferon-γ [IFNγ]).[22] Placental vesicles have also been shown to directly stimulate neutrophils, resulting in increased superoxide production[38] and the formation of extracellular lattices (neutrophil extracellular traps [NETs]), which, *ex vivo*, are increased in the intervillous space of pre-eclampsia placentas.[30]

15.3.3 Effects of Placental Vesicles on Endothelial Cells

As well as having effects on innate immune cells, placental vesicles have also been shown to affect the function of endothelial cells, inhibiting their proliferation and growth as a monolayer *in vitro*[20,39,40] and inhibiting the relaxation of preconstricted blood vessels *ex vivo*.[39–41] Furthermore, when human umbilical vein endothelial cells (HUVECs) are cultured with placental vesicles the culture supernatants can secondarily activate neutrophils, demonstrating the potential for a vicious cycle of inflammatory activation.[42] These effects may contribute to the endothelial dysfunction of pre-eclampsia *in vivo*, which is a characteristic of the maternal syndrome.

15.3.4 Procoagulant Effects of Placental Vesicles

Pregnancy is associated with a physiological increase in many procoagulant factors and inhibitors of fibrinolysis such that the mother is in a hypercoagulable state. This is an important mechanism in normal pregnancy, which prevents postpartum hemorrhage. Placental vesicles may play a role in this process, being known to express tissue factor and phosphatidylserine, both of which initiate the coagulation cascade[43,44] (see Chapter 1).

15.3.5 Differential Effects of Placental Vesicles from Normal and Pre-Eclampsia Placentas on Immune and Endothelial Cells and the Coagulation System

A key question is whether, in addition to the quantitative differences already described, there are qualitative differences between vesicles shed from normal and pre-eclampsia placentas that could alter their function. There is currently little information on this as the majority of studies have used vesicles from normal placentas due to the practical difficulties of preparing vesicles from pre-eclampsia placentas. However, a recent report has shown that vesicles from pre-eclampsia placentas caused significantly higher activation of peripheral blood mononuclear cells (PBMCs) to produce a range of cytokines and chemokines, including IL-1β, compared to normal placental vesicles.[45] It has also been shown that higher levels of superoxide production are induced in neutrophils by vesicles prepared from pre-eclampsia placentas compared to controls.[38] This may be due to the higher content of peroxidized lipids and greater susceptibility to oxidation of vesicles prepared from pre-eclampsia placentas.[46]

In pre-eclampsia there is excessive activation of the coagulation system with increased platelet activation.[47] Our recent work has shown that vesicles prepared from pre-eclampsia placentas express higher levels of functional tissue factor than those from normal placentas, which may account for the excessive activation of the clotting system seen in this disorder.[43,44] However, no differential effects of vesicles prepared from normal and pre-eclampsia placentas on endothelial cells have been found.[39]

15.4 Characterization of Placental Vesicles

As discussed above, placental vesicles have been shown to have a wide range of functional activities, suggesting that they carry a variety of bioactive molecules into the maternal circulation. A crucial step is to define which molecules are present and whether there are differences in these between vesicles from normal and pre-eclampsia placentas that could explain their role in this disorder. A summary of some candidate molecules and their potential activities is shown in Fig. 15.1.

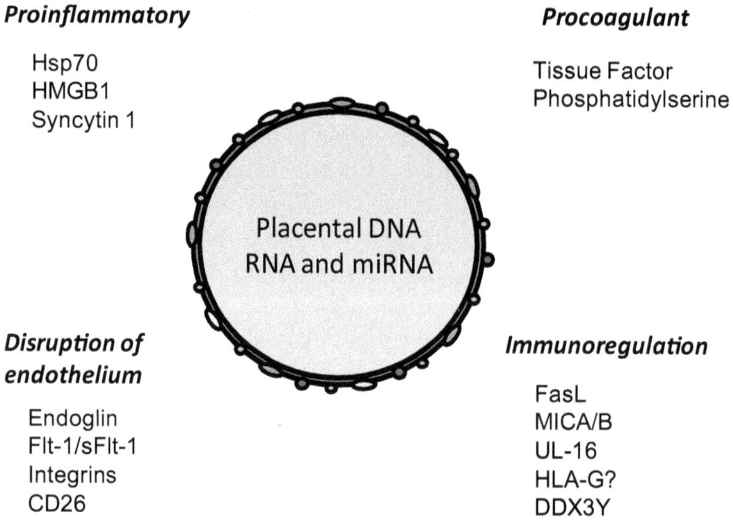

Figure 15.1 The functional activities of placental vesicles and the relevant molecular cargos they carry. Proinflammatory effects of placental vesicles may be due to their expression of danger molecules, including Hsp70, HMGB1, and Syncytin 1, and their procoagulant activity may result from their expression of tissue factor and phosphatidylserine. Flt-1 (and/or sFlt-1), endoglin, integrins, and CD26 carried by placental vesicles may contribute to endothelial dysfunction. Placental vesicles have also been shown to be immunoregulatory, suppressing NK and T-cell responses *in vitro*, possibly due to their expression of MICA/B and UL-16, and Fas ligand, HLA-G, and the minor histocompatibility antigen DDX3Y, respectively. Adapted from Ref. 12, Copyright (2012), with permission from Elsevier.

15.4.1 Placental Vesicle Protein Expression

In our initial attempts to characterize the active factors responsible for the functional effects of placental vesicles on endothelial cells, we carried out sequential ion exchange and gel filtration chromatography followed by mass spectrometry. The major protein bands identified were integrins α_5 (CD49e) and α_V (CD51), dipeptidyl peptidase IV (DPP IV, CD26), α-actinin, transferrin, transferrin receptor, PLAP, and monamine oxidase A.[48] Antibodies against integrins α_5 and α_V and DPP IV all reduced the placental vesicle inhibition of HUVEC proliferation, which was also reversed by added fibronectin, suggesting that adhesion molecules are a major source of the inhibitory activity.[49] In a more recent study it has been shown that vesicles prepared by perfusion of normal placentas expressed the soluble forms of the fms-like tyrosine kinase (sFlt-1) and endoglin, which exert antiangiogenic actions and could thereby cause endothelial dysfunction. Plasminogen activator inhibitors (PAI-1/PAI-2), which regulate fibrinolysis, were also found, and these could be responsible for the very high levels of fibrin deposition in the intervillous space and placental infarction observed in these pregnancies.[50] Flt-1 expression by placental as well as platelet vesicles in the maternal plasma has also been demonstrated by flow cytometry, but this only accounted for 5% of the total Flt-1 present.[51]

In terms of molecules that interact with the maternal immune system, the syncytiotrophoblast vesicles are devoid of all classic (HLA-A, B, C) and nonclassic (HLA-G, E, F) MHC class 1 antigens (Tannetta et al., unpublished observation) and are therefore unable to elicit a classic T-cell response. Vesicles derived from extravillous cytotrophoblasts may express HLA-G. However, this molecule does not stimulate a T-cell response either but rather inhibits both CD4-positive and CD8-positive T-cells.[52] A recent report has shown the presence of the Y chromosome–linked minor histocompatibility antigen DDX3Y in shed trophoblast debris, and the authors speculate that this may lead to the induction of antigen-specific regulatory CD8-positive T-cells, conferring immunological tolerance to the mother.[53] As mentioned above, placental vesicles produced by first-trimester explant culture have been shown to express other MHC class I–related antigens, including MIC proteins and ULBPs, which bind to and down-regulate the NKG2D-activating receptor on NK cells, thereby preventing them from attacking the placenta.[32,33] The immunosuppressive effects of placental vesicles may also be mediated through their expression of the Fas ligand, which acts by inducing apoptosis of maternal T-cells.[31]

In contrast, little is known about the proinflammatory molecules carried by placental vesicles. Prime candidates are "danger molecules" or "alarmins." These are molecules that have important intracellular functions but become intensely proinflammatory when released from cells. Examples of these are heat shock protein 70 (HSP70) and high mobility group box 1 (HMGB1), both of which are expressed on placental vesicles.[12] A further candidate is syncytin 1, a protein encoded by the human endogenous retrovirus-W (HERV-W), which has been shown to be present on placental vesicles and stimulates IL-1α and a range of other proinflammatory cytokine production by peripheral blood mononuclear cells.[23]

15.4.2 Proteomic Analysis of Placental Vesicles

The alternative to investigating individual candidate proteins is to carry out a full proteomic analysis of placental vesicles. While proteomic analyses of normal and pre-eclampsia placentas,[54,55] isolated trophoblasts,[56] serum,[57,58] and plasmas[59,60] have previously been reported, we have performed, we believe, the first such comparison on perfused placental vesicle pools from normal and pre-eclampsia placentas using multidimensional protein identification technology (MudPIT).[12] The analysis found 538 proteins unique to pre-eclamptic placental vesicles, 604 unique to normal placental vesicles and 1,421 common to both preparations. The complex protein cargo carried by placental vesicles was demonstrated by preliminary analysis showing the presence of alarmins (HSP70 and galectin 3), exosomal proteins (CD63, CD9, CD81), immunoregulatory molecules (CD26, CD200, CD47, galectin 1), complement and complement regulatory molecules (C1q, C3, CD55, CD59, and vitronectin), amino acid transporters (CD98), and antiangiogenic molecules (endoglin) in both normal and pre-eclampsia placental vesicles.

The potential for circulating placental vesicles to act as biopsies of the placenta is shown by the altered expression of proteins, reported previously in pre-eclampsia placenta samples, being mirrored in perfused placental vesicles. This includes the up-regulated expression of apolipoprotein AI, protein disulphide isomerase, myosin, and centrosome protein.[30,54] Several proteins shown to circulate at increased levels in pre-eclampsia were also highly expressed in perfused placental vesicles, including ceruloplasmin, fibrinogen, hemopexin, and vitronectin.[58–60] This raises the issue of whether molecules assumed to be "soluble" are in fact carried by vesicles, as most protocols for plasma preparation do not include an ultracentrifugation step to deplete samples of microvesicles and exosomes. To date, there are no published reports of proteomic analysis of placenta-derived exosomes. However, vesicles released by the telomerase-immortalized first-trimester cytotrophoblast cell line Swan 71, which resembles the extravillous cytotrophoblast, have been characterized and shown to have the size and density range of exosomes. Ion trap mass spectrometry demonstrated the presence of exosomal proteins, as well as proteins involved in diverse functions such as cell structure, adhesion, coagulation, protein synthesis, deoxyribonucleic acid (DNA) replication, channels, and metabolism, thus demonstrating that despite their small size, exosomes carry a complex protein cargo.[61]

15.4.3 Placental Vesicle DNA, mRNA, and miRNA

The discovery of circulating fetal genetic material in maternal blood[62,63] has fueled interest in its use as a noninvasive alternative to amniocentesis and chorionic villus sampling—invasive prenatal diagnosis techniques that carry a significant risk of miscarriage. With the growing number of studies reporting associations between circulating extracellular nucleic acids and obstetric complications, their use as biomarkers of placental insufficiency is gaining credence. A general overview of placental vesicle DNA, messenger ribonucleic acid (mRNA), and micro-RNA (miRNA) is outlined here. For a more comprehensive review refer to Hromadnokova.[64]

15.4.3.1 Fetal (placental) DNA

Initial efforts to recover fetal DNA from maternal blood concentrated on the isolation of fetal cells. This proved to be challenging as they circulate in very low numbers. Cell-free fetal DNA, however, is far more abundant and can be detected from as early as 5 weeks of gestation.[65,66] Circulating levels of cell-free fetal DNA also fall rapidly following delivery, indicating that the placenta is the predominant source.[5] Increases in the levels of circulating cell-free fetal DNA have been demonstrated in conditions involving placental impairment such as intrauterine growth restriction[67] and pre-eclampsia[68,69] with the latter showing a correlation with the severity of the disease,[69] and elevated circulating cell-free fetal DNA prior to the onset of maternal symptoms.[70] However, in the clinical setting, analysis of cell-free fetal DNA has been limited to mothers carrying male fetuses or rhesus-negative mothers carrying rhesus-positive fetuses,[71] to enable the distinction of fetal DNA from circulating maternal DNA.

The association of fetal DNA with placental microvesicles, however, would allow the isolation of fetal-derived genetic material using placental specific markers such as PLAP, making possible the screening of DNA from both male and female fetuses. Placental vesicles prepared *in vitro* contain fetal DNA,[7,17] yet no characterization of the size of the vesicles containing DNA has been reported. It is likely that these vesicles are apoptotic bodies, released by cells undergoing apoptosis, which contain DNA fragments. These tend to be >1um in size and are therefore easily detectable by flow cytometry. The trophoblast cell line JEG-3, when induced to undergo apoptosis, released DNA containing vesicles detectable by flow cytometry using picogreen, a fluorescent dye that selectively binds double-stranded DNA.[72,73] The mode of *in vitro* placental vesicle preparation has been shown to affect the concentrations of fetal DNA carried by the placenta-derived vesicles, with those from explants containing higher levels of DNA.[17] Explant culture was also shown to be associated with much higher levels of fetal DNA, which could not be pelleted by ultracentrifugation, suggesting that placental vesicle–free DNA could be released mostly by the apoptotic/necrotic syncytiotrophoblast layer during the initial culture period.[17] Indeed, hypoxia/reoxygenation of explants significantly increased markers of apoptosis and necrosis and the release of cell-free DNA—effects that were significantly reduced by the addition of antioxidants.[74] Circulating placental vesicles containing DNA have also been detected in the maternal circulation using double labeling with picogreen and either anti-PLAP or anti-HLA-G antibody. Increased levels were found in pre-eclampsia,[7] perhaps reflecting the placental oxidative stress associated with this disorder.

15.4.3.2 Fetal (placental) mRNA

As with DNA, mRNA also circulates at high levels in maternal blood. mRNA transcripts from placenta-expressed genes, such as corticotrophin-releasing hormone (CRH), are readily detectable in maternal plasma, providing a gender-independent target for noninvasive prenatal gene expression profiling.[75,76] Circulating mRNA levels increase with gestational age in uncomplicated pregnancies.[77] Pre-eclampsia, however, caused a marked increase in CRH mRNA concentrations, which showed a positive correlation with disease severity.[69,75,77] mRNA expression profiling has also revealed differential expression of multiple mRNAs in placental tissue from pre-eclamptic pregnancies.[78] As plasma contains ribonuclease (RNase) activity, the mRNAs must circulate in a protected form, perhaps encased in EVs. In support of this, placental vesicles prepared *in vitro* have been shown to contain fetal mRNA.[17] To date, no characterization of the size of vesicles containing mRNA has been reported; and as such the mechanisms resulting in the release of vesicle-packaged mRNA are poorly understood. As with DNA, the mode of preparation has been shown to affect the concentrations of fetal mRNA carried by the placenta-derived vesicles, with perfusion-derived vesicles containing high mRNA levels, as assessed by measuring CRH mRNA.[17] Also placental vesicle mRNA integrity may reflect levels of placental oxidative stress, as demonstrated by a reduction in the integrity of mRNA transcripts released in vesicles from cultured explants exposed to hypoxia/reoxygenation.[79] However, the assumption that circulating placental vesicles reflect the state of the placenta at the time of vesicle formation and release assumes that there is no enrichment of particular mRNA into vesicles. This requires further investigation in the light of studies such as those of Kumpel et al. (2008), which showed markedly elevated vesicular mRNA for human placental lactogen and human chorionic gonadotrophin compared to isolated trophoblast cells, suggesting sorting of certain mRNA into placental vesicles.[80]

15.4.3.3 Fetal (placental) miRNA

miRNAs are small RNA species that regulate gene expression post-transcriptionally and are implicated in many diseases, including several human cancers (see Chapter 13). The placenta expresses several unique miRNAs alongside abundant ubiquitous species.[81,82–84] Placenta-specific miRNAs are found in the chromosome 19 miRNA cluster. This is the largest human cluster and is primate specific, encoding 46 miRNA species and appearing relatively recently in evolution.[85] Placenta-specific miRNAs have been shown to be the predominant forms of miRNA in the placenta, expressed by the syncytiotrophoblast layer, primary human term trophoblasts,[86] BeWo cells,[24,86] and JEG-3 cells.[86] Interestingly, syncytialization of isolated trophoblasts did not affect the expression of chromosome 19 microRNA cluster miRNAs, while hypoxia selectively suppressed miR 520c-3p.[86] Placenta-specific miRNAs are also abundant in the plasma of pregnant women, showing increasing circulating levels as pregnancy progresses, and are cleared rapidly after delivery.[24,84,87,88]

Very few studies have specifically investigated placental vesicle–associated miRNA. Microinterfering RNAs have been found in various bodily fluids (e.g., plasma, urine, saliva, and breast milk) as part of a novel mechanism for intercellular exchange of genetic material.[89,90] Investigations to date into miRNA release mechanisms have demonstrated circulating miRNAs associated with exosomes, microvesicles, argonaut2 protein, and high-density lipoprotein cholesterol, which may all contribute to stabilizing secreted miRNAs.[89,91–94] Several placenta-specific miRNAs are released in exosomes by the trophoblast cell line BeWo.[24] Likewise, isolated human term trophoblasts release exosomes containing predominantly placenta-specific miRNAs whose profile closely resembles that of the parent cells.[86]

Investigation of the involvement of miRNAs in the pathogenesis of pre-eclampsia is in its infancy. However, several studies have reported altered miRNA expression profiles in pre-ecamplsia placentas, although with little consistency.[78,83,95–98] Hypoxia-associated miR210 has been shown to be significantly up-regulated in pre-ecamplsia placentas and plasma.[83,95,99] Possible involvement of miR210 in the poor spiral artery remodeling associated with pre-eclampsia was demonstrated by Zhang et al. (2012). Here, hypoxia was shown to up-regulate the expression of miR210 in both first-trimester cytotrophoblast cells and the trophoblast cell line JAR, while overexpression of miR210 significantly attenuated first-trimester cytotrophoblast migration and invasion,[99] processes evidently impaired in pre-eclampsia. Zhang et al. determined that mRNA of *homeobox-A9* and *ephrin-A3*, genes involved in migration and invasion processes, were targets for miR210 and protein levels of both were significantly reduced in pre-eclampsia-derived placentas. Overexpression of the angiogenesis-targeting miRNAs miR-17, -20a, and -20b in pre-eclampsia placentas has also been implicated in impaired early placental development.[97] In addition modeling miR29b overexpression, present in pre-eclampsia placentas,[98] in trophoblast cell lines, induced apoptosis and impaired invasion and angiogenesis through the regulation of myeloid cell leukaemia sequence 1, matrix metalloprotease 2, VEGF A, and integrin β1 expression.[100] Analysis of circulating miRNAs in severe pre-eclampsia has shown them to be both up- and down-regulated,[101] with some significantly elevated between 12 to 16 weeks' gestation in pregnancies that subsequently developed pre-eclampsia, thus providing evidence of their potential use as predictive markers.[102]

15.5 Placental Vesicles: Exosomes, Microvesicles, and Apoptotic Bodies

So far in this chapter the term "placental vesicle" has been used to describe material shed from the placenta that is small enough to pass through the capillaries of the lungs and enter the maternal blood. However, this material is undoubtedly comprised of a mixture of different vesicle types, including exosomes, microvesicles, and apoptotic vesicles, which are primarily distinguished on the basis of their size (30–100 nm, 100–1 μm and 1–4 μm, respectively), together with necrotic debris (Fig. 15.2). However, many researchers do not attempt to define the vesicle populations they are studying. This is confounded by the use of different centrifugation speeds to pellet the vesicles, resulting in some groups only studying the larger vesicles,[103,104] while others focus on smaller exosomes to the exclusion of others.[33] Until recently vesicle size could only be accurately determined by electron microscopy, which is a low-throughput, subjective technique. However, the advent of techniques such as nanoparticle tracking analysis (NTA) has allowed, for the first time, the accurate sizing and counting of vesicles in biological fluids[105] (see Chapter 11). We have used this technique to determine the size of vesicles prepared by placental perfusion. Sizes ranged from approximately 40 nm upward, with the majority (>90%) being less than 1 μm and 70% being less than 300 nm in diameter, suggesting that they are predominantly a mixture of exosomes and microvesicles. Distinct peaks at approximately 100 nm (exosomes) and 200 nm (microvesicles) can often be seen by this technique[12] (Fig. 15.3).

The composition of placental vesicles in terms of the proportions of exosomes, microvesicles, and apoptotic bodies present may have an important bearing on their functional characteristics, with exosomes being predominantly involved in immunoregulation and intercellular communication and microvesicles being more proinflammatory. This has led us to speculate whether changes in the balance between exosomes and microvesicles in vesicles shed from

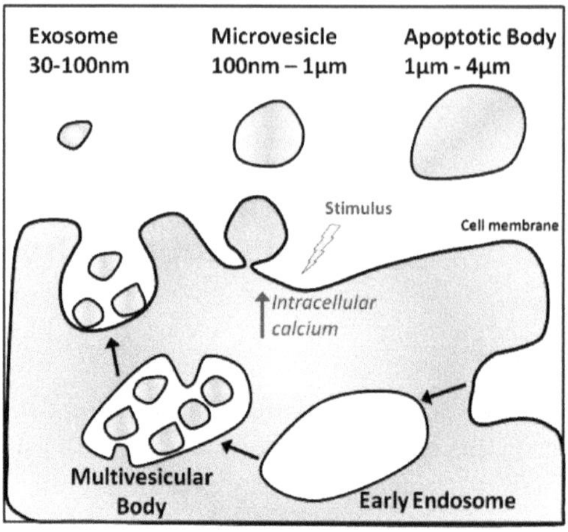

Figure 15.2 Three vesicle subtypes: exosomes, microvesicles, and apoptotic bodies. Exosomes (30–100 nm) are generated from reverse budding of the endosome membrane, resulting in a multivesicular body, which fuses with the plasma membrane of the cell, and exosomes are released by exocytosis. Microvesicles (100 nm–1 μm) are produced by direct budding or shedding of the plasma membrane. They are released in response to stimuli causing an increase in intracellular calcium levels and remodeling of the plasma cell membrane. Apoptotic bodies (1–4 μm) are released from cells undergoing apoptosis. Reprinted from Ref. 117, Copyright (2009), with permission from Elsevier.

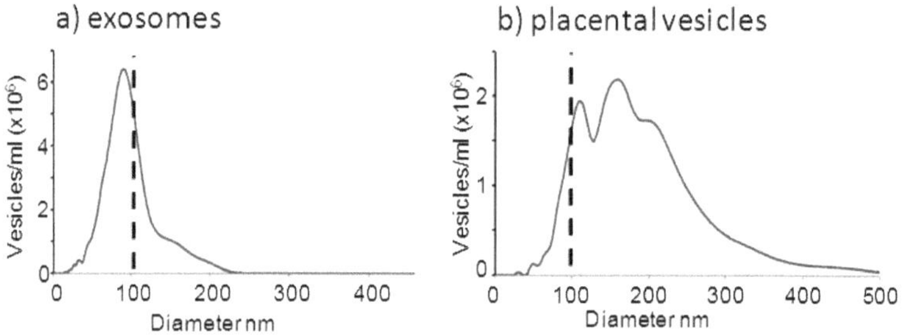

Figure 15.3 Measurement of cellular microvesicles and exosomes using NTA. (a) Exosomes from a neuroblastoma cell line measured by NTA and (b) perfused placental vesicles measured by NTA, showing a larger size distribution than (a), possibly due to the presence of a mixture of exosomes and microvesicles. Reprinted from Ref. 12, Copyright (2012), with permission from Elsevier.

pre-eclampsia placentas could alter their overall functional effects to a more proinflammatory, antiangiogenic, and procoagulant state[106] (Fig. 15.4). Future studies separating exosomes from microvesicles by immunoaffinity beads or filtration through nanomembranes and subsequent proteomic and DNA/RNA analysis of their contents will be important to determine their relative functional importance.

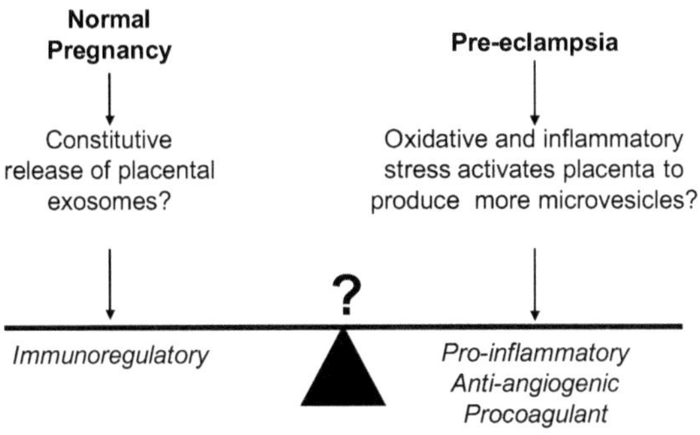

Figure 15.4 The role of exosomes and microvesicles in normal pregnancy and pre-eclampsia. We speculate that in normal pregnancy, vesicles shed from the placenta are predominantly constitutively released exosomes, which are immunoregulatory in nature. In contrast, in pre-eclampsia oxidative and inflammatory stress activates the syncytiotrophoblast, causing it to release more microvesicles, which have proinflammatory, antiangiogenic, and procoagulant activity and trigger the maternal syndrome.

15.6 Extracellular Vesicles in Pregnancy and Pre-Eclampsia Plasma from Other Cell Types

We have proposed that increased numbers and an altered phenotype of placental vesicles in the maternal circulation in pre-eclampsia may cause an inflammatory response, endothelial dysfunction, and activation of the coagulation system characteristic of the disorder. A downstream consequence of this would be the activation and release of cellular vesicles from other vascular cell types, including leukocytes, endothelium, platelets, and red blood cells (RBCs) which further exacerbate the disease. Thus it might be possible that by measuring placental vesicles (the cause) in parallel with circulating vesicles derived from other cell types (the consequence), the prediction and monitoring of the condition could be improved (Fig. 15.5).

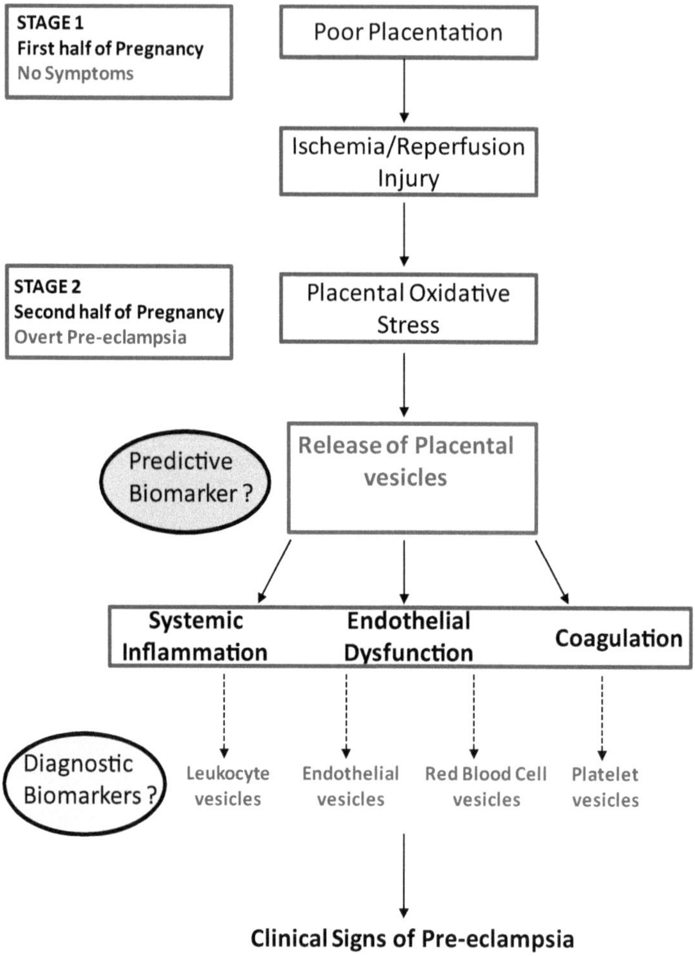

Figure 15.5 Proposed pathogenesis of pre-eclampsia. The release of placental vesicles into the maternal circulation may cause an inflammatory response, endothelial dysfunction, and activation of the coagulation system, characteristic of pre-eclampsia. This in turn leads to the activation and release of cellular vesicles from other cell types, including leukocytes, the endothelium, RBCs, and platelets, to further exacerbate the disease. Placental vesicles have the potential to be used as predictive biomarkers, whereas the other cellular-derived vesicles could be used as diagnostic biomarkers.

There have been a large number of studies investigating vesicles shed from cells in the vascular compartment in normal pregnancy and pre-eclampsia. These include measures of total vesicle numbers and those derived from platelets, RBCs, endothelial cells, and leukocytes and their subpopulations (monocytes, granulocytes, and B- and T-lymphocytes). While some studies suggest that particular vesicle types are altered in pregnant compared to nonpregnant women[107–111] and further changed in women with pre-eclampsia, consistent with leukocyte activation, endothelial injury, and platelet activation,[6,16,47,107–110,112–114] there is little consensus, as others have found no change.[6,51,107,109–111,113] This makes interpretation of these experiments extremely difficult. These discrepant findings probably result from differences in the methods for isolating the vesicles (e.g., centrifugation speeds), resulting in different populations being analyzed, the use of different antibody panels to identify the vesicles by flow cytometry, and, in some studies, the use of inappropriate unmatched controls. This analysis is further confounded by the observation that antigens commonly used to identify the parent cells are not necessarily expressed on the vesicles derived from those cells.[111] These issues highlight the need for standardized protocols for microvesicle and exosome analysis, as discussed in Chapter 5.

15.6.1 Functional Effects of Maternal Circulating Vesicles

Notwithstanding the lack of consensus on the numbers of different vesicle types in the maternal circulation in normal pregnancy and pre-eclampsia, there have been studies of the functional effects of total vesicles derived from the maternal plasma on monocytes and the endothelium. It has been shown that vesicles isolated from pre-eclampsia plasma stimulated a significant increase in intracellular adhesion molecule-1 (ICAM-1) (a cell activation marker) expression in the "Monomac-6" monocyte cell line compared to vesicles from normal pregnant and nonpregnant women, but only when they were cocultured in the presence of endothelial cells (HUVECs), suggesting an interaction between the monocytes and endothelial cells in this process, possibly via cytokine release.[115] In another study, vesicles prepared from the plasma of pre-eclamptic but not normal pregnant women were found to induce vascular hyporeactivity to serotonin in *ex vivo* preparations of human omental arteries and mouse aortas, This hyporeactivity was reversed by a nitric oxide (NO) synthase inhibitor with associated increase in NO production.[112] However, a study of complement activation on the surface of plasma-derived vesicles, measuring bound C1q and activated C4 and/or C3, showed no difference between normal pregnant and pre-eclamptic women. The levels of the complement activator molecules CRP, SAP, immunoglobulin M (IgM), and immunoglobulin G (IgG) binding to the vesicles were also studied, and higher concentrations of vesicles binding CRP were found in pre-eclampsia compared to normal pregnant and nonpregnant controls, but this did not correlate with increased activation of the classical complement pathway as determined by C1q binding.[116] These preparations will have been composed of vesicles derived from all the cells in the vascular compartment discussed above, together with any placental vesicles present, so it cannot be determined which populations were responsible for these effects.

15.7 Future Perspectives

The ability to measure, isolate, and phenotype placental and other cellular vesicles in the maternal circulation has great potential for improving the detection and management of pre-eclampsia. Firstly, there is growing interest in biomarkers that could be used in early pregnancy to predict which women are at risk of developing the disorder. Tests currently being trialled use candidate markers such as sFlt-1, Endoglin, and placental growth factor (PLGF).[57,60] However, as we have shown, placental vesicles carry hundreds of proteins, many of which are unique to pre-eclampsia and may therefore provide better markers, either alone or in multiple arrays. Secondly, by understanding the way that the placenta signals to the mother through microvesicles and exosomes in normal pregnancy and how these signals change in pre-eclampsia, it may be possible to find ways of neutralizing the activity of damaging vesicles. Dampening down the maternal inflammatory response would allow the pregnancy to continue long enough to avoid the need for very early preterm delivery and all the problems this brings both for the baby and the mother.

Acknowledgments

This work was supported by a Wellcome Trust Programme Grant (GR079862), Wellcome Trust Technology Development Grant (GR087730), MRC Programme Grant (MR/J003360/1), and the Oxford Partnership Comprehensive Biomedical Research Centre (http://www.oxfordbrc.org), with funding from the Department of Health's National Institute for Health Research Biomedical Research Centres funding scheme. The views expressed in this publication are those of the authors and not necessarily those of the Department of Health.

References

1. Burton GJ, Jones CJ (2009) Syncytial knots, sprouts, apoptosis, and trophoblast deportation from the human placenta, *Taiwan J. Obstet. Gynecol.*, **48**, 28–37.
2. Johansen M, Redman CW, Wilkins T, Sargent IL (1999) Trophoblast deportation in human pregnancy—its relevance for pre-eclampsia, *Placenta*, **20**, 531–539.
3. Knight M, Redman CW, Linton EA, Sargent IL (1998) Shedding of syncytiotrophoblast microvilli into the maternal circulation in pre-eclamptic pregnancies, *BJOG*, **105**, 632–640.
4. Germain SJ, Sacks GP, Sooranna SR, Sargent IL, Redman CW (2007) Systemic inflammatory priming in normal pregnancy and preeclampsia: the role of circulating syncytiotrophoblast microparticles, *J. Immunol.*, **178**, 5949–5956.
5. Reddy A, Zhong XY, Rusterholz C, Hahn S, Holzgreve W, Redman CW, Sargent IL (2008) The effect of labour and placental separation on the shedding of syncytiotrophoblast microparticles, cell-free DNA and mRNA in normal pregnancy and pre-eclampsia, *Placenta*, **29**, 942–949.
6. Lok CA, Van Der Post JA, Sargent IL, Hau CM, Sturk A, Boer K, Nieuwland R (2008) Changes in microparticle numbers and cellular origin during pregnancy and preeclampsia, *Hypertens Pregnancy*, **27**, 344–360.
7. Orozco AF, Jorgez CJ, Ramos-Perez WD, Popek EJ, Yu X, Kozinetz CA, Bischoff FZ, Lewis DE (2009) Placental release of distinct DNA-associated micro-particles into maternal circulation: reflective of gestation time and preeclampsia, *Placenta*, **30**, 891–897.

8. Pap E, Pallinger E, Falus A, Kiss AA, Kittel A, Kovacs P, Buzas EI (2008) T lymphocytes are targets for platelet- and trophoblast-derived microvesicles during pregnancy, *Placenta,* **29**, 826–832.
9. Sabapatha A, Gercel-Taylor C, Taylor DD (2006) Specific isolation of placenta-derived exosomes from the circulation of pregnant women and their immunoregulatory consequences, *Am. J. Reprod. Immunol.* **56**, 345–355.
10. Burton GJ, Jauniaux E (2004) Placental oxidative stress: from miscarriage to preeclampsia, *J. Soc. Gynecol. Invest.,* **11**, 342–352.
11. Redman CW, Sargent IL (2005) Latest advances in understanding preeclampsia, *Science,* **308**, 1592–1594.
12. Redman CW, Tannetta DS, Dragovic RA, Gardiner C, Southcombe JH, Collett GP, Sargent IL (2012) Review: does size matter? Placental debris and the pathophysiology of pre-eclampsia, *Placenta,* **33**(Suppl), S48–S54.
13. Goswami D, Tannetta DS, Magee LA, Fuchisawa A, Redman CW, Sargent IL, von Dadelszen P (2006) Excess syncytiotrophoblast microparticle shedding is a feature of early-onset pre-eclampsia, but not normotensive intrauterine growth restriction, *Placenta,* **27**, 56–61.
14. Chen Y, Huang Y, Jiang R, Teng Y (2012) Syncytiotrophoblast-derived microparticle shedding in early-onset and late-onset severe pre-eclampsia, *IJGO,* **119**, 234–238.
15. Marques F, Campos FM, Filho OA, Carvalho AT, Dusse LM, Gomes KB (2012) Circulating microparticles in severe preeclampsia, *Clin. Chim. Acta,* **414**, 253–258.
16. Lok CA, Nieuwland R, Sturk A, Hau CM, Boer K, Vanbavel E, Vanderpost JA (2007) Microparticle-associated P-selectin reflects platelet activation in preeclampsia, *Platelets,* **18**, 68–72.
17. Gupta AK, Holzgreve W, Huppertz B, Malek A, Schneider H, Hahn S (2004) Detection of fetal DNA and RNA in placenta-derived syncytiotrophoblast microparticles generated *in vitro, Clin. Chem.,* **50**, 2187–2190.
18. Siman CM, Sibley CP, Jones CJ, Turner MA, Greenwood SL (2001) The functional regeneration of syncytiotrophoblast in cultured explants of term placenta, *Am. J. Physiol. Regul. Integr. Comp. Physiol.,* **280**, R1116–R1122.
19. Schneider H, Huch A (1985) Dual *in vitro* perfusion of an isolated lobe of human placenta: method and instrumentation, *Contrib. Gynecol. Obstet.,* **13**, 40–47.

20. Gupta AK, Rusterholz C, Huppertz B, Malek A, Schneider H, Holzgreve W, Hahn S (2005) A comparative study of the effect of three different syncytiotrophoblast micro-particles preparations on endothelial cells, *Placenta,* **26**, 59–66.

21. Gupta AK, Rusterholz C, Holzgreve W, Hahn S (2005) Syncytiotrophoblast micro-particles do not induce apoptosis in peripheral T lymphocytes, but differ in their activity depending on the mode of preparation, *J. Reprod. Immunol.,* **68**, 15–26.

22. Southcombe J, Tannetta D, Redman C, Sargent I (2011) The immunomodulatory role of syncytiotrophoblast microvesicles, *PLOS ONE,* **6**, e20245.

23. Holder BS, Tower CL, Forbes K, Mulla MJ, Aplin JD, Abrahams VM (2012) Immune cell activation by trophoblast-derived microvesicles is mediated by syncytin 1, *Immunology,* **136**, 184–191.

24. Luo SS, Ishibashi O, Ishikawa G, Ishikawa T, Katayama A, Mishima T, Takizawa T, Shigihara T, Goto T, Izumi A, Ohkuchi A, Matsubara S, Takeshita T (2009) Human villous trophoblasts express and secrete placenta-specific microRNAs into maternal circulation via exosomes, *Biol. Reprod.,* **81**, 717–729.

25. Collett GP, Goh XF, Linton EA, Redman CW, Sargent IL (2012) RhoE is regulated by cyclic AMP and promotes fusion of human BeWo choriocarcinoma cells, *PLOS ONE,* **7**, e30453.

26. Collett GP, Linton EA, Redman CW, Sargent IL (2010) Downregulation of caveolin-1 enhances fusion of human BeWo choriocarcinoma cells, *PLOS ONE,* **5**, e10529.

27. Tannetta DS, Sargent IL, Linton EA, Redman CW (2008) Vitamins C and E inhibit apoptosis of cultured human term placenta trophoblast, *Placenta,* **29**, 680–690.

28. Thibault G, Degenne D, Girard AC, Guillaumin JM, Lacord M, Bardos P (1991) The inhibitory effect of human syncytiotrophoblast plasma membrane vesicles on *in vitro* lymphocyte proliferation is associated with reduced interleukin 2 receptor expression, *Cell. Immunol.,* **138**, 165–174.

29. Arkwright PD, Rademacher TW, Boutignon F, Dwek RA, Redman CW (1994) Suppression of allogeneic reactivity *in vitro* by the syncytiotrophoblast membrane glycocalyx of the human term placenta is carbohydrate dependent, *Glycobiology,* **4**, 39–47.

30. Gupta AK, Hasler P, Holzgreve W, Gebhardt S, Hahn S (2005) Induction of neutrophil extracellular DNA lattices by placental microparticles

and IL-8 and their presence in preeclampsia, *Hum. Immunol.*, **66**, 1146–1154.

31. Taylor DD, Akyol S, Gercel-Taylor C (2006) Pregnancy-associated exosomes and their modulation of T cell signaling, *J. Immunol.*, **176**, 1534–1542.

32. Mincheva-Nilsson L, Nagaeva O, Chen T, Stendahl U, Antsiferova J, Mogren I, Hernestal J, Baranov V (2006) Placenta-derived soluble MHC class I chain-related molecules down-regulate NKG2D receptor on peripheral blood mononuclear cells during human pregnancy: a possible novel immune escape mechanism for fetal survival, *J. Immunol.*, **176**, 3585–3592.

33. Hedlund M, Stenqvist AC, Nagaeva O, Kjellberg L, Wulff M, Baranov V, Mincheva-Nilsson L (2009) Human placenta expresses and secretes NKG2D ligands via exosomes that down-modulate the cognate receptor expression: evidence for immunosuppressive function, *J. Immunol.*, **183**, 340–351.

34. Sacks GP, Studena K, Sargent K, Redman CW (1998) Normal pregnancy and preeclampsia both produce inflammatory changes in peripheral blood leukocytes akin to those of sepsis, *Am. J. Obstet. Gynecol.*, **179**, 80–86.

35. Messerli M, May K, Hansson SR, Schneider H, Holzgreve W, Hahn S, Rusterholz C (2010) Feto-maternal interactions in pregnancies: placental microparticles activate peripheral blood monocytes, *Placenta*, **31**, 106–112.

36. Atay S, Gercel-Taylor C, Suttles J, Mor G, Taylor DD (2011) Trophoblast-derived exosomes mediate monocyte recruitment and differentiation, *Am. J. Reprod. Immunol.* **65**, 65–77.

37. Atay S, Gercel-Taylor C, Taylor DD (2011) Human trophoblast-derived exosomal fibronectin induces pro-inflammatory IL-1beta production by macrophages, *Am. J. Reprod. Immunol.* **66**, 259–269.

38. Aly AS, Khandelwal M, Zhao J, Mehmet AH, Sammel MD, Parry S (2004) Neutrophils are stimulated by syncytiotrophoblast microvillous membranes to generate superoxide radicals in women with preeclampsia, *Am. J. Obstet. Gynecol.*, **190**, 252–258.

39. Smarason AK, Sargent IL, Starkey PM, Redman CW (1993) The effect of placental syncytiotrophoblast microvillous membranes from normal and pre-eclamptic women on the growth of endothelial cells *in vitro*, *BJOG*, **100**, 943–949.

40. Hoegh AM, Tannetta D, Sargent I, Borup R, Nielsen FC, Redman C, Sorensen S, Hviid TV (2006) Effect of syncytiotrophoblast microvillous

membrane treatment on gene expression in human umbilical vein endothelial cells, *BJOG,* **113**, 1270–1279.

41. Cockell AP, Learmont JG, Smarason AK, Redman CW, Sargent IL, Poston L (1997) Human placental syncytiotrophoblast microvillous membranes impair maternal vascular endothelial function, *BJOG,* **104**, 235–240.

42. von Dadelszen P, Hurst G, Redman CW (1999) Supernatants from co-cultured endothelial cells and syncytiotrophoblast microvillous membranes activate peripheral blood leukocytes *in vitro*, *Hum. Reprod.,* **14**, 919–924.

43. Gardiner C, Tannetta DS, Simms CA, Harrison P, Redman CW, Sargent IL (2011) Syncytiotrophoblast microvesicles released from pre-eclampsia placentae exhibit increased tissue factor activity, *PLOS ONE,* **6**, e26313.

44. Reverdiau P, Jarousseau AC, Thibault G, Khalfoun B, Watier H, Lebranchu Y, Bardos P, Gruel Y (1995) Tissue factor activity of syncytiotrophoblast plasma membranes and tumoral trophoblast cells in culture, *Thromb. Haemost.* **73**, 49–54.

45. Holder BS, Tower CL, Jones CJ, Aplin JD, Abrahams VM (2012) Heightened pro-inflammatory effect of preeclamptic placental microvesicles on peripheral blood immune cells in humans, *Biol. Reprod.,* **86**, 103.

46. Cester N, Staffolani R, Rabini RA, Magnanelli R, Salvolini E, Galassi R, Mazzanti L, Romanini C (1994) Pregnancy induced hypertension: a role for peroxidation in microvillus plasma membranes, *Mol. Cell. Biochem.,* **131**, 151–155.

47. Macey MG, Bevan S, Alam S, Verghese L, Agrawal S, Beski S, Thuraisingham R, MacCallum PK (2010) Platelet activation and endogenous thrombin potential in pre-eclampsia, *Thromb. Res.,* **125**, e76–e81.

48. Kertesz Z, Hurst G, Ward M, Willis AC, Caro H, Linton EA, Sargent IL, Redman CW (1999) Purification and characterization of a complex from placental syncytiotrophoblast microvillous membranes which inhibits the proliferation of human umbilical vein endothelial cells, *Placenta,* **20**, 71–79.

49. Kertesz Z, Linton EA, Redman CW (2000) Adhesion molecules of syncytiotrophoblast microvillous membranes inhibit proliferation of human umbilical vein endothelial cells, *Placenta,* **21**, 150–159.

50. Guller S, Tang Z, Ma YY, Di Santo S, Sager R, Schneider H (2011) Protein composition of microparticles shed from human placenta during

placental perfusion: potential role in angiogenesis and fibrinolysis in preeclampsia, *Placenta,* **32**, 63–69.

51. Lok CA, Boing AN, Sargent IL, Sooranna SR, van der Post JA, Nieuwland R, Sturk A (2008) Circulating platelet-derived and placenta-derived microparticles expose Flt-1 in preeclampsia, *Reprod. Sci.,* **15**, 1002–1010.
52. Bainbridge DR, Ellis SA, Sargent IL (2000) HLA-G suppresses proliferation of CD4(+) T-lymphocytes, *J. Reprod. Immunol.,* **48**, 17–26.
53. Holland OJ, Linscheid C, Hodes HC, Nauser TL, Gilliam M, Stone P, Chamley LW, Petroff MG (2012) Minor histocompatibility antigens are expressed in syncytiotrophoblast and trophoblast debris: implications for maternal alloreactivity to the fetus, *Am. J. Pathol.,* **180**, 256–266.
54. Gharesi-Fard B, Zolghadri J, Kamali-Sarvestani E (2010) Proteome differences of placenta between pre-eclampsia and normal pregnancy, *Placenta,* **31**, 121–125.
55. Jin H, Ma KD, Hu R, Chen Y, Yang F, Yao J, Li XT, Yang PY (2008) Analysis of expression and comparative profile of normal placental tissue proteins and those in preeclampsia patients using proteomic approaches, *Anal. Chim. Acta,* **629**, 158–164.
56. Sun LZ, Yang NN, De W, Xiao YS (2007) Proteomic analysis of proteins differentially expressed in preeclamptic trophoblasts, *Gynecol. Obstet. Invest.,* **64**, 17–23.
57. Rasanen J, Girsen A, Lu X, Lapidus JA, Standley M, Reddy A, Dasari S, Thomas A, Jacob T, Pouta A, Surcel HM, Tolosa JE, Gravett MG, Nagalla SR (2010) Comprehensive maternal serum proteomic profiles of preclinical and clinical preeclampsia, *J. Proteome Res.,* **9**, 4274–4281.
58. Liu C, Zhang N, Yu H, Chen Y, Liang Y, Deng H, Zhang Z (2011) Proteomic analysis of human serum for finding pathogenic factors and potential biomarkers in preeclampsia, *Placenta,* **32**, 168–174.
59. Auer J, Camoin L, Guillonneau F, Rigourd V, Chelbi ST, Leduc M, Laparre J, Mignot TM, Vaiman D (2010) Serum profile in preeclampsia and intra-uterine growth restriction revealed by iTRAQ technology, *J. Proteomics,* **73**, 1004–1017.
60. Blumenstein M, McMaster MT, Black MA, Wu S, Prakash R, Cooney J, McCowan LM, Cooper GJ, North RA (2009) A proteomic approach identifies early pregnancy biomarkers for preeclampsia: novel linkages between a predisposition to preeclampsia and cardiovascular disease, *Proteomics,* **9**, 2929–2945.

61. Atay S, Gercel-Taylor C, Kesimer M, Taylor DD (2011) Morphologic and proteomic characterization of exosomes released by cultured extravillous trophoblast cells, *Exp. Cell Res.*, **317**, 1192–1202.

62. Lo YM, Corbetta N, Chamberlain PF, Rai V, Sargent IL, Redman CW, Wainscoat JS (1997) Presence of fetal DNA in maternal plasma and serum, *Lancet*, **350**, 485–487.

63. Simpson JL, Elias S (1993) Isolating fetal cells from maternal blood. Advances in prenatal diagnosis through molecular technology, *JAMA*, **270**, 2357–2361.

64. Hromadnikova I (2012) Extracellular nucleic acids in maternal circulation as potential biomarkers for placental insufficiency, *DNA Cell Biol.*, **31**, 1221–1232.

65. Lo YM, Tein MS, Lau TK, Haines CJ, Leung TN, Poon PM, Wainscoat JS, Johnson PJ, Chang AM, Hjelm NM (1998) Quantitative analysis of fetal DNA in maternal plasma and serum: implications for noninvasive prenatal diagnosis, *Am. J. Hum. Genet.*, **62**, 768–775.

66. Birch L, English CA, O'Donoghue K, Barigye O, Fisk NM, Keer JT (2005) Accurate and robust quantification of circulating fetal and total DNA in maternal plasma from 5 to 41 weeks of gestation, *Clin. Chem.*, **51**, 312–320.

67. Al Nakib M, Desbriere R, Bonello N, Bretelle F, Boubli L, Gabert J, Levy-Mozziconacci A (2009) Total and fetal cell-free DNA analysis in maternal blood as markers of placental insufficiency in intrauterine growth restriction, *Fetal Diagn. Ther.*, **26**, 24–28.

68. Zhong XY, Laivuori H, Livingston JC, Ylikorkala O, Sibai BM, Holzgreve W, Hahn S (2001) Elevation of both maternal and fetal extracellular circulating deoxyribonucleic acid concentrations in the plasma of pregnant women with preeclampsia, *Am. J. Obstet. Gynecol.*, **184**, 414–419.

69. Zhong XY, Gebhardt S, Hillermann R, Tofa KC, Holzgreve W, Hahn S (2005) Parallel assessment of circulatory fetal DNA and corticotropin-releasing hormone mRNA in early- and late-onset preeclampsia, *Clin. Chem.*, **51**, 1730–1733.

70. Zhong XY, Holzgreve W, Hahn S (2002) The levels of circulatory cell free fetal DNA in maternal plasma are elevated prior to the onset of preeclampsia, *Hypertens. Pregnancy*, **21**, 77–83.

71. Muller SP, Bartels I, Stein W, Emons G, Gutensohn K, Kohler M, Legler TJ (2008) The determination of the fetal D status from maternal plasma for decision making on Rh prophylaxis is feasible, *Transfusion*, **48**, 2292–2301.

72. Orozco AF, Jorgez CJ, Horne C, Marquez-Do DA, Chapman MR, Rodgers JR, Bischoff FZ, Lewis DE (2008) Membrane protected apoptotic trophoblast microparticles contain nucleic acids: relevance to preeclampsia, *Am. J. Pathol.*, **173**, 1595–1608.

73. Orozco AF, Bischoff FZ, Horne C, Popek E, Simpson JL, Lewis DE (2006) Hypoxia-induced membrane-bound apoptotic DNA particles: potential mechanism of fetal DNA in maternal plasma, *Ann. NY Acad. Sci.*, **1075**, 57–62.

74. Tjoa ML, Cindrova-Davies T, Spasic-Boskovic O, Bianchi DW, Burton GJ (2006) Trophoblastic oxidative stress and the release of cell-free feto-placental DNA, *Am. J. Pathol.*, **169**, 400–404.

75. Ng EK, Tsui NB, Lau TK, Leung TN, Chiu RW, Panesar NS, Lit LC, Chan KW, Lo YM (2003) mRNA of placental origin is readily detectable in maternal plasma, *Proc. Natl. Acad. Sci. U. S. A.* **100**, 4748–4753.

76. Poon LL, Leung TN, Lau TK, Lo YM (2000) Presence of fetal RNA in maternal plasma, *Clin. Chem.*, **46**, 1832–1834.

77. Farina A, Chan CW, Chiu RW, Tsui NB, Carinci P, Concu M, Banzola I, Rizzo N, Lo YM (2004) Circulating corticotropin-releasing hormone mRNA in maternal plasma: relationship with gestational age and severity of preeclampsia, *Clin. Chem.*, **50**, 1851–1854.

78. Mayor-Lynn K, Toloubeydokhti T, Cruz AC, Chegini N (2011) Expression profile of microRNAs and mRNAs in human placentas from pregnancies complicated by preeclampsia and preterm labor, *Reprod. Sci.*, **18**, 46–56.

79. Rusterholz C, Holzgreve W, Hahn S (2007) Oxidative stress alters the integrity of cell-free mRNA fragments associated with placenta-derived syncytiotrophoblast microparticles, *Fetal Diagn. Ther.*, **22**, 313–317.

80. Kumpel B, King MJ, Sooranna S, Jackson D, Eastlake J, Cheng R, Johnson M (2008) Phenotype and mRNA expression of syncytiotrophoblast microparticles isolated from human placenta, *Ann. NY Acad. Sci.*, **1137**, 144–147.

81. Mouillet JF, Chu T, Sadovsky Y (2011) Expression patterns of placental microRNAs, *Birth Defects Res. A: Clin. Mol. Teratol.*, **91**, 737–743.

82. Liang Y, Ridzon D, Wong L, Chen C (2007) Characterization of microRNA expression profiles in normal human tissues, *BMC Genomics,* **8**, 166.

83. Zhu XM, Han T, Sargent IL, Yin GW, Yao YQ (2009) Differential expression profile of microRNAs in human placentas from preeclamptic

pregnancies vs normal pregnancies, *Am. J. Obstet. Gynecol.*, **200**, 661 e661–667.

84. Miura K, Miura S, Yamasaki K, Higashijima A, Kinoshita A, Yoshiura K, Masuzaki H (2010) Identification of pregnancy-associated microRNAs in maternal plasma, *Clin. Chem.*, **56**, 1767–1771.

85. Morales Prieto DM, Markert UR (2011) MicroRNAs in pregnancy, *J. Reprod. Immunol.*, **88**, 106–111.

86. Donker RB, Mouillet JF, Chu T, Hubel CA, Stolz DB, Morelli AE, Sadovsky Y (2012) The expression profile of C19MC microRNAs in primary human trophoblast cells and exosomes, *Mol. Hum. Reprod.*, **18**, 417–424.

87. Chim SS, Shing TK, Hung EC, Leung TY, Lau TK, Chiu RW, Lo YM (2008) Detection and characterization of placental microRNAs in maternal plasma, *Clin. Chem.*, **54**, 482–490.

88. Kotlabova K, Doucha J, Hromadnikova I (2011) Placental-specific microRNA in maternal circulation—identification of appropriate pregnancy-associated microRNAs with diagnostic potential, *J. Reprod. Immunol.*, **89**, 185–191.

89. Valadi H, Ekstrom K, Bossios A, Sjostrand M, Lee JJ, Lotvall JO (2007) Exosome-mediated transfer of mRNAs and microRNAs is a novel mechanism of genetic exchange between cells, *Nat. Cell Biol.*, **9**, 654–659.

90. Chen X, Liang H, Zhang J, Zen K, Zhang CY (2012) Horizontal transfer of microRNAs: molecular mechanisms and clinical applications, *Protein Cell*, **3**, 28–37.

91. Turchinovich A, Weiz L, Langheinz A, Burwinkel B (2011) Characterization of extracellular circulating microRNA, *Nucleic Acids Res.*, **39**, 7223–7233.

92. Arroyo JD, Chevillet JR, Kroh EM, Ruf IK, Pritchard CC, Gibson DF, Mitchell PS, Bennett CF, Pogosova-Agadjanyan EL, Stirewalt DL, Tait JF, Tewari M (2011) Argonaute2 complexes carry a population of circulating microRNAs independent of vesicles in human plasma, *Proc. Natl. Acad. Sci. U. S. A.* **108**, 5003–5008.

93. Vickers KC, Palmisano BT, Shoucri BM, Shamburek RD, Remaley AT (2011) MicroRNAs are transported in plasma and delivered to recipient cells by high-density lipoproteins, *Nat. Cell Biol.*, **13**, 423–433.

94. Li L, Zhu D, Huang L, Zhang J, Bian Z, Chen X, Liu Y, Zhang CY, Zen K (2012) Argonaute 2 Complexes Selectively Protect the Circulating MicroRNAs in Cell-Secreted Microvesicles, *PLOS ONE,* **7**, e46957.

95. Pineles BL, Romero R, Montenegro D, Tarca AL, Han YM, Kim YM, Draghici S, Espinoza J, Kusanovic JP, Mittal P, Hassan SS, Kim CJ (2007) Distinct subsets of microRNAs are expressed differentially in the human placentas of patients with preeclampsia, *Am. J. Obstet. Gynecol.,* **196**(261), e261–266.

96. Enquobahrie DA, Abetew DF, Sorensen TK, Willoughby D, Chidambaram K, Williams MA (2011) Placental microRNA expression in pregnancies complicated by preeclampsia, *Am. J. Obstet. Gynecol.,* **204**(178), e112–e121.

97. Wang W, Feng L, Zhang H, Hachy S, Satohisa S, Laurent LC, Parast M, Zheng J, Chen DB (2012) Preeclampsia up-regulates angiogenesis-associated microRNA (i.e., miR-17, -20a, and -20b) that target ephrin-B2 and EPHB4 in human placenta, *J. Clin. Endocrinol. Metab.,* **97**, E1051–E1059.

98. Hu Y, Li P, Hao S, Liu L, Zhao J, Hou Y (2009) Differential expression of microRNAs in the placentae of Chinese patients with severe pre-eclampsia, *CCLM/FESCC,* **47**, 923–929.

99. Zhang Y, Fei M, Xue G, Zhou Q, Jia Y, Li L, Xin H, Sun S (2012) Elevated levels of hypoxia-inducible microRNA-210 in pre-eclampsia: new insights into molecular mechanisms for the disease, *J. Cell. Mol. Med.,* **16**, 249–259.

100. Li P, Guo W, Du L, Zhao J, Wang Y, Liu L, Hu Y, Hou Y (2012) MicroRNA-29b contributes to pre-eclampsia through its effects on apoptosis, invasion and angiogenesis of trophoblast cells, *Clin. Sci. (Lond.),* **124**, 27–40.

101. Wu L, Zhou H, Lin H, Qi J, Zhu C, Gao Z, Wang H (2012) Circulating microRNAs are elevated in plasma from severe preeclamptic pregnancies, *Reproduction,* **143**, 389–397.

102. Hromadnikova I, Kotlabova K, Doucha J, Dlouha K, Krofta L (2012) Absolute and relative quantification of placenta-specific micrornas in maternal circulation with placental insufficiency-related complications, *JMD,* **14**, 160–167.

103. Chamley LW, Chen Q, Ding J, Stone PR, Abumaree M (2011) Trophoblast deportation: just a waste disposal system or antigen sharing? *J. Reprod. Immunol.,* **88**, 99–105.

104. Abumaree MH, Chamley LW, Badri M, El-Muzaini MF (2012) Trophoblast debris modulates the expression of immune proteins in macrophages: a key to maternal tolerance of the fetal allograft? *J. Reprod. Immunol.*, **94**, 131–141.

105. Dragovic RA, Gardiner C, Brooks AS, Tannetta DS, Ferguson DJ, Hole P, Carr B, Redman CW, Harris AL, Dobson PJ, Harrison P, Sargent IL (2011) Sizing and phenotyping of cellular vesicles using nanoparticle tracking analysis, *Nanomed.: Nanotechnol. Biol. Med.*, **7**, 780–788.

106. Redman CW (2011) Preeclampsia: a multi-stress disorder, *La Revue de medecine interne/fondee. par la Societe nationale francaise de medecine interne*, **32**(Suppl 1), S41–S44.

107. VanWijk MJ, Nieuwland R, Boer K, van der Post JA, VanBavel E, Sturk A (2002) Microparticle subpopulations are increased in preeclampsia: possible involvement in vascular dysfunction? *Am. J. Obstet. Gynecol.*, **187**, 450–456.

108. Bretelle F, Sabatier F, Desprez D, Camoin L, Grunebaum L, Combes V, D'Ercole C, Dignat-George F (2003) Circulating microparticles: a marker of procoagulant state in normal pregnancy and pregnancy complicated by preeclampsia or intrauterine growth restriction, *Thromb. Haemost.*, **89**, 486–492.

109. Lok CA, Jebbink J, Nieuwland R, Faas MM, Boer K, Sturk A, Van Der Post JA (2009) Leukocyte activation and circulating leukocyte-derived microparticles in preeclampsia, *Am. J. Reprod. Immunol.*, **61**, 346–359.

110. Alijotas-Reig J, Palacio-Garcia C, Farran-Codina I, Ruiz-Romance M, Llurba E, Vilardell-Tarres M (2011) Circulating cell-derived microparticles in severe preeclampsia and in fetal growth restriction, *Am. J. Reprod. Immunol.*, **67**, 140–151.

111. Dragovic RA, Southcombe JII, Tannetta DS, Redman CW, Sargent, IL (2013) Multicolor flow cytometry and nanoparticle tracking analysis of extracellular vesicles in the plasma of normal pregnant and pre-eclamptic women, *Biol Reprod.*, [Epub ahead of print].

112. Meziani F, Tesse A, David E, Martinez MC, Wangesteen R, Schneider F, Andriantsitohaina R (2006) Shed membrane particles from preeclamptic women generate vascular wall inflammation and blunt vascular contractility, *Am. J. Pathol.*, **169**, 1473–1483.

113. Gonzalez-Quintero VH, Jimenez JJ, Jy W, Mauro LM, Hortman L, O'Sullivan MJ, Ahn Y (2003) Elevated plasma endothelial microparticles in preeclampsia, *Am. J. Obstet. Gynecol.*, **189**, 589–593.

114. Gonzalez-Quintero VH, Smarkusky LP, Jimenez JJ, Mauro LM, Jy W, Hortsman LL, O'Sullivan MJ, Ahn YS (2004) Elevated plasma endothelial microparticles: preeclampsia versus gestational hypertension, *Am. J. Obstet. Gynecol.,* **191**, 1418–1424.

115. Lok CA, Snijder KS, Nieuwland R, Van Der Post JA, de Vos P, Faas MM (2012) Microparticles of pregnant women and preeclamptic patients activate endothelial cells in the presence of monocytes, *Am. J. Reprod. Immunol.,* **67**, 206–215.

116. Biro E, Lok CA, Hack CE, van der Post JA, Schaap MC, Sturk A, Nieuwland R (2007) Cell-derived microparticles and complement activation in preeclampsia versus normal pregnancy, *Placenta,* **28**, 928–935.

117. Cocucci E, Racchetti G, Meldolesi J (2009) Shedding microvesicles: artefacts no more, *Trends Cell Biol.,* **19**, 43–51.

Chapter 16

Extracellular Vesicles in Cardiovascular Disease

Anne-Clémence Vion[a,b] **and Chantal M. Boulanger**[a,b]
[a]*U970, INSERM, Paris Cardiovascular Research Center (PARCC), 56 rue leblanc, 75015 Paris, France*
[b]*UMR-S970, Université Paris Descartes, Sorbonne Paris Cité, Paris, France*
Chantal.boulanger@inserm.fr

16.1 Introduction

Cardiovascular diseases (CVDs), including coronary artery disease, stroke, and peripheral arterial disease, are responsible for most of the cardiovascular morbidity and mortality in the world. These diseases are mainly the clinical consequences of atherosclerosis.

Epidemiological studies indicate that the prevalence of atherosclerosis is increasing because of the adoption of a Western lifestyle and will reach epidemic proportions in the coming decades.

Atherosclerosis was first described as a simple proliferative process, with passive deposition of lipid debris on the arterial wall. Within the past three decades, atherosclerosis emerged as a chronic inflammatory disease, involving increased endothelial cell

Extracellular Vesicles in Health and Disease
Edited by Paul Harrison, Christopher Gardiner, and Ian L. Sargent

ISBN 978-981-4411-98-1 (Hardcover), 978-981-4411-99-8 (eBook)
www.panstanford.com

permeability and accumulation of low-density lipoproteins (LDLs) in the subendothelial space, followed by the diapedesis of leukocytes and formation of foam cells, migration and proliferation of smooth muscle cells, production of connective tissue, and neovascularization. The culminating point of this slowly developing disease is plaque rupture or erosion, resulting in thrombosis and arterial occlusion.

Activated cells release extracellular vesicles (EVs). This review summarizes their involvement in different pathophysiological processes implicated in CVDs, in particular atherosclerosis.

16.2 Extracellular Vesicles: A Consequence of CVDs

16.2.1 Circulating EV Levels in CVDs

EVs are detectable in the plasma of healthy subject, and their levels are modified in several CVDs.

In patients with cardiovascular risk factors (diabetes, hypertension, metabolic syndrome, hypertriglyceridemia) endothelial and platelet microvesicle (MV) levels are augmented.[1-5] In patients with subclinical atherosclerosis, levels of EVs originating from leukocytes are augmented,[6] and in more advanced stages, such as coronary calcification, acute coronary syndrome, or stable coronary artery diseases, platelet vesicles and endothelial EVs are also elevated.[7-15] Several reports demonstrate that plasma EV levels are an emerging marker of endothelial dysfunction in various diseases like chronic renal failure, diabetes with endothelial dysfunction, obesity, coronary artery disease, or acute endothelial dysfunction.[16-20]

Heart failure and valvular disease modulate endothelial vesicle, platelet vesicle, and leukocyte vesicle levels but also increase the release of tissue factor (TF)-positive EVs. EV levels are also modulated in other vascular diseases such as pre-eclampsia[21,22] (see Chapter 15), rheumatoid arthritis,[23] inflammatory vasculitis,[24,25] and antiphospholipid syndrome.[26,27]

It is important to point out that the plasma levels of these vesicles reflect the balance between their production and their clearance. In CVDs the contribution to each phenomenon to plasma vesicle levels has not been fully investigated. In healthy animals,

EVs are cleared from the circulation by not only the liver and the spleen but also the lung.[28–31] These studies have also identified the major role of lactadherin and Del-1 in the clearance of MVs by macrophages and endothelial cells, respectively.[28,30,32] So far, there is no information on exosome *in vivo* clearance by specific organs, but macropinocytosis appears to be one way of exosome integration within cells.[33] Alternatively, the lactadherin pathway may also contribute to exosome clearance, although this has not been proven *in vivo* yet. Despite knowing the exact contribution of each process on plasma vesicle levels, modification in the plasma levels remains important information about the severity of the diseases in patients suffering from CVDs.

16.2.2 Mechanisms Implicated

Several factors involved in the development of atherosclerotic lesions, such as lipoproteins, cytokines, oxidative stress, or shear stress level, increase *in vitro* the release of EVs from vascular and/or circulating cells (Table 16.1).

On the basis of the results obtained from experiments on platelets, EV formation at the plasma membrane of the cell appears to require some specific modifications. First, intracellular calcium- and caspase-dependent mechanisms are major determinants of the loss of membrane asymmetry.[34] Intracellular calcium levels are modulated by mitochondrial permeability and plasma membrane calcium channel activities. This leads to calcium-dependent up-regulation of scramblase or floppase-/adenosine triphosphate (ATP)-binding cassette A1 and inhibition of translocase/flippase activity–induced exposure of phosphatidylserine (PS) on the outer leaflet.[34] Store-operated calcium entry (SOCE) and $P2X_1$ activation by calcium also contributes to PS exposure. Second, blebbing requires cytoskeletal reorganization. Calcium could contribute to this reorganization in activating calpains and proteases. During apoptosis, bleb formation depends on the actin cytoskeleton and actin-myosin contraction, which is regulated by caspase-3-dependent Rho kinase I and II activation.[35,36] Interestingly, cytoskeletal reorganization and PS exposure could interact as RhoA could modulate SOCE activity and in turn increase PS exposure.[34] A recent study identified TMEM16F (a calcium-dependent scramblase) as a potent regulator of PS exposure in Scott syndrome patients,[37] as these patients have a defect in platelet activation and release of EVs, suggesting a potential role of TMEM16F.

Endothelial vesicle formation and release have received significant attention over the past recent years, and different signaling pathways have been identified depending on the stimuli (Table 16.1).[38] Endothelial vesicle shedding can occur independently of endothelial apoptosis.[39] Curtis et al.[40] identified p38 mitogen–activated protein kinase as a key factor for the shedding of endothelial cells under tumor necrosis factor-α (TNFα) stimulation. In opposition, thrombin stimulation of endothelial cells induces a complex biphasic release of endothelial vesicles. Several different mechanisms concur to vesiculation. First, thrombin binds to its protease-activated receptor-1, followed by Rho kinase II activation. Second, a later pathway involves TRAIL/Apo2L, a cytokine that belongs to the TNFα superfamily,[41] followed by interleukin (IL)-1 release and IL-1 receptor activation.[42] The second phase is characterized by an amplification loop based on the release by endothelial cells stimulated by thrombin of soluble forms of TRAIL and of IL-1 that act in an autocrine or paracrine manner on endothelial cells and stimulate EV shedding. A recent study also demonstrated that costimulation of endothelial cells with CD40 ligand (CD40L) and thrombin leads to an increase in endothelial vesicles 10+ EVs via p38 activation.[43] Interestingly, these findings demonstrate that thrombin-induced activation of endothelial cells leads to the release of EVs of different compositions. Angiotensin II stimulation *in vitro* increase endothelial vesicle release and this is mediated by Rho kinase activation and involves a cholesterol-rich domain such as lipid rafts and caveolaes.[44] Endogenous nitric oxide (NO) appears to play a protective role against endothelial vesicle formation by a mechanism involving tetrahydro-biopterin, as observed after C-reactive protein (CRP) endothelial activation.[45] So far, no other study has addressed the potential effects of NO on endothelial vesicle formation and release. Monocyte macrophages also release MVs under activation (Table 16.1). Endotoxin activates macrophage MV formation via a pathway requiring inducible nitric oxide synthase (NOS) activation.[46] Furthermore, tobacco smoke stimulates the generation of highly procoagulant monocytic MVs in a process requiring extracellular signal-regulated kinase (ERK1/2) activation and caspase-3-dependent apoptosis.[47]

Contrary to platelet MV release, little information is available on the molecular mechanisms leading to the PS exposure and EV formation in erythrocytes. Increases in intracellular calcium and oxidative stress promote erythrocyte MV release.[48,49] In addition, senescence of erythrocytes, as well as of platelets, leads to PS exposure on the cell membrane outer leaflet and release of EVs.[50] In senescent platelets, the release of MVs depends on cytochrome C release and subsequent activation of caspase-3 and Rho kinase I.[50]

Table 16.1 Relevant stimuli for atherosclerosis, leading to EV release from circulating or vascular cells

	EVs from			
Stimuli	**Endothelial cell**	**Platelet**	**Smooth muscle cell**	**Monocyte/ Macrophage**
Cigarette extract	–	–	–	Li et al.[47]
Modified LDL	Nomura et al.[118]	–	Llorente-Cortes et al.	–
HDL cholesterol	Liu et al.[119]	–	–	Liu et al.[119]
Uremic toxin	Faure et al.[120]	–	–	–
Flow conditions	Vion et al.[121]	Nomura et al.[122]	Stampfuss et al.[123]	–
Angiotensin II	Burger et al.[44]	–	–	–
Thrombin	Sapet et al.[39] Simoncini et al.[41] Martinez de Lizarrondo et al.[43]	Barry et al.[124] Dale et al.[125] Chang et al.[126]	–	–
Collagen	–	Barry et al.[124] Boilard et al.[23] Chang et al.[126]	–	–
Homocysteine	Sekula et al.[127]	Olas et al.[128]	–	–

Stimuli	EVs from Endothelial cell	Platelet	Smooth muscle cell	Monocyte/ Macrophage
Activated protein C	Pérez-casal et al.[129]	–	–	Pérez-casal et al.[129]
PAI-1	Brodsky et al.[130]	–	–	–
Proinflammatory cytokines (TNFα–IL-1b) and CRP	Combes et al.[26] Curtis et al.[40] Abid hussein et al.[92] Wang et al.[45]	Nomura et al.[122] Piguet et al.[131]	Schecter et al.[132]	Jungel et al.[133]
Oxidative stress	Vince et al.[134] Szotowski et al.[135]	–	–	–
Fas ligand	–	–	Essayagh et al.[136]	–
PDGF	–	–	Schecter et al.[137]	–

Source: Adapted from Rautou et al., 2010

Note: Only those studies that showed a significant increase in EV release compared to the basal condition have been included in this table.

Abbreviations: HDL, high-density lipoprotein; PDGF, platelet-derived growth factor; PAI, plasminogen activator inhibitor.

16.3 Extracellular Vesicles: Effectors of CVDs

EVs carry different molecules able to modulate cell activation. It was well demonstrated now that these EVs carry different proteins, depending on the cell type they originate from but also depending on signaling events that lead to their release. These proteins could be present inside EVs or at their surface and contribute to specific interactions between cells and EVs. EVs could also carry specific lipids and messenger ribonucleic acid (mRNA), and recent studies demonstrated that they serve as micro-RNA cargos (Fig. 16.1).

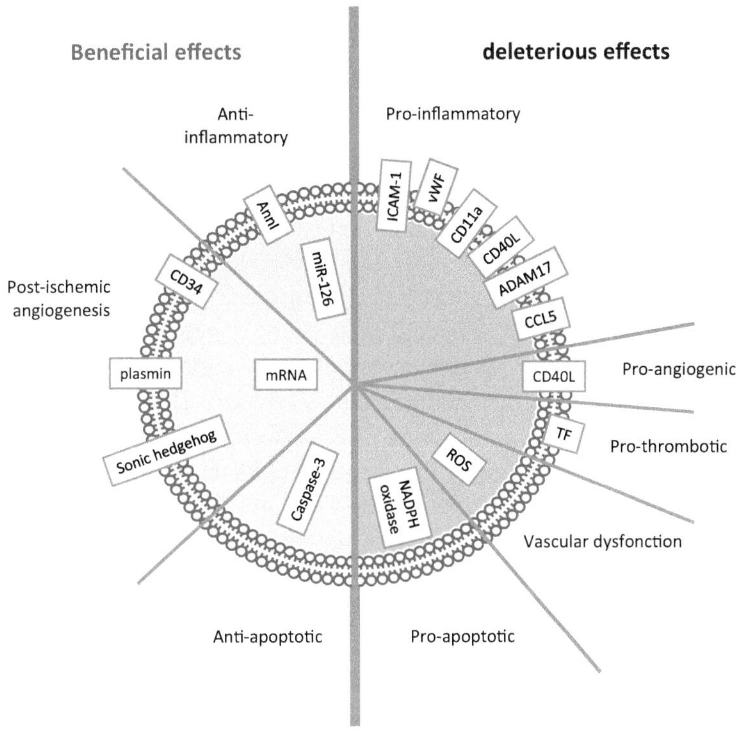

Figure 16.1 Mediators of the biological effect carried by EVs. EVs could have beneficial and deleterious effects, depending on the molecules they carry. *Abbreviations*: AnnI, annexin 1; ROS, reactive oxygen species. (Illustration realized using Servier Medical Art bank of images.)

More than simply "shed membranes," these EVs appear to be able to transfer active biological information to targeted cells and at last modify their activation and orientation in a beneficial or deleterious manner.

16.3.1 Deleterious Effects

In most CVDs, inflammation plays a crucial role. So far, no study has assessed the effect of EVs isolated from CVD patients on inflammatory pathways. Only *in vitro* studies provide us information on the EV effect on inflammation. Circulating EVs from various origins may contribute to inflammation via their influence on cell–cell interactions and cytokine release. They increased the release of IL-6 or IL-8 by endothelial cells and leukocytes.[51–54] They also stimulate monocyte chemotaxis and the expression of adhesion molecules (especially intracellular adhesion molecule [ICAM]-1) at the surface of endothelial cells and CD11a on monocytes.[51,54,55] Increased adhesiveness was partially mediated by arachidonic acid and oxidized phospholipids present in EVs as well as by transfer of the proatherogenic chemokine regulated upon activation, normal T-cells expressed and secreted (RANTES/CCL5).[56,57] Platelet vesicles from apoptotic platelets also promote resident macrophage differentiation to professional phagocytes.[58] Activated endothelial cells could also activate monocytes via the release of von Willebrand factor (VWF)-positive endothelial EV release (Fig. 16.2). These studies essentially used platelet EVs generated *in vitro*, and the *in vivo* relevance of obtained results needs to be investigated. Indeed, a study using a supernatant of activated platelets (containing platelet EVs) demonstrated that this supernatant had no effect on leukocyte adhesion *in vivo* in apolipoprotein E (ApoE) −/− mice and did not increase development of atherosclerotic plaques.[59]

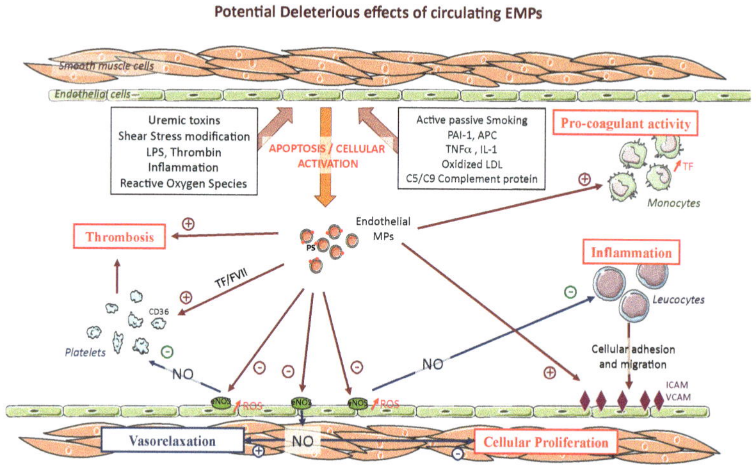

Figure 16.2 Potential deleterious effects of circulating endothelial EVs on CVDs. *Abbreviations*: VCAM, vascular adhesion molecule, PAI: plasminogen activator inhibitor, APC, activated protein C; LPS, lipopolysaccaride (Illustration realized using Servier Medical Art bank of images). Adapted from Amabile et al. (2010), *Sem. Thromb. Hemost.*, **36**, 907–916.

More is known about the effect of EVs accumulating in atherosclerotic plaques. These plaques contained a large amount of EVs, essentially of a leukocyte origin, reflecting the local inflammation.[60] Plaque EVs express major histocompatibility complex class II and costimulation molecules such as CD40L. They are able to stimulate T-lymphocytes that in turn might activate B-lymphocytes to produce specific immunoglobulins against plaque antigens (phosphatydilcholine on oxidized LDL and apoptotic cell membrane, for example).[61,62] Exosomes could also take part in the process of antigen presentation because of the ability of dendritic cell–derived exosomes to stimulate antigen-specific T-cell activation.[63]

Endothelial EVs are also found in atherosclerotic lesions; these EVs possibly originate from microvessels within plaque rather than endothelial cells from the luminal wall of arteries. *In vitro*–generated endothelial EV are capable of inducing plasmacytoid dendritic cell maturation with the production of IL-6 and IL-8; moreover CD4-positive T-cells primed with these dendritic cells produced Th1 cytokines.[64] Altogether, these findings support the hypothesis that local high endothelial EV concentration in atherosclerotic plaques might enhance the local proinflammatory responses. Finally, plaque EVs also carry catalytically active TNFα-converting enzyme (TACE/ADAM17), which could increase the release of TNFα.[65] This proinflammatory effect of plaque EVs is further enhanced by the transfer of functionally integrated ICAM-1 from EVs to the target endothelial membrane, leading to increased monocyte adhesion and recruitment.[66] Cholesterol-induced monocytic EVs, as well as starvation-induced monocytic EVs, have also a proinflammatory effect by increasing leukocyte rolling and adherence to postcapillary venules or increasing monocyte and T-cell infiltration in plaque *in vivo*.[67,68]

Angiogenesis is also a crucial point in atherosclerosis, contributing to plaque instability by promoting neovascularization. Indeed, vulnerable plaques are characterized by an increase in the number of vasa vasorum and more frequent intraplaque hemorrhage. Leroyer et al. demonstrated that CD40L-positive EVs isolated from human plaque promote *in vivo* angiogenesis, and this might contribute to increased plaque vulnerability.[61] Proteolysis of basement membrane matrix cellular components is necessary to promote endothelial invasion into the surrounding interstitial matrix. Human plaque EVs harbor active proteases,[65] and protease activity has also been detected on EVs generated *in vitro* from various cell types.[69–71] Especially, endothelial vesicles carry the proform of several matrix metalloproteinases (MMPs), including MMP-2, MMP-9, and MT1-MMP.[71] Exosomes could also contribute to the evolution of atherosclerotic plaque as exosomes originate from T-cell-induced cholesterol accumulation in human monocytes via activation of phosphatidylserine receptors;[72] and exosomes could also carry proteases.[73]

CVDs are often associated with increased thrombosis,[63] as detailed in other sections of this revue (see Chapter 14). EVs play an important role in this event, providing the anionic phospholipid necessary for the assembly of clotting enzymes and thus promoting thrombin generation. EVs (especially those originated from monocytes) could express TF, a major initiator of blood coagulation. Actions of EVs were not only observed *in vitro* but also *in vivo*, where they favored thrombus propagation.[74,75]

Circulating EVs modulate vascular function in CVDs. Especially, EVs generated *in vitro* or isolated from patients with acute coronary syndrome, pre-eclampsia, chronic renal failure, or metabolic syndrome can induce acute endothelial dysfunction in rat aortic rings, reducing the acetylcholine-induced endothelium-dependent relaxing response[8,22] (Fig. 16.2). In these studies, endothelial EVs appear to be the major effector, but smooth muscle cell EVs, lymphocyte-derived EVs, or endothelial EVs generated *in vitro* appear to have the same effect. This acute endothelial dysfunction could be due to a decrease in NO synthesis and endothelial nitric oxide synthase (eNOS) function or an increase in caveolin-1 (Fig. 16.2). *In vivo*– and *in vitro*–generated EVs can also induce the production of superoxide anion (O^{2-}) by aortic rings or human umbilical vein endothelial cells (HUVECs); this O^{2-} can uncouple eNOS and induce transformation of NO to peroxynitrite. Interestingly this deleterious effect seems to be abolished by antioxidant such as superoxide dismutase. Erythrocyte EVs also have deleterious effect: blood storage increases hemoglobin and erythrocyte EV release that react with and scavenge NO, increasing endothelial dysfunction.[76]

EVs isolated from patients with severe sepsis (highly enriched in endothelial and platelet EVs) have been reported to increase vascular contraction in mice by enhancing thromboxane A2 (TXA2) production. *In vitro* findings confirm this result, demonstrating that plasma EVs could synthesize TXA2 and increase rabbit pulmonary artery contraction. However, in patients with cirrhosis, transfer of phospholipids from plasma EVs to the endothelial layer stimulates cyclooxygenase-1 activity and the release of vasodilatory prostanoids, which in turn impair vasoconstrictive responses to catecholamines, contributing to the vascular hyporeactivity observed in cirrhotic patients.[77]

Finally, both EVs and exosomes are capable of inducing apoptosis, which certainly would contribute to their deleterious effects.[78] For instance, platelet-derived exosomes are implicated in septic vascular dysfunction by inducing endothelial cell apoptosis through generation of peroxynitrite due to their reduced nicotinamide adenine dinucleotide phosphate (NADPH) oxidase activity.[79]

16.3.2 Beneficial Effects

Despite all the deleterious effects of MVs and exosomes observed in various CVDs, numerous studies have demonstrated that certain types of EVs could provide benefic and protective effects against diseases.

First, EVs could have an anti-inflammatory action. Indeed, leukocyte-derived EVs hamper the inflammatory response of leukocytes to lipopolysaccharides and increase the secretion of anti-inflammatory cytokine-transforming growth factor β1.[80] This effect appears to be mediated by the exposure of annexin1 (an anti-inflammatory protein) at the surface of these EVs.[81] *In vitro*, EVs are also taken up by B-cells and monocytes by different mechanisms and modulate the activation of monocytes and B-cells toward an anti-inflammatory phenotype.[82] In CVDs, neutrophils could also release potent anti-inflammatory effectors, called ectosomes, at the earliest stage of inflammation.[80,83]

Secondly, EVs promote postischemic vascularization. Leroyer et al. have demonstrated that ischemic muscles from mice with induced hind-limb ischemia contain more EVs than nonischemic muscles. These EVs mainly originate from endothelial cells and are able to induce *in vitro* differentiation of bone marrow–derived progenitor cells into endothelial cells, increasing their *in vivo* proangiogenic potential.[84] Moreover, in ischemic tissues, EVs could activate endothelial cells by mRNA transfer from endothelial progenitor cells to quiescent endothelial cells, thus stimulating angiogenesis.[85] Degradation of the interstitial matrix is a necessary step for cell migration and vessel formation. EVs may play a role in this process by carrying proteolytic factors, such as plasmin, that could activate MMPs or proteases that belong to the MMP family.[70,71] Finally, the presence of a morphogen sonic hedgehog on EVs originated from human lymphoid cells is able to stimulate *in vitro* capillary formation by increasing cell adhesion and up-regulating proangiogenic factors.[55] Exosomes play an important role in angiogenesis. Exosomes secreted by mesenchymal stem cells reduce myocardial ischemia/reperfusion injury, exosomes derived from cardiac progenitor cells increase the migratory capacity of endothelial cells *in vitro*, and this may contribute to myocardial angiogenesis *in vivo*.[86,87] Moreover, CD34-positive exosomes could promote angiogenesis both *in vivo* and *in vitro*[88] by carrying a sonic hedgehog on their surface.[89]

Interestingly, EV release might be a way by which cells escape from apoptosis. Indeed, in endothelial cells, endothelial vesicle secretion diminishes the caspase-3 level within cells by trapping caspase-3 in EVs. EVs could also carry different antiapoptotic factors, such as protein C receptor or activated protein C (APC), and promote cell survival by induction of the cytoprotective effect.[90–92] Furthermore, endothelial EVs generated under starvation prevent apoptosis of target endothelial cells by inhibiting p38 activation.[93] EVs from monocytes could also have a beneficial effect on vascular function, as they promote endothelial NO production *in vitro* by decreasing caveolin-1 expression.[94]

Finally, it has been reported that EVs carry and shuttle a broad range of premature and mature micro-RNAs.[95,96] These EVs are also enriched in certain types of micro-RNA compared to the micro-RNA profile of the cell they originate from.[97] Some of these micro-RNAs could have atheroprotective effects. In inhibiting the regulator of G-protein signaling 16 (RGS16) protein expression miR-126 increases CXCR4 and CXCL12. Injection of EVs enriched in miR-126 conferred beneficial effects on atherosclerosis in limiting plaque size, increasing plaque stability, and enhancing Sca-1+ progenitor cell recruitment.[98] EVs secreted by KLF2-transduced or shear-stress-stimulated HUVECs are enriched in miR-143/145 and control target gene expression in cocultured smooth muscle cells; these EVs also reduce atherosclerotic lesion formation in the aorta of ApoE −/− mice.[99]

In summary, EVs and exosomes have deleterious effects as well as benefic effects in cardiovascular pathologies. Release of EVs greatly depends of the initial stimuli to generate them, and it was important to point out that most of the studies demonstrating the effects of EVs mainly depend of the model used. Moreover, *in vitro* or *in vivo* generation, cellular origin, and the site of production affect the EV molecular composition (proteins, mRNA, micro-RNA, or phospholipids), therefore modulating their biological effects.

16.4 Microvesicles and Exosomes: Prognostic Value and Therapeutic Target?

16.4.1 Impact of Cardiovascular Therapeutics

Today, most patients suffering from CVDs or presenting risk factors are under treatment. Different studies have demonstrated the effect of therapeutics on the EV level. For example antiplatelet agents (glycoprotein IIb-IIIa inhibitor or thienopyridines), decreasing platelet activation, reduce platelet vesicles released in blood circulation; during acute coronary syndrome, these drugs could also have an effect on leukocyte vesicle generation.[11,100–102] Antihypertensive agents such as β-blockers or calcium channel inhibitors demonstrated an impact on platelet vesicle levels.[103,104] Furthermore, in patients with hypertension and diabetes, losartan (angiotensin II receptor antagonist) plus statins decrease the monocyte derived EV level;[105] in comparison, the same treatment in patients with diabetes and dyslipidemia decreases platelet vesicle levels in plasma.[106] Iserbastan, another angiotensin II receptor antagonist, also decreases the EV level.[107] *In vitro*, statins (inhibiting 3-hydroxy-3-methyl-glutaryl-coenzyme A [HMG-CoA] reductase) also decrease endothelial vesicle release by a mechanism dependent on the Rho kinase pathway.[39,108,109] In diabetic patients, acarbose treatment reduces platelet vesicle levels.[110] Insulin could be implicated in reducing TF at the surface of monocyte-derived EVs, as suggested by *in vitro* findings.[111] Antioxidant agents have also a benefic effect on plasma EV levels: in patients with congestive heart failure, vitamin C prevents procoagulant EV release by decreasing cell apoptosis.[112] Finally, erythropoietin therapy, decreasing blood viscosity and subsequently increasing shear stress, lowered the level of circulating endothelial EVs in patients with end-stage renal failure.[113]

Some studies have clearly identified direct effects of drugs on EV release (statins, vitamin C), but often it was difficult to determine whether or not these molecules act directly on mechanisms governing EV release and/or clearance or act more globally on risk factors and in turn reduce EV release.

16.4.2 Predictive Value of Circulating EVs

As EVs carry different information, depending on the cellular origin and production site, they might be used as biological markers of CVDs. A few studies investigated the potential prognosis potential of these EVs. For the first time, Simak et al. found a correlation between the endothelial EV level, lesion volume, and the clinical outcome in patients with acute ischemic stroke.[12] Amabile et al. reported that CD62e+ endothelial vesicle levels could predict the 1 year outcome of patients suffering from pulmonary hypertension.[114] Nozaki et al. also observed that endothelial vesicle levels predicted the outcome of coronary heart disease in a stable population independently of the C-reactive protein level, the brain natriuretic peptide level, and the Framingham score. Similar results were obtained from chronic renal failure patients where CD31+CD41-endothelial EVs were independent predictors of cardiovascular mortality.[78] Moreover, elevated levels of circulating miR-133a, carried by myocardiac exosomes, can be used as a marker for cardiomyocyte death, and it may have functions in CVDs.[115] Altogether, these data suggest that EVs, and more specifically EV measurement, could be used in the future as biomarkers, contributing to a better stratification of patients with intermediate risk and better identification and treatment of those with a higher risk in this population. EV measurement might also be used to monitor the efficacy of a therapy in clinical practice. However, major advances in this field are currently hampered by the lack of standardized methods of measurement before EV determination can provide useful clinical information for patients.

As biomarkers, EVs allow access to usually inaccessible tissues such as the endothelium. Further research might lead to interventions targeting EV release and function *in vivo* or *in vitro*. *In vivo*, increasing proangiogenic EV release could be beneficial during ischemic events; augmented levels of CD34-positive exosomes could also increase progenitor cell transplantation, thus contributing to therapeutic angiogenesis. *In vitro* generation of specific EVs, containing different types of micro-RNAs or expressing molecules to target a specific cell type, might be used in the future as therapeutic vectors or drug delivery systems.[116,117]

16.5 Conclusion

Increasing numbers of studies highlight the contribution of EVs in different stages of CVDs. However, the extent of their contribution remains uncertain because animal models of atherosclerosis with selective defect in EV generation or uptake are unfortunately lacking. Clearly, this will be possible once the molecular mechanisms governing EV formation and release are dissected out. Then, one could envisage therapeutic avenues designed to either prevent their deleterious effects or promote their repair capacity.

References

1. Sabatier F, Darmon P, Hugel B, Combes V, Sanmarco M, Velut JG, Arnoux D, Charpiot P, Freyssinet JM, Oliver C, Sampol J, Dignat-George F (2002) Type 1 and type 2 diabetic patients display different patterns of cellular microparticles, *Diabetes*, **51**, 2840–2845.
2. Diamant M, Nieuwland R, Pablo RF, Sturk A, Smit JW, Radder JK (2002) Elevated numbers of tissue-factor exposing microparticles correlate with components of the metabolic syndrome in uncomplicated type 2 diabetes mellitus, *Circulation*, **106**, 2442–2447.
3. Preston RA, Jy W, Jimenez JJ, Mauro LM, Horstman LL, Valle M, Aime G, Ahn YS (2003) Effects of severe hypertension on endothelial and platelet microparticles, *Hypertension*, **41**, 211–217.
4. Arteaga RB, Chirinos JA, Soriano AO, Jy W, Horstman L, Jimenez JJ, Mendez A, Ferreira A, de Marchena E, Ahn YS (2006) Endothelial microparticles and platelet and leukocyte activation in patients with the metabolic syndrome, *Am. J. Cardiol.*, **98**, 70–74.
5. Ferreira AC, Peter AA, Mendez AJ, Jimenez JJ, Mauro LM, Chirinos JA, Ghany R, Virani S, Garcia S, Horstman LL, Purow J, Jy W, Ahn YS, de Marchena E (2004) Postprandial hypertriglyceridemia increases circulating levels of endothelial cell microparticles, *Circulation*, **110**, 3599–3603.

6. Chironi G, Simon A, Hugel B, Del Pino M, Gariepy J, Freyssinet JM, Tedgui A (2006) Circulating leukocyte-derived microparticles predict subclinical atherosclerosis burden in asymptomatic subjects, *Arterioscler. Thromb. Vasc. Biol.*, **26**, 2775–2780.
7. Jayachandran M, Litwiller RD, Owen WG, Heit JA, Behrenbeck T, Mulvagh SL, Araoz PA, Budoff MJ, Harman SM, Miller VM (2008) Characterization of blood borne microparticles as markers of premature coronary calcification in newly menopausal women, *Am. J. Physiol.*, **295**, H931–H938.
8. Boulanger CM, Scoazec A, Ebrahimian T, Henry P, Mathieu E, Tedgui A, Mallat Z (2001) Circulating microparticles from patients with myocardial infarction cause endothelial dysfunction, *Circulation*, **104**, 2649–2652.
9. Mallat Z, Benamer H, Hugel B, Benessiano J, Steg PG, Freyssinet JM, Tedgui A (2000) Elevated levels of shed membrane microparticles with procoagulant potential in the peripheral circulating blood of patients with acute coronary syndromes, *Circulation*, **101**, 841–843.
10. Bernal-Mizrachi L, Jy W, Jimenez JJ, Pastor J, Mauro LM, Horstman LL, de Marchena E, Ahn YS (2003) High levels of circulating endothelial microparticles in patients with acute coronary syndromes, *Am. Heart J.*, **145**, 962–970.
11. Morel O, Hugel B, Jesel L, Mallat Z, Lanza F, Douchet MP, Zupan M, Chauvin M, Cazenave JP, Tedgui A, Freyssinet JM, Toti F (2004) Circulating procoagulant microparticles and soluble GPV in myocardial infarction treated by primary percutaneous transluminal coronary angioplasty. A possible role for GPIIb-IIIa antagonists, *J. Thromb. Haemost.*, **2**, 1118–1126.
12. Simak J, Gelderman MP, Yu H, Wright V, Baird AE (2006) Circulating endothelial microparticles in acute ischemic stroke: a link to severity, lesion volume and outcome, *J. Thromb. Haemost.*, **4**, 1296–1302.
13. Jung KH, Chu K, Lee ST, Park HK, Bahn JJ, Kim DH, Kim JH, Kim M, Kun Lee S, Roh JK (2009) Circulating endothelial microparticles as a marker of cerebrovascular disease, *Ann. Neurol.*, **66**, 191–199.
14. Nomura S, Imamura A, Okuno M, Kamiyama Y, Fujimura Y, Ikeda Y, Fukuhara S (2000) Platelet-derived microparticles in patients with arteriosclerosis obliterans: enhancement of high shear-induced microparticle generation by cytokines, *Thromb. Res.*, **98**, 257–268.
15. Tan KT, Tayebjee MH, Lynd C, Blann AD, Lip GY (2005) Platelet microparticles and soluble P selectin in peripheral artery disease: relationship to extent of disease and platelet activation markers, *Ann. Med.*, **37**, 61–66.

16. Heiss C, Amabile N, Lee AC, Real WM, Schick SF, Lao D, Wong ML, Jahn S, Angeli FS, Minasi P, Springer ML, Hammond SK, Glantz SA, Grossman W, Balmes JR, Yeghiazarians Y (2008) Brief secondhand smoke exposure depresses endothelial progenitor cells activity and endothelial function: sustained vascular injury and blunted nitric oxide production, *J. Am. Coll. Cardiol.*, **51**, 1760–1771.
17. Amabile N, Guerin AP, Leroyer A, Mallat Z, Nguyen C, Boddaert J, London GM, Tedgui A, Boulanger CM (2005) Circulating endothelial microparticles are associated with vascular dysfunction in patients with end-stage renal failure, *J. Am. Soc. Nephrol.*, **16**, 3381–3388.
18. Werner N, Wassmann S, Ahlers P, Kosiol S, Nickenig G (2006) Circulating CD31+/annexin V+ apoptotic microparticles correlate with coronary endothelial function in patients with coronary artery disease, *Arterioscler. Thromb. Vasc. Biol.*, **26**, 112–116.
19. Koga H, Sugiyama S, Kugiyama K, Watanabe K, Fukushima H, Tanaka T, Sakamoto T, Yoshimura M, Jinnouchi H, Ogawa H (2005) Elevated levels of VE-cadherin-positive endothelial microparticles in patients with type 2 diabetes mellitus and coronary artery disease, *J. Am. Coll. Cardiol.*, **45**, 1622–1630.
20. Esposito K, Ciotola M, Schisano B, Gualdiero R, Sardelli L, Misso L, Giannetti G, Giugliano D (2006) Endothelial microparticles correlate with endothelial dysfunction in obese women, *J. Clin. Endocrinol. Metab.*, **91**, 3676–3679.
21. Gonzalez-Quintero VH, Jimenez JJ, Jy W, Mauro LM, Hortman L, O'Sullivan MJ, Ahn Y (2003) Elevated plasma endothelial microparticles in preeclampsia, *Am. J. Obstet. Gynecol.*, **189**, 589–593.
22. VanWijk MJ, Nieuwland R, Boer K, van der Post JA, VanBavel E, Sturk A (2002) Microparticle subpopulations are increased in preeclampsia: possible involvement in vascular dysfunction? *Am. J. Obstet. Gynecol.*, **187**, 450–456.
23. Boilard E, Nigrovic PA, Larabee K, Watts GF, Coblyn JS, Weinblatt ME, Massarotti EM, Remold-O'Donnell E, Farndale RW, Ware J, Lee DM (2010) Platelets amplify inflammation in arthritis via collagen-dependent microparticle production, *Science*, **327**, 580–583.
24. Brogan PA, Shah V, Brachet C, Harnden A, Mant D, Klein N, Dillon MJ (2004) Endothelial and platelet microparticles in vasculitis of the young, *Arthritis Rheum.*, **50**, 927–936.
25. Sellam J, Proulle V, Jungel A, Ittah M, Miceli Richard C, Gottenberg JE, Toti F, Benessiano J, Gay S, Freyssinet JM, Mariette X (2009) Increased

levels of circulating microparticles in primary Sjogren's syndrome, systemic lupus erythematosus and rheumatoid arthritis and relation with disease activity, *Arthritis Res. Ther.*, **11**, R156.

26. Combes V, Simon AC, Grau GE, Arnoux D, Camoin L, Sabatier F, Mutin M, Sanmarco M, Sampol J, Dignat-George F (1999) *In vitro* generation of endothelial microparticles and possible prothrombotic activity in patients with lupus anticoagulant, *J. Clin. Invest.*, **104**, 93–102.
27. Dignat-George F, Camoin-Jau L, Sabatier F, Arnoux D, Anfosso F, Bardin N, Veit V, Combes V, Gentile S, Moal V, Sanmarco M, Sampol J (2004) Endothelial microparticles: a potential contribution to the thrombotic complications of the antiphospholipid syndrome, *Thromb. Haemost.*, **91**, 667–673.
28. Dasgupta SK, Abdel-Monem H, Niravath P, Le A, Bellera RV, Langlois K, Nagata S, Rumbaut RE, Thiagarajan P (2009) Lactadherin and clearance of platelet-derived microvesicles, *Blood*, **113**, 1332–1339.
29. Dasgupta SK, Le A, Chavakis T, Rumbaut RE, Thiagarajan P (2012) Developmental endothelial locus-1 (Del-1) mediates clearance of platelet microparticles by the endothelium, *Circulation*, **125**, 1664–1672.
30. Willekens FL, Werre JM, Kruijt JK, Roerdinkholder-Stoelwinder B, Groenen-Dopp YA, van den Bos AG, Bosman GJ, van Berkel TJ Liver (2005) Kupffer cells rapidly remove red blood cell-derived vesicles from the circulation by scavenger receptors, *Blood*, **105**, 2141–2145.
31. Al Faraj A, Gazeau F, Wilhelm C, Devue C, Guerin CL, Pechoux C, Paradis V, Clement O, Boulanger CM, Rautou PE (2012) Endothelial cell-derived microparticles loaded with iron oxide nanoparticles: feasibility of MR imaging monitoring in mice, *Radiology*, **263**, 169–178.
32. Ait-Oufella H, Kinugawa K, Zoll J, Simon T, Boddaert J, Heeneman S, Blanc-Brude O, Barateau V, Potteaux S, Merval R, Esposito B, Teissier E, Daemen MJ, Leseche G, Boulanger C, Tedgui A, Mallat Z (2007) Lactadherin deficiency leads to apoptotic cell accumulation and accelerated atherosclerosis in mice, *Circulation*, **115**, 2168–2177.
33. Feng D, Zhao WL, Ye YY, Bai XC, Liu RQ, Chang LF, Zhou Q, Sui SF (2010) Cellular internalization of exosomes occurs through phagocytosis, *Traffic*, **11**, 675–687.
34. Morel O, Jesel L, Freyssinet JM, Toti F (2011) Cellular mechanisms underlying the formation of circulating microparticles, *Arterioscler. Thromb. Vasc. Biol.*, **31**, 15–26.

35. Coleman ML, Sahai EA, Yeo M, Bosch M, Dewar A, Olson MF (2001) Membrane blebbing during apoptosis results from caspase-mediated activation of ROCK I, *Nat. Cell Biol.*, **3**, 339–345.

36. Sebbagh M, Renvoize C, Hamelin J, Riche N, Bertoglio J, Breard J (2001) Caspase-3-mediated cleavage of ROCK I induces MLC phosphorylation and apoptotic membrane blebbing, *Nat. Cell Biol.*, **3**, 346–352.

37. Suzuki J, Umeda M, Sims PJ, Nagata S (2010) Calcium-dependent phospholipid scrambling by TMEM16F, *Nature*, **468**, 834–838.

38. Dignat-George F, Boulanger CM (2011) The many faces of endothelial microparticles, *Arterioscler. Thromb. Vasc. Biol.*, **31**, 27–33.

39. Sapet C, Simoncini S, Loriod B, Puthier D, Sampol J, Nguyen C, Dignat-George F, Anfosso F (2006) Thrombin-induced endothelial microparticle generation: identification of a novel pathway involving ROCK-II activation by caspase-2, *Blood*, **108**, 1868–1876.

40. Curtis AM, Wilkinson PF, Gui M, Gales TL, Hu E, Edelberg JM (2009) p38 mitogen-activated protein kinase targets the production of proinflammatory endothelial microparticles, *J. Thromb. Haemost.*, **7**, 701–709.

41. Simoncini S, Njock MS, Robert S, Camoin-Jau L, Sampol J, Harle JR, Nguyen C, Dignat-George F, Anfosso F (2009) TRAIL/Apo2L mediates the release of procoagulant endothelial microparticles induced by thrombin *in vitro*: a potential mechanism linking inflammation and coagulation, *Circ. Res.*, **104**, 943–951.

42. Leroyer AS, Anfosso F, Lacroix R, Sabatier F, Simoncini S, Njock SM, Jourde N, Brunet P, Camoin-Jau L, Sampol J, Dignat-George F (2010) Endothelial-derived microparticles: biological conveyors at the crossroad of inflammation, thrombosis and angiogenesis, *Thromb. Haemost.*, **104**, 456–463.

43. Martinez de Lizarrondo S, Roncal C, Calvayrac O, Rodriguez C, Varo N, Purroy A, Lorente L, Rodriguez JA, Doeuvre L, Hervas-Stubbs S, Angles-Cano E, Paramo JA, Martinez-Gonzalez J, Orbe J (2012) Synergistic effect of thrombin and CD40 ligand on endothelial matrix metalloproteinase-10 expression and microparticle generation *in vitro* and *in vivo*, *Arterioscler. Thromb. Vasc. Biol.*, **32**, 1477–1487.

44. Burger D, Montezano AC, Nishigaki N, He Y, Carter A, Touyz RM (2011) Endothelial microparticle formation by angiotensin II is mediated via Ang II receptor type I/NADPH oxidase/ Rho kinase pathways targeted to lipid rafts, *Arterioscler. Thromb. Vasc. Biol.*, **31**, 1898–1907.

45. Wang JM, Wang Y, Huang JY, Yang Z, Chen L, Wang LC, Tang AL, Lou ZF, Tao J (2007) C-reactive protein-induced endothelial microparticle

generation in HUVECs is related to BH4-dependent NO formation, *J. Vasc. Res.*, **44**, 241–248.

46. Gauley J, Pisetsky DS (2010) The release of microparticles by RAW 264.7 macrophage cells stimulated with TLR ligands, *J. Leukocyte Biol.*, **87**, 1115–1123.
47. Li M, Yu D, Williams KJ, Liu ML (2010) Tobacco smoke induces the generation of procoagulant microvesicles from human monocytes/macrophages, *Arterioscler. Thromb. Vasc. Biol.*, **30**, 1818–1824.
48. Comfurius P, Senden JM, Tilly RH, Schroit AJ, Bevers EM, Zwaal RF (1990) Loss of membrane phospholipid asymmetry in platelets and red cells may be associated with calcium-induced shedding of plasma membrane and inhibition of aminophospholipid translocase, *Biochim. Biophys. Acta*, **1026**, 153–160.
49. Freikman I, Amer J, Cohen JS, Ringel I, Fibach E (2008) Oxidative stress causes membrane phospholipid rearrangement and shedding from RBC membranes—an NMR study, *Biochim. Biophys. Acta*, **1778**, 2388–2394.
50. Dasgupta SK, Argaiz ER, Mercado JE, Maul HO, Garza J, Enriquez AB, Abdel-Monem H, Prakasam A, Andreeff M, Thiagarajan P (2010) Platelet senescence and phosphatidylserine exposure, *Transfusion*, **50**, 2167–2175.
51. Barry OP, Pratico D, Savani RC, FitzGerald GA (1998) Modulation of monocyte-endothelial cell interactions by platelet microparticles, *J. Clin. Invest.*, **102**, 136–144.
52. Mesri M, Altieri DC (1998) Endothelial cell activation by leukocyte microparticles, *J. Immunol.*, **161**, 4382–4387.
53. Mesri M, Altieri DC (1999) Leukocyte microparticles stimulate endothelial cell cytokine release and tissue factor induction in a JNK1 signaling pathway, *J. Biol. Chem*, **274**, 23111–23118.
54. Nomura S, Tandon NN, Nakamura T, Cone J, Fukuhara S, Kambayashi J (2001) High-shear-stress-induced activation of platelets and microparticles enhances expression of cell adhesion molecules in THP-1 and endothelial cells, *Atherosclerosis*, **158**, 277–287.
55. Soleti R, Benameur T, Porro C, Panaro MA, Andriantsitohaina R, Martinez MC (2009) Microparticles harboring Sonic Hedgehog promote angiogenesis through the upregulation of adhesion proteins and proangiogenic factors, *Carcinogenesis*, **30**, 580–588.
56. Huber J, Vales A, Mitulovic G, Blumer M, Schmid R, Witztum JL, Binder BR, Leitinger N (2002) Oxidized membrane vesicles and blebs from

apoptotic cells contain biologically active oxidized phospholipids that induce monocyte-endothelial interactions, *Arterioscler. Thromb. Vasc. Biol.*, **22**, 101–107.

57. Mause SF, von Hundelshausen P, Zernecke A, Koenen RR, Weber C (2005) Platelet microparticles: a transcellular delivery system for RANTES promoting monocyte recruitment on endothelium, *Arterioscler. Thromb. Vasc. Biol.*, **25**, 1512–1518.

58. Vasina EM, Cauwenberghs S, Feijge MA, Heemskerk JW, Weber C, Koenen RR (2011) Microparticles from apoptotic platelets promote resident macrophage differentiation, *Cell Death Dis.*, **2**, e211.

59. Huo Y, Schober A, Forlow SB, Smith DF, Hyman MC, Jung S, Littman DR, Weber C, Ley K (2003) Circulating activated platelets exacerbate atherosclerosis in mice deficient in apolipoprotein E, *Nat. Med.*, **9**, 61–67.

60. Leroyer AS, Isobe H, Leseche G, Castier Y, Wassef M, Mallat Z, Binder BR, Tedgui A, Boulanger CM (2007) Cellular origins and thrombogenic activity of microparticles isolated from human atherosclerotic plaques, *J. Am. Coll. Cardiol.*, **49**, 772–777.

61. Leroyer AS, Rautou PE, Silvestre JS, Castier Y, Leseche G, Devue C, Duriez M, Brandes RP, Lutgens E, Tedgui A, Boulanger CM (2008) CD40 ligand+ microparticles from human atherosclerotic plaques stimulate endothelial proliferation and angiogenesis a potential mechanism for intraplaque neovascularization, *J. Am. Coll. Cardiol.*, **52**, 1302–1311.

62. Mayr M, Grainger D, Mayr U, Leroyer AS, Leseche G, Sidibe A, Herbin O, Yin X, Gomes A, Madhu B, Griffiths JR, Xu Q, Tedgui A, Boulanger CM (2009) Proteomics, metabolomics, immunomics on microparticles derived from human atherosclerotic plaques, *Circ. Cardiovasc. Genet.*, **2**, 379–388.

63. Thery C, Duban L, Segura E, Veron P, Lantz O, Amigorena S (2002) Indirect activation of naive CD4+ T cells by dendritic cell-derived exosomes, *Nat. Immunol.*, **3**, 1156–1162.

64. Angelot F, Seilles E, Biichle S, Berda Y, Gaugler B, Plumas J, Chaperot L, Dignat-George F, Tiberghien P, Saas P, Garnache-Ottou F (2009) Endothelial cell-derived microparticles induce plasmacytoid dendritic cell maturation: potential implications in inflammatory diseases, *Haematologica*, **94**, 1502–1512.

65. Canault M, Leroyer AS, Peiretti F, Leseche G, Tedgui A, Bonardo B, Alessi MC, Boulanger CM, Nalbone G (2007) Microparticles of human atherosclerotic plaques enhance the shedding of the tumor necrosis factor-alpha converting enzyme/ADAM17 substrates, tumor necrosis

factor and tumor necrosis factor receptor-1, *Am. J. Pathol.*, **171**, 1713–1723.

66. Rautou PE, Leroyer AS, Ramkhelawon B, Devue C, Duflaut D, Vion AC, Nalbone G, Castier Y, Leseche G, Lehoux S, Tedgui A, Boulanger CM (2011) Microparticles from human atherosclerotic plaques promote endothelial ICAM-1-dependent monocyte adhesion and transendothelial migration, *Circ. Res.*, **108**, 335–343.
67. Liu ML, Scalia R, Mehta JL, Williams KJ (2012) Cholesterol-induced membrane microvesicles as novel carriers of damage-associated molecular patterns: mechanisms of formation, action, detoxification, *Arterioscler. Thromb. Vasc. Biol.*, **32**(9), 2113–2121.
68. Hoyer FF, Giesen MK, Franca C, Lutjohann D, Nickenig G, Werner N (2012) Monocytic microparticles promote atherogenesis by modulating inflammatory cells in mice, *J. Cell Mol. Med.*, **16**(11), 2777–2788.
69. Dashevsky O, Varon D, Brill A (2009) Platelet-derived microparticles promote invasiveness of prostate cancer cells via upregulation of MMP-2 production, *Int. J. Cancer*, **124**, 1773–1777.
70. Lacroix R, Sabatier F, Mialhe A, Basire A, Pannell R, Borghi H, Robert S, Lamy E, Plawinski L, Camoin-Jau L, Gurewich V, Angles-Cano E, Dignat-George F (2007) Activation of plasminogen into plasmin at the surface of endothelial microparticles: a mechanism that modulates angiogenic properties of endothelial progenitor cells *in vitro*, *Blood*, **110**, 2432–2439.
71. Taraboletti G, D'Ascenzo S, Borsotti P, Giavazzi R, Pavan A, Dolo V (2002) Shedding of the matrix metalloproteinases MMP-2, MMP-9, and MT1-MMP as membrane vesicle-associated components by endothelial cells, *Am. J. Pathol.*, **160**, 673–680.
72. Zakharova L, Svetlova M, Fomina AF (2007) T cell exosomes induce cholesterol accumulation in human monocytes via phosphatidylserine receptor, *J. Cell Physiol.*, **212**, 174–181.
73. Escrevente C, Morais VA, Keller S, Soares CM, Altevogt P, Costa J (2008) Functional role of N-glycosylation from ADAM10 in processing, localization and activity of the enzyme, *Biochim. Biophys. Acta*, **1780**, 905–913.
74. Biro E, Sturk-Maquelin KN, Vogel GM, Meuleman DG, Smit MJ, Hack CE, Sturk A, Nieuwland R (2003) Human cell-derived microparticles promote thrombus formation *in vivo* in a tissue factor-dependent manner, *J. Thromb. Haemost.*, **1**, 2561–2568.
75. Chou J, Mackman N, Merrill-Skoloff G, Pedersen B, Furie BC, Furie B (2004) Hematopoietic cell-derived microparticle tissue factor

contributes to fibrin formation during thrombus propagation, *Blood*, **104**, 3190–3197.

76. Donadee C, Raat NJ, Kanias T, Tejero J, Lee JS, Kelley EE, Zhao X, Liu C, Reynolds H, Azarov I, Frizzell S, Meyer EM, Donnenberg AD, Qu L, Triulzi D, Kim-Shapiro DB, Gladwin MT (2011) Nitric oxide scavenging by red blood cell microparticles and cell-free hemoglobin as a mechanism for the red cell storage lesion, *Circulation*, **124**, 465–476.
77. Rautou PE, Bresson J, Sainte-Marie Y, Vion AC, Paradis V, Renard JM, Devue C, Heymes C, Letteron P, Elkrief L, Lebrec D, Valla D, Tedgui A, Moreau R, Boulanger CM (2012) Abnormal plasma microparticles impair vasoconstrictor responses in patients with cirrhosis, *Gastroenterology*, **143**, 166–176, e166.
78. Amabile N, Guerin AP, Tedgui A, Boulanger CM, London GM (2012) Predictive value of circulating endothelial microparticles for cardiovascular mortality in end-stage renal failure: a pilot study, *Nephrol. Dial. Transplant.*, **27**, 1873–1880.
79. Gambim MH, do Carmo Ade O, Marti L, Verissimo-Filho S, Lopes LR, Janiszewski M (2007) Platelet-derived exosomes induce endothelial cell apoptosis through peroxynitrite generation: experimental evidence for a novel mechanism of septic vascular dysfunction, *Crit Care*, **11**, R107.
80. Gasser O, Schifferli JA (2004) Activated polymorphonuclear neutrophils disseminate anti-inflammatory microparticles by ectocytosis, *Blood*, **104**, 2543–2548.
81. Dalli J, Norling LV, Renshaw D, Cooper D, Leung KY, Perretti M (2008) Annexin 1 mediates the rapid anti-inflammatory effects of neutrophil-derived microparticles, *Blood*, **112**, 2512–2519.
82. Koppler B, Cohen C, Schlondorff D, Mack M (2006) Differential mechanisms of microparticle transfer toB cells and monocytes: anti-inflammatory propertiesof microparticles, *Eur J. Immunol.*, **36**, 648–660.
83. Angelillo-Scherrer A (2012) Leukocyte-derived microparticles in vascular homeostasis, *Circ. Res.*, **110**, 356–369.
84. Leroyer AS, Ebrahimian TG, Cochain C, Recalde A, Blanc-Brude O, Mees B, Vilar J, Tedgui A, Levy BI, Chimini G, Boulanger CM, Silvestre JS (2009) Microparticles from ischemic muscle promotes postnatal vasculogenesis, *Circulation*, **119**, 2808–2817.
85. Deregibus MC, Cantaluppi V, Calogero R, Lo Iacono M, Tetta C, Biancone L, Bruno S, Bussolati B, Camussi G (2007) Endothelial progenitor cell

derived microvesicles activate an angiogenic program in endothelial cells by a horizontal transfer of mRNA, *Blood*, **110**, 2440–2448.

86. Lai RC, Arslan F, Lee MM, Sze NS, Choo A, Chen TS, Salto-Tellez M, Timmers L, Lee CN, El Oakley RM, Pasterkamp G, de Kleijn DP, Lim SK (2010) Exosome secreted by MSC reduces myocardial ischemia/reperfusion injury, *Stem Cell Res.*, **4**, 214–222.
87. Vrijsen KR, Sluijter JP, Schuchardt MW, van Balkom BW, Noort WA, Chamuleau SA, Doevendans PA (2010) Cardiomyocyte progenitor cell-derived exosomes stimulate migration of endothelial cells, *J. Cell Mol. Med*, **14**, 1064–1070.
88. Sahoo S, Klychko E, Thorne T, Misener S, Schultz KM, Millay M, Ito A, Liu T, Kamide C, Agrawal H, Perlman H, Qin G, Kishore R, Losordo DW (2011) Exosomes from human CD34(+) stem cells mediate their proangiogenic paracrine activity, *Circ. Res.*, **109**, 724–728.
89. Mackie AR, Klyachko E, Thorne T, Schultz KM, Millay M, Ito A, Kamide CE, Liu T, Gupta R, Sahoo S, Misener S, Kishore R, Losordo DW (2012) Sonic Hedgehog-Modified Human CD34+ Cells Preserve Cardiac Function After Acute Myocardial Infarction, *Circ. Res.*, **111**, 312–321.
90. Boing AN, Hau CM, Sturk A, Nieuwland R (2008) Platelet microparticles contain active caspase 3, *Platelets*, **19**, 96–103.
91. Abid Hussein MN, Boing AN, Sturk A, Hau CM, Nieuwland R (2007) Inhibition of microparticle release triggers endothelial cell apoptosis and detachment, *Thromb. Haemost.*, **98**, 1096–1107.
92. Abid Hussein MN, Nieuwland R, Hau CM, Evers LM, Meesters EW, Sturk A (2005) Cell-derived microparticles contain caspase 3 *in vitro* and *in vivo*, *J. Thromb. Haemost.*, **3**, 888–896.
93. Jansen F, Yang X, Hoyer FF, Paul K, Heiermann N, Becher MU, Hussein NA, Kebschull M, Bedorf J, Franklin BS, Latz E, Nickenig G, Werner N (2012) Endothelial microparticle uptake in target cells is annexin i/phosphatidylserine receptor dependent and prevents apoptosis, *Arterioscler. Thromb. Vasc. Biol.*, **32**, 1925–1935.
94. Mastronardi ML, Mostefai HA, Soleti R, Agouni A, Martinez MC, Andriantsitohaina R (2011) Microparticles from apoptotic monocytes enhance nitrosative stress in human endothelial cells, *Fundam. Clin. Pharmacol.*, **25**, 653–660.
95. Hunter MP, Ismail N, Zhang X, Aguda BD, Lee EJ, Yu L, Xiao T, Schafer J, Lee ML, Schmittgen TD, Nana-Sinkam SP, Jarjoura D, Marsh CB (2008) Detection of microRNA expression in human peripheral blood microvesicles, *PLOS ONE*, **3**, e3694.
96. Eldh M, Ekstrom K, Valadi H, Sjostrand M, Olsson B, Jernas M, Lotvall J (2010) Exosomes communicate protective messages during oxidative stress, possible role of exosomal shuttle RNA, *PLOS ONE*, **5**, e15353.

97. Diehl P, Fricke A, Sander L, Stamm J, Bassler N, Htun N, Ziemann M, Helbing T, El-Osta A, Jowett JB, Peter K (2012) Microparticles: major transport vehicles for distinct microRNAs in circulation, *Cardiovasc. Res.*, **93**, 633–644.

98. Zernecke A, Bidzhekov K, Noels H, Shagdarsuren E, Gan L, Denecke B, Hristov M, Koppel T, Jahantigh MN, Lutgens E, Wang S, Olson EN, Schober A, Weber C (2009) Delivery of microRNA-126 by apoptotic bodies induces CXCL12-dependent vascular protection, *Sci. Signal*, **2**, ra81.

99. Hergenreider E, Heydt S, Treguer K, Boettger T, Horrevoets AJ, Zeiher AM, Scheffer MP, Frangakis AS, Yin X, Mayr M, Braun T, Urbich C, Boon RA, Dimmeler S (2012) Atheroprotective communication between endothelial cells and smooth muscle cells through miRNAs, *Nat. Cell Biol.*, **14**, 249–256.

100. Kagawa H, Nomura S, Nagahama M, Ozaki Y, Fukuhara S (1999) Effect of ticlopidine on platelet-derived microparticles in patients with connective tissue diseases, *Haemostasis*, **29**, 255–261.

101. Shouzu A, Nomura S, Omoto S, Hayakawa T, Nishikawa M, Iwasaka T (2004) Effect of ticlopidine on monocyte-derived microparticles and activated platelet markers in diabetes mellitus, *Clin. Appl. Thromb. Hemost.*, **10**, 167–173.

102. Goto S, Tamura N, Li M, Handa M, Ikeda Y, Handa S, Ruggeri ZM (2003) Different effects of various anti-GPIIb-IIIa agents on shear-induced platelet activation and expression of procoagulant activity, *J. Thromb. Haemost.*, **1**, 2022–2030.

103. Nomura S, Shouzu A, Omoto S, Nishikawa M, Iwasaka T (2005) Long-term treatment with nifedipine modulates procoagulant marker and C-C chemokine in hypertensive patients with type 2 diabetes mellitus, *Thromb. Res.*, **115**, 277–285.

104. Nomura S, Shouzu A, Omoto S, Nishikawa M, Iwasaka T (2005) Benidipine improves oxidized LDL-dependent monocyte and endothelial dysfunction in hypertensive patients with type 2 diabetes mellitus, *J. Hum. Hypertens.*, **19**, 551–557.

105. Nomura S, Shouzu A, Omoto S, Nishikawa M, Iwasaka T (2004) Effects of losartan and simvastatin on monocyte-derived microparticles in hypertensive patients with and without type 2 diabetes mellitus, *Clin. Appl. Thromb. Hemost.*, **10**, 133–141.

106. Nomura S, Inami N, Shouzu A, Omoto S, Kimura Y, Takahashi N, Tanaka A, Urase F, Maeda Y, Ohtani H, Iwasaka T (2009) The effects of pitavastatin, eicosapentaenoic acid and combined therapy on platelet-

derived microparticles and adiponectin in hyperlipidemic, diabetic patients, *Platelets*, **20**, 16–22.

107. Georgescu A, Alexandru N, Andrei E, Titorencu I, Dragan E, Tarziu C, Ghiorghe S, Badila E, Bartos D, Popov D (2012) Circulating microparticles and endothelial progenitor cells in atherosclerosis: pharmacological effects of irbesartan, *J. Thromb. Haemost.*, **10**, 680–691.
108. Tramontano AF, O'Leary J, Black AD, Muniyappa R, Cutaia MV, El-Sherif N (2004) Statin decreases endothelial microparticle release from human coronary artery endothelial cells: implication for the Rho-kinase pathway, *Biochem. Biophys. Res. Commun.*, **320**, 34–38.
109. Diamant M, Tushuizen ME, Abid-Hussein MN, Hau CM, Boing AN, Sturk A, Nieuwland R (2008) Simvastatin-induced endothelial cell detachment and microparticle release are prenylation dependent, *Thromb. Haemost.*, **100**, 489–497.
110. Shimazu T, Inami N, Satoh D, Kajiura T, Yamada K, Iwasaka T, Nomura S (2009) Effect of acarbose on platelet-derived microparticles, soluble selectins, and adiponectin in diabetic patients, *J. Thromb. Thrombolysis*, **28**, 429–435.
111. Gerrits AJ, Koekman CA, Yildirim C, Nieuwland R, Akkerman JW (2009) Insulin inhibits tissue factor expression in monocytes, *J. Thromb. Haemost.*, **7**, 198–205.
112. Rossig L, Hoffmann J, Hugel B, Mallat Z, Haase A, Freyssinet JM, Tedgui A, Aicher A, Zeiher AM, Dimmeler S (2001) Vitamin C inhibits endothelial cell apoptosis in congestive heart failure, *Circulation*, **104**, 2182–2187.
113. Boulanger CM, Amabile N, Guerin AP, Pannier B, Leroyer AS, Mallat CN, Tedgui A, London GM (2007) *In vivo* shear stress determines circulating levels of endothelial microparticles in end-stage renal disease, *Hypertension*, **49**, 902–908.
114. Amabile N, Heiss C, Chang V, Angeli FS, Damon L, Rame EJ, McGlothlin D, Grossman W, De Marco T, Yeghiazarians Y (2009) Increased CD62e(+) endothelial microparticle levels predict poor outcome in pulmonary hypertension patients, *J. Heart Lung Transplant.*, **28**, 1081–1086.
115. Kuwabara Y, Ono K, Horie T, Nishi H, Nagao K, Kinoshita M, Watanabe S, Baba O, Kojima Y, Shizuta S, Imai M, Tamura T, Kita T, Kimura T (2011) Increased microRNA-1 and microRNA-133a levels in serum of patients with cardiovascular disease indicate myocardial damage, *Circ. Cardiovasc. Genet.*, **4**, 446–454.
116. Lai RC, Tan SS, Teh BJ, Sze SK, Arslan F, de Kleijn DP, Choo A, Lim SK (2012) Proteolytic Potential of the MSC exosome proteome:

implications for an exosome-mediated delivery of therapeutic proteasome, *Int. J. Proteomics*, **2012**, 971907.

117. Chen TS, Arslan F, Yin Y, Tan SS, Lai RC, Choo AB, Padmanabhan J, Lee CN, de Kleijn DP, Lim SK (2011) Enabling a robust scalable manufacturing process for therapeutic exosomes through oncogenic immortalization of human ESC-derived MSCs, *J. Transl. Med.*, **9**, 47.
118. Nomura S, Shouzu A, Omoto S, Nishikawa M, Iwasaka T, Fukuhara S (2004) Activated platelet and oxidized LDL induce endothelial membrane vesiculation: clinical significance of endothelial cell-derived microparticles in patients with type 2 diabetes, *Clin. Appl. Thromb. Hemost.*, **10**, 205–215.
119. Liu ML, Reilly MP, Casasanto P, McKenzie SE, Williams KJ (2007) Cholesterol enrichment of human monocyte/macrophages induces surface exposure of phosphatidylserine and the release of biologically-active tissue factor-positive microvesicles, *Arterioscler. Thromb. Vasc. Biol.*, **27**, 430–435.
120. Faure V, Dou L, Sabatier F, Cerini C, Sampol J, Berland Y, Brunet P, Dignat-George F (2006) Elevation of circulating endothelial microparticles in patients with chronic renal failure, *J. Thromb. Haemost.*, **4**, 566–573.
121. Vion AC, Ramkelawon B, Loyer X, Chironi G, Devue C, Loirand G, Tedgui A, Lehoux S, Boulanger CM (2012) Laminar shear stress decreases endothelial microparticles release by a nitric oxide- and ABCA1-dependent mechanism, *Arterioscler. Thromb. Vasc. Biol.*, **32**, Abstract No. 208.
122. Nomura S, Nakamura T, Cone J, Tandon NN, Kambayashi J (2000) Cytometric analysis of high shear-induced platelet microparticles and effect of cytokines on microparticle generation, *Cytometry*, **40**, 173–181.
123. Stampfuss JJ, Censarek P, Fischer JW, Schror K, Weber AA (2006) Rapid release of active tissue factor from human arterial smooth muscle cells under flow conditions, *Arterioscler. Thromb. Vasc. Biol.*, **26**, e34–e37.
124. Barry OP, Pratico D, Lawson JA, FitzGerald GA (1997) Transcellular activation of platelets and endothelial cells by bioactive lipids in platelet microparticles. *J. Clin. Invest.*, **99**, 2118–2127.
125. Dale GL, Remenyi G, Friese P (2009) Tetraspanin CD9 is required for microparticle release from coated-platelets, *Platelets*, **20**, 361–366.
126. Chang CP, Zhao J, Wiedmer T, Sims PJ (1993) Contribution of platelet microparticle formation and granule secretion to the transmembrane migration of phosphatidylserine, *J. Biol. Chem.*, **268**, 7171–7178.
127. Sekula M, Janawa G, Stankiewicz E, Stepien E (2011) Endothelial microparticle formation in moderate concentrations of homocysteine and methionine *in vitro*, *Cell Mol. Biol. Lett.*, **16**, 69–78.

128. Olas B, Malinowska J, Rywaniak J (2010) Homocysteine and its thiolactone may promote apoptotic events in blood platelets *in vitro*, *Platelets*, **21**, 533–540.

129. Perez-Casal M, Downey C, Fukudome K, Marx G, Toh CH (2005) Activated protein C induces the release of microparticle-associated endothelial protein C receptor, *Blood*, **105**, 1515–1522.

130. Brodsky SV, Malinowski K, Golightly M, Jesty J, Goligorsky MS (2002) Plasminogen activator inhibitor-1 promotes formation of endothelial microparticles with procoagulant potential, *Circulation*, **106**, 2372–2378.

131. Piguet PF, Vesin C, Da Kan C (2002) Activation of platelet caspases by TNF and its consequences for kinetics, *Cytokine*, **18**, 222–230.

132. Schecter AD, Spirn B, Rossikhina M, Giesen PL, Bogdanov V, Fallon JT, Fisher EA, Schnapp LM, Nemerson Y, Taubman MB (2000) Release of active tissue factor by human arterial smooth muscle cells, *Circ. Res.*, **87**, 126–132.

133. Jungel A, Distler O, Schulze-Horsel U, Huber LC, Ha HR, Simmen B, Kalden JR, Pisetsky DS, Gay S, Distler JH (2007) Microparticles stimulate the synthesis of prostaglandin E(2) via induction of cyclooxygenase 2 and microsomal prostaglandin E synthase 1, *Arthritis Rheum.*, **56**, 3564–3574.

134. Vince RV, Chrismas B, Midgley AW, McNaughton LR, Madden LA (2009) Hypoxia mediated release of endothelial microparticles and increased association of S100A12 with circulating neutrophils, *Oxid. Med., Cell Longev.*, **2**, 2–6.

135. Szotowski B, Antoniak S, Goldin-Lang P, Tran QV, Pels K, Rosenthal P, Bogdanov VY, Borchert HH, Schultheiss HP, Rauch U (2007) Antioxidative treatment inhibits the release of thrombogenic tissue factor from irradiation- and cytokine-induced endothelial cells, *Cardiovasc. Res.*, **73**, 806–812.

136. Essayagh S, Brisset AC, Terrisse AD, Dupouy D, Tellier L, Navarro C, Arnal JF, Sie P (2005) Microparticles from apoptotic vascular smooth muscle cells induce endothelial dysfunction, a phenomenon prevented by beta3-integrin antagonists, *Thromb. Haemost.*, **94**, 853–858.

137. Schecter AD, Giesen PL, Taby O, Rosenfield CL, Rossikhina M, Fyfe BS, Kohtz DS, Fallon JT, Nemerson Y, Taubman MB (1997) Tissue factor expression in human arterial smooth muscle cells. TF is present in three cellular pools after growth factor stimulation, *J. Clin. Invest.*, **100**, 2276–2285.

Chapter 17

Extracellular Vesicles in Immunology

Els J. van der Vlist, Esther N.M Nolte-'t Hoen, and Marca H. M. Wauben
Department of Biochemistry & Cell Biology, Faculty of Veterinary Medicine, Utrecht University, Utrecht, The Netherlands
M.H.M.Wauben@uu.nl

17.1 General Introduction

Intercellular communication via vesicles has several clear advantages. First, the selective sorting of proteins, lipids, and ribonucleic acids (RNAs) into vesicles allows the transfer of unique combinations of multiple signaling molecules within one vehicle.[1] Second, vesicles uniquely allow the intercellular transfer of transmembrane proteins. These proteins can be involved in adhesion, targeting, and/or signaling of the vesicle to the target cell. Third, cytosolic proteins and RNA present in the lumen of the vesicle are sequestered and thereby protected against degradation and shielded for unwanted collateral effects. Fourth, after their release into the extracellular space, vesicles can bind soluble factors from their environment, which can also be transferred to target cells. Finally, cell-derived vesicles not only can be targeted to cells in their direct environment

Extracellular Vesicles in Health and Disease
Edited by Paul Harrison, Christopher Gardiner, and Ian L. Sargent

ISBN 978-981-4411-98-1 (Hardcover), 978-981-4411-99-8 (eBook)
www.panstanford.com

but can also reach target cells over longer distances via circulation in body fluids. Hence, the selective incorporation of different molecules into vesicles and the regulated release and targeting of vesicles ensure the timely spread of custom-made vehicles for intercellular communication.

Many, if not all, cell types can release vesicles, either constitutively or upon specific triggers.[1,2] Vesicles derived from viable cells range in size between 50 nm and 1,000 nm and can be roughly divided into two groups, vesicles that are released into their environment after shedding from the plasma membrane and vesicles derived from endosomal compartments. Multivesicular bodies (MVBs) are late endosomal compartments that contain multiple intraluminal vesicles, which are formed through the inward budding of the limiting membrane of the MVBs. Once proteins are incorporated into intraluminal vesicles they can either be degraded, when the MVB fuses with lysosomes, or be released into the extracellular space when the MVB fuses with the plasma membrane.[3] Both MVB pathways can be simultaneously operational in one cell.[4] The intraluminal vesicles released in the extracellular space are generally referred to as exosomes.[3] Exosomes have been described to be 50–150 nm in size and have a buoyant density in sucrose ranging between 1.10 g/mL and 1.19 g/mL.[1,3] They contain endosomal proteins such as Tsg101 and Alix and tetraspanins such as CD9 and CD63.[1,3] Depending on their parental cell, exosomes can also carry cell-type-specific proteins such as major histocompatibility complex (MHC) II for dendritic cell (DC)-derived exosomes and the T-cell receptor (TCR) for T-cell-derived exosomes.[5,6] DC- and T-cell-derived exosomes also carry a selective set of small RNAs, which differs from the RNA composition of their parental cells.[7,8]

Vesicles that are shed from the plasma membrane, which are often referred to as microvesicles, microparticles, or ectosomes, are very heterogeneous in size (50–1,000 nm).[2,9] Plasma membrane–derived vesicles are formed by budding at particular regions of the plasma membrane. Specific cytosolic and plasma membrane–derived proteins can be sorted into or excluded from these buds.[2] In addition to the formation at MVBs, exosome biogenesis can also occur at specialized plasma membrane microdomains enriched in endosomal proteins, as was shown for T-cells.[10] Although the exact mechanisms of protein sorting into different types of cell-derived vesicles have yet to be elucidated, cholesterol-rich microdomains are believed to be involved in this process.[2,11]

The currently used classification of vesicles is mainly based on the cell type of origin or the presumed subcellular origin of the vesicle. This has resulted in a rather confusing nomenclature including terms such as "prostasomes," "exosomes," "microvesicles," "microparticles," "ectosomes," and "shedding vesicles." Although in theory these different vesicle types can be quite well defined on the basis of their (sub)cellular origin, once released in the extracellular space this classification is very difficult. First, endosomally derived vesicles and plasma membrane–derived vesicles can overlap in size and in buoyant density. Second, up to now there are no markers described that are exclusively present on one vesicle type. Consequently, it is not yet possible to phenotypically classify a released vesicle on the basis of its subcellular origin. Classification based on function is also complicated. The molecular composition of a released vesicle is dynamic and depends not only on the subcellular origin of the vesicle but also on the producing cell type and the activation status of these cells. Hence, the total population of released vesicles is heterogeneous and probably consists of a mixture of different vesicle types.[12,13]

Extracellular vesicles can affect their target cells by signaling via receptor–ligand interactions,[14] via modulatory lipids,[15] or through transfer of regulatory (small) RNAs.[16] Besides the molecular composition of the vesicles, the status of the target cell determines the functional outcome of the crosstalk between vesicles and target cells. Altogether, these different factors might explain the pleiotropic biological functions that have been described for extracellular vesicles.

17.1.1 Methods to Characterize Individual Vesicles and Vesicle Populations

The majority of the vesicles released by cells is smaller than 300 nm.[17,18] As a result, high-resolution imaging techniques, such as electron microscopy (EM) or atomic force microscopy (AFM) (see Chapter 9), are required to visualize individual nanosized vesicles. These high-resolution imaging techniques, however, are not suitable to study large numbers of vesicles needed to reliably characterize vesicle populations that are largely heterogeneous.

Techniques such as western blotting, proteomics, lipidomics, and transcriptomics have been used to characterize bulk isolates of entire vesicle populations. These techniques yield valuable information on the molecular composition at the level of the total vesicle population but are less suited to study the heterogeneity of vesicle populations. Detection of a certain molecule, for instance, will not reveal whether this molecule is present in all vesicles or uniquely expressed in a certain vesicle subset. More importantly, molecules characteristic for a small vesicle subset may be missed in such bulk-based approaches. Furthermore, the lack of "household" markers, which are constitutively sorted into vesicles, precludes reliable quantitative analysis of vesicles using bulk-based analysis techniques.

To analyze different vesicle types or subpopulations within a heterogeneous vesicle population, multiparameter, high-throughput analysis of individual vesicles is required. Nanoparticle tracking analysis (NTA) (see Chapter 11) allows accurate size determination of individual vesicles on the basis of their Brownian motion.[19] However, our recent data indicate that heterogeneously sized vesicle populations are difficult to quantify and characterize using NTA.

Flow cytometry is a powerful technique that can be used to combine high-throughput and multiparameter analysis of cells. However, since most conventional flow cytometers have a lower limit for light scatter detection of 300–500 nm, they are unable to detect nanosized vesicles.[20] We have therefore developed a fluorescence-based high-resolution flow cytometric method that can be used for quantitative and qualitative analysis of individual nanosized vesicles smaller than 300 nm.[21,22] By using bright fluorescent labeling of vesicles, combined with reduced wide-angle light scattering on a high-end flow cytometer, this method is suitable for the analysis of vesicles down to ~70 nm in size. This high-resolution flow cytometric method allows multiparameter analysis of individual vesicles and can thus be used for the analysis of vesicle subsets within heterogeneous vesicle populations. NTA can complement the flow cytometry–based analysis of nanosized vesicles since with the latter method only relative and approximate size information can be obtained, whereas NTA measurements provide more accurate information on the absolute size of vesicles. The recent and future development of methods to analyze large numbers of individual vesicles will allow more insight into the physiological role of changes in the composition of both individual vesicles and vesicle subsets.

17.2 T-Cell-Derived Vesicles

T-cells are key effector cells of the adaptive immune response and are also important in the regulation of B-cell and T-cell responses. $CD4^+$ T-cells recognize their cognate peptide presented in the context of MHC II and provide help to enhance both humoral responses and cellular immunity mediated by $CD8^+$ cytotoxic T-cells (CTL). Depending on numerous polarizing signals $CD4^+$ T-cells can be skewed to many different functional subsets, such as Th1 cells, Th2 cells, Th17 cells, or regulatory T-cells.[23] Each subset secretes different effector molecules and can actively regulate other T-cell subsets. Vesicles derived from T-cells might therefore differ in their molecular composition, depending on the T-cell subset they originate from. Remarkably, only a limited number of studies have focused on T-cell-derived vesicles. Due to large variations in the source of T-cells and exogenously applied stimuli in these studies, the molecular composition of the released vesicles is likely to vary substantially. Additionally, the use of different vesicle isolation procedures further complicates interstudy comparisons. Nevertheless, these studies show that vesicles derived from T-cells can be targeted to many different cell types and may induce a large range of immune modulatory effects (Fig. 17.1).

Table 17.1 gives an overview of the variation in parental T-cell sources, type, and duration of activation stimuli, vesicle isolation protocols, target cells, and functional assays that were used to study the role of T-cell-derived vesicles. In addition, the morphological and molecular characteristics of the different vesicle populations, as well as the methods used to characterize these vesicles, are listed in this table. As cellular sources of vesicles, primary T-cells, different T-cell clones and cell lines, such as Jurkat cells, were used. These cells were triggered to release vesicles by mitogens, such as phorbol myristate acetate (PMA) or phytohemagglutinin (PHA),[24–30] or (antibody-mimicked) cognate interactions with antigen-presenting cells (APCs).[6,12,26,29,31–33] T-cells were allowed to release vesicles for different time periods ranging from 1 h up to 4 days.[6,24,25,30,34] However, the release of T-cell-derived vesicles was mostly limited to 10–24 h of culture.[6,12,26,29,31,33,35,36] In most studies, vesicles were isolated from cell culture supernatants by differential steps of centrifugation. Vesicle sedimentation was most often performed at 100,000 g, but in some studies lower g-forces, such as 15,000 g or 20,000 g, were used.[27–29,32] Sedimentation at lower g-forces yields a vesicle population of larger or aggregated vesicles. Only three studies used density gradient floatation to further characterize the T-cell-derived vesicles.[10,12,32] The size and composition of the isolated vesicles were determined by electron microscopy, western blotting, or flow cytometric analysis (of bead-associated vesicles). TCR/CD3 complexes, MHC I, CD2, and lymphocyte function-associated antigen 1 (LFA-1) were frequently detected on the surface of T-cell-derived vesicles. The proapoptotic protein Fas ligand (FasL) was also detected in multiple studies. For *in vitro* functional studies, isolated vesicles were incubated with different target cells, after which proliferation, cytokine production, and/or apoptosis induction were measured. For *in vivo* functional studies, mice were injected intravenously with isolated T-cell-derived vesicles, after which tumor growth, killing of injected target cells, and the onset of diabetes were assessed (Table 17.1).

17.3 CD8$^+$ T-Cell-Derived Vesicles

Activated CD8$^+$ T-cells kill infected cells or tumor cells by releasing the contents of their secretory lysosomes in the synaptic cleft formed between the CD8$^+$ T-cell and the target cell. Secretory lysosomes are late endosomal compartments that contain lethal components, such as perforin and granzymes. Interestingly, these secretory lysosomes also contain multiple nanosized intraluminal vesicles that carry TCR, CD3, CD8, and MHC I.[34] Peters et al. were the first to describe the intracellular presence and release of CD8$^+$ T-cell-derived vesicles on the basis of elaborate EM studies.[34] At that time, it was hypothesized that perforin and granzymes were released in a vesicle-mediated or vesicle-associated manner.[37] The presence of CD3/TCR, CD8, and possibly other proteins on these nanosized vesicles was proposed to ensure unidirectional delivery of the lethal compounds to the target cell to avoid bystander damage.[38] Later, the proapoptotic transmembrane protein FasL was identified on both the limiting membrane and nanosized intraluminal vesicles.[39] On the basis of this subcellular origin FasL-bearing vesicles are often referred to as exosomes. The receptor of FasL, named Fas (or Fas receptor), is expressed on various immune and nonimmune cells. Binding of FasL to Fas triggers apoptosis of Fas-bearing cells.[40] In this perspective, it is likely that also FasL, present on CD8+ T-cell-derived vesicles, is involved in the vesicle-mediated killing in the lytic synapse. It is unclear whether membrane-bound FasL can also be released via plasma membrane–derived vesicles. However, the sorting of FasL to late endosomal compartments avoids inactivation of FasL, which normally occurs at the cell surface through cleavage of FasL by metalloproteases.[41] Hence, the sorting to the endosomal compartment and the subsequent release of exosomes ensure the delivery of active membrane-bound FasL to target cells.

After the initial description of CD8+ T-cell-derived exosomes, the function of CD8+ T-cell-derived vesicles has been addressed by a few other groups. These studies focused on the total pool of vesicles released by CD8+ T-cells, which may be a mixture of plasma membrane–derived vesicles and exosomes.[6,32,36] More importantly, the vesicles studied were isolated from cell culture supernatants and might therefore be different from the population of CD8+ effector vesicles released in the lytic synapse between CD8+ T-cells and target cells. Indeed, these studies demonstrated that CD8+ T-cell-derived vesicles released in culture supernatants have immune regulatory properties.

17.3.1 Immune-Suppressive Effect of CD8+ T-Cell-Derived Vesicles

The immune-suppressive effect of CD8+ T-cell-derived vesicles became clear from a study that described that vesicles derived from ovalbumin (OVA)-OT-I CD8+ T-cells could inhibit CD8+ cytotoxic responses both *in vitro* and *in vivo*[36] (Fig. 17.1). These vesicles blocked MHC I–OVA–peptide complexes on DCs, thereby reducing *in vitro* CD8+ proliferation without affecting CD4+ proliferation. Besides shielding of MHC I, these CD8+ T-cell-derived vesicles induced apoptosis of OVA-pulsed DCs, which could be prevented by incubating the vesicles with FasL-blocking antibodies before incubation with the DCs. Remarkably, despite the fact that the CD8+ T-cell-derived vesicles induced apoptosis of OVA-pulsed DCs, *in vitro* CD4+ T-cell proliferation remained unaffected in this study.[36] *In vivo* administration of the CD8+ T-cell-derived vesicles inhibited cytotoxic T-cell responses against OVA-expressing tumor cells. Furthermore, administration of vesicles from OT-I CD8+ T-cells were shown to be beneficial in a DC_{OVA}-induced diabetes model, since they inhibited the $CD8^+$ T-cell-mediated killing of pancreatic beta cells.[36] Blocking of MHC I–peptide complexes and apoptosis induction via the Fas/FasL pathway have both been suggested to be involved in this process. It is, however, unclear whether one vesicle population released by activated OVA-specific $CD8^+$ T-cells can perform either functions or that different subpopulations of vesicles are responsible for these effects.

17.3.2 Immune-Activating Effects of $CD8^+$ T-Cell-Derived Vesicles

The immune-activating potential of $CD8^+$ T-cell-derived vesicles became clear from a study describing increased antiviral properties of $CD4^+$ T-cells after binding of $CD8^+$ T-cell-derived vesicles[32] (Fig. 17.1). Both acute and chronically infected $CD4^+$ T-cells showed increased suppression of human immunodeficiency virus-1 (HIV-1) replication *in vitro* upon binding of the $CD8^+$ T-cell-derived vesicles. The poorly understood antiviral mechanism displayed by $CD8^+$ T-cells remained unsolved for a long time, since it was attributed to a soluble protein (e.g., cytokine or chemokine). However, recently Tumne et al. proved that 15,000 g sedimented vesicles, but also isolated homogenized $CD8^+$ T-cell membranes, were able to suppress HIV-1 replication.[32] $CD3^+$ MHC II^+ vesicles enriched in tetraspanins were shown to be responsible for this effect. Although the cell culture supernatant from $CD8^+$ T-cells was able to suppress HIV replication, cell-to-cell contact between $CD8^+$ T-cells and HIV-1-infected $CD4^+$ T-cells was necessary for maximal inhibition of virus replication. This direct cell–cell contact may be a trigger for the $CD8^+$ T-cell to release vesicles or allow more efficient targeting of vesicles to the $CD4^+$ T-cell.

17.4 $CD4^+$ T-Cell-Derived Vesicles

Similar to $CD8^+$ T-cell-derived vesicles released in the lytic synapse between a $CD8^+$ T-cell and a target cell,[34] $CD4^+$ T-cell-derived vesicles can be released in the immune synapse formed between a $CD4^+$ T-cell and an APC.[7] It has been demonstrated that cognate interaction between T-cells and APCs induces the directional release of micro-RNA (miRNA)-containing vesicles from the T-cell toward the interacting APC.[7] These T-cell-derived vesicle-associated miRNAs were functionally transferred to the interacting APC.[7] Besides the direct targeting of T-cell-derived vesicles to the interacting APC, $CD4^+$ T-cell-derived vesicles can also be isolated from the cell culture supernatant. In analogy with $CD8^+$ T-cell-derived vesicles, these nondirectionally released vesicles can play a role in immune regulation via the targeting of bystander cells.

17.4.1 Immune-Suppressive Effects of $CD4^+$ T-Cell-Derived Vesicles

Vesicles released by either suppressive $CD4^+$ T-cells or effector $CD4^+$ T-cells can be involved in the down-regulation of immune responses, via interaction with other T-cells or DCs (Fig. 17.1). Similar to $CD8^+$ T-cells, FasL and another apoptosis-inducing ligand, APO2L, are sorted to the MVBs of $CD4^+$ T-cells.[39] Upon activation, $CD4^+$ T-cells can release APO2L/FasL-bearing vesicles through fusion of MVBs with the plasma membrane.[24,25] These FasL-bearing vesicles can be targeted to nonactivated $CD4^+$ T-cells and induce apoptosis in these cells,[24] which may be a mechanism for dampening of immune responses. The fact that these vesicles were released within 1 h, well before their parental cells showed signs of activation-induced cell death, clearly distinguishes them from apoptotic blebs.

Another type of immune cell that is targeted by $CD4^+$ T-cell-derived vesicles is the APC. Our group was the first to show that proteins derived from rat T-cells, such as TCR and MHC II (present on activated human and rat T-cells), were transferred to APCs in a vesicle-mediated manner.[35] We showed that vesicles derived from anergic $CD4^+$ T-cells, which can suppress T-cell responses similar to regulatory T-cells, could endow APCs with immune-suppressive properties. APCs incubated with these anergic T-cell-derived vesicles down-modulated $CD4^+$ T-cell responses during subsequent antigen exposure.[35] In contrast, vesicles from nonanergic T-cells did not change the T-cell stimulatory capacity of these APCs. The vesicles derived from anergic T-cells exhibited higher levels of MHC II, interleukin (IL)-2-Rα, and intracellular adhesion molecule-1 (ICAM-1). It is, however, unclear whether these molecules are responsible for the immune-suppressive effect or for improved targeting of other immune-suppressive entities, for example, miRNAs.

More recently, it was shown that OVA-specific (OT-II) $CD4^+$ T-cells release FasL-positive vesicles that can bind to OVA-pulsed DCs (DC_{OVA}) in an MHC IIp/TCR- and ICAM-1/LFA-1-dependent manner.[33] DC_{OVA} incubated with vesicles derived from OT-II CD4$^+$ T-cells that had been activated via cognate interactions showed decreased capacity to induce proliferation of naive OT-II CD4$^+$ T-cells *in vitro*. Immunization of mice with DC_{OVA} in combination with OT-II CD4$^+$ T-cell vesicles resulted in severely reduced numbers of OVA-specific CD8$^+$ T-cells as compared to immunization with DC_{OVA} alone or DC_{OVA} together with vesicles from control ConA-stimulated polyclonal CD4$^+$ T-cells. Subsequent cytotoxicity assays demonstrated that *in vivo* administration of vesicles derived from OT-II CD4$^+$ T-cells inhibit DC_{OVA}-stimulated effector CD8$^+$ cytotoxic responses.[33] This effect could be either due to masking of MHC II–peptide complexes on the DC_{OVA} by vesicle-associated OVA-specific TCR, resulting in less efficient CD4$^+$ T-cell-mediated help for CD8$^+$ T-cells. Alternatively, vesicle-associated FasL could have led to apoptosis induction of Fas-expressing DC_{OVA}. Although *in vitro* binding of the T-cell-derived vesicles to DC_{OVA} could be blocked by anti-MHC II antibodies, it was not investigated whether either anti-MHC II or anti-FasL blocking could inhibit the described *in vivo* effects. Yet another study indicates that CD4$^+$ T-cell-derived vesicles could be involved in suppression of immune responses via blocking of MHC–peptide complexes on DCs.[31] DCs that acquired TCR molecules from OVA-specific CD4$^+$ T-cells were shown to be less efficient in activating naive OVA-specific CD4$^+$ T-cells, whereas their ability to prime OVA-specific CD8$^+$ T-cells was unaffected. Addition of an excess of OVA peptide restored the capacity of TCR-bearing DCs to activate OVA-specific CD4$^+$ T-cells. This suggests that transferred TCR complexes had masked the OVA-peptide-bearing MHC II molecules, thereby reducing their accessibility for CD4$^+$ T-cells.[31] Hence, these findings indicate that CD4$^+$ T-cell derived vesicles could play a role in the antigen-specific dampening of immune responses.

17.4.2 Enhancement of Immune Responses by CD4$^+$ T-Cell-Derived Vesicles

Vesicles released by CD4$^+$ T-cells have also been described to be involved in enhancement of immune responses, for example, via activation of innate monocytes and mast cells[27–30] (Fig. 17.1). Incubation of human monocytes with vesicles derived from mitogen-activated CD4$^+$ T-cells induced the production of proinflammatory cytokines such as tumor necrosis factor-α (TNFα) and IL-1β.[27,28,30]

The vesicle-induced production of these cytokines could be inhibited by high-density lipoprotein (HDL) from human serum, whereas the release of other factors such as sIL-1Ra and CCL2 remained unaffected.[27,28] Incubation of monocytes with 100–800 nm sized T-cell-derived vesicles was also shown to lead to the accumulation of cholesterol in the cytosol of these cells. This accumulation was thought to occur through the uptake of the cholesterol-rich vesicles by the monocytes and could be inhibited by blocking the phosphatidylserine (PS) receptor.[30] Since both activated T-cells and monocytes are present in atherosclerotic plaques, it has been suggested that activated T-cell-derived vesicles may promote atherogenesis by inducing intracellular cholesterol accumulation in monocytes and creating a proinflammatory cytokine-rich environment.[30] Whether HDL could plays a role in inhibiting this vesicle-mediated process remains to be elucidated.

When targeted to mast cells, vesicles derived from anti-CD3/anti-CD28-activated T-cells caused degranulation and release of IL-8 and oncostatin M. These effects were also observed after incubation of mast cells with crude membrane fragments derived from these T-cells.[29] Degranulation induced by activated T-cell-derived vesicles was dose dependent and increased over time to almost 20% degranulation after 20 h, whereas incubation with vesicles from nonactivated T-cells led to less than 5% degranulation. This effect may be extracellular signal-regulated kinase (ERK) dependent, since vesicles from activated T-cells were shown to induce phosphorylation of ERK in mast cells and pretreatment of mast cells with an ERK inhibitor significantly reduced the ability of activated T-cell vesicles to induce mast cell degranulation.[29] Since in these studies the T-cell-derived vesicles displayed similar activity as solubilized membranes and membrane fragments,[27–30] it is not clear whether the described effects were mediated by a specifically released population of vesicles or by nonspecifically shed plasma membrane fragments.

17.5 Assessing the Quality and Quantity of T-Cell-Derived Vesicles

The studies described above show that T-cell-derived vesicles are targeted to various types of immune cells and can regulate immune responses at distinct levels (Fig. 17.1). The broad spectrum of target cells and functions of these vesicles may be caused by differences in the molecular composition of vesicles released by functionally distinct T-cell subsets. Few studies have addressed the changes in the population of vesicles released by T-cells upon activation. On the basis of analyses of bulk isolates of T-cell-derived vesicles, it was suggested that TCR engagement is an important trigger for CD4⁺, and possibly also CD8⁺, T-cells to increase vesicle release.[6,26] Using our flow cytometric method for detection and characterization of individual vesicles, we showed that CD4⁺ T-cells indeed released higher numbers of vesicles upon TCR triggering. Moreover, additional costimulatory signals via CD28 further increased the number of released vesicles.[12] Using this technique, we also showed that the population of vesicles released by T-cells is heterogeneous and that T-cells differentially regulate the release of distinct vesicle subpopulations, depending on their activation status.[12]

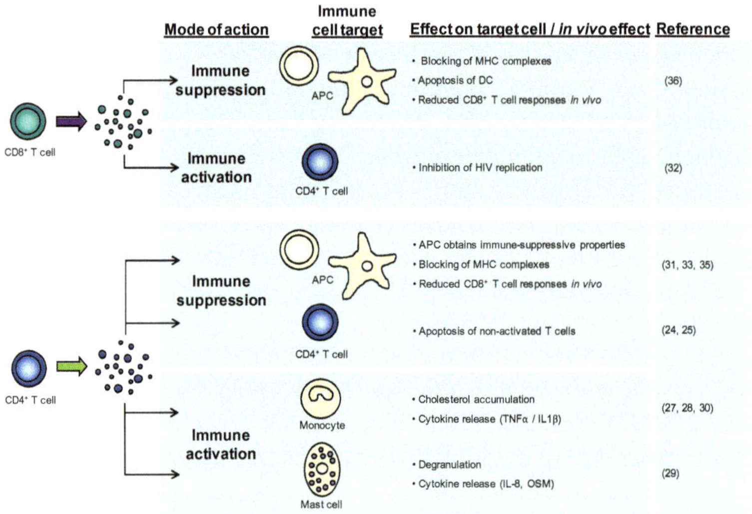

Figure 17.1 Targets and pleiotropic effects of T-cell-derived vesicles. This figure summarizes the immune cell targets and effects that have been described for vesicles derived from CD4⁺ and CD8⁺ T-cells. The effects summarized in this figure are observed *in vitro* unless indicated otherwise.

Hence, different vesicle populations within the total pool of vesicles might be targeted to distinct cell types and exert different functions. To understand the pleiotropic effects of T-cell-derived vesicles it is therefore important to further unravel the triggers for the release of different vesicle subsets and to study their molecular composition and target cells.

17.6 Dendritic Cell-Derived Vesicles

A key-player in immune regulation is the DC. In all tissues DCs constantly sample their environment and present peptides on MHC molecules, which can be recognized by T-cells bearing a TCR specific for this complex (cognate DC–T-cell interaction). DCs that are not activated by danger signals, for example, pathogens or tissue damage, are called immature DCs. Immature DCs do not induce effector T-cell responses and can play an important role in the maintenance of tolerance by inhibition of effector T-cells and/or expansion of regulatory T-cells.[42] Danger signals, such as the presence of bacterial-derived lipopolysaccharide (LPS), can induce DCs to mature and become efficient inducers of effector T- and B-cells.[42,43] The interaction between mature DCs and $CD4^+$ T-cells lies at the heart of the adaptive immune response and is crucial for efficient eradication of invading pathogens and maintenance of homeostasis.[43] Consequently, the functional outcome of cognate DC–T-cell interactions is tightly regulated and depends on the subtype and activation status of the cells engaged in this interaction. The balance between activation and tolerance is regulated via direct cell–cell interactions, secreted soluble factors, and released vesicles.

The first report on vesicle release by APCs came from Raposo et al.[14] who described that B-cells could secrete vesicles containing MHC II. These vesicles were thought to amplify the antigen-presenting capacity of APCs and were quickly recognized as potential therapeutic agents. As a consequence, both the triggers for release and the molecular composition of vesicles released by APCs have been studied far more extensively than T-cell-derived vesicles, and many excellent reviews have been written (see Refs. 1 and 44 for examples). In this section we focus on vesicles derived from DCs, since these professional APCs are key to the initiation of immune responses. The proposed functions and immune target cells of DC-derived vesicles are summarized in Fig. 17.2. Various studies indicate that vesicles derived from DCs can be targeted to $CD4^+$ and $CD8^+$ T-cells, DCs, and natural killer (NK) cells. The effects of DC-derived vesicles have been studied both *in vitro* and *in vivo*. Importantly, the composition and number of vesicles released by DCs depend on the activation (maturation) status of the parental cell and in turn determine how these vesicles modify the function of target cells.

Both immature DCs and mature DCs release vesicles. It was originally postulated that LPS-matured DCs release two- to threefold less vesicles than immature DCs.[5,45] This suggestion was based on Western blot detection of specific proteins and analysis of the total protein content of vesicle populations. However, a recent study and our own unpublished data, in which high-resolution flow cytometry and NTA were used for quantification of vesicles, indicate that LPS-matured DCs release approximate twofold more vesicles than nonactivated DCs.[46] Importantly, we found that also during cognate DC–$CD4^+$ T-cell interactions the release of DC-derived vesicles strongly increased, demonstrating a link between vesicle production and productive T-cell–DC interactions.[4,21]

Besides the number of released vesicles, also the molecular composition of DC-derived vesicles differs depending on the activation status of the parental cell. Vesicles derived from mature DCs were shown to contain higher levels of MHC class II, CD86, and ICAM-1 but lower levels of MFG-E8 compared to vesicles from immature DCs[21,45] (unpublished data). Vesicles derived from immature and mature DCs also differ in their miRNA content.[8] These data indicate that the release and cargo composition of DC-derived vesicles are regulated, which allows a timely spread of tailor-made messengers for intercellular communication.

Vesicles derived from either immature or mature DCs were shown to be efficiently targeted to other DCs[47,48] and to $CD4^+$ T-cells.[48,49] However, the fate of the bound vesicles differs between these target cells. $CD4^+$ T-cells capture DC vesicles via high-affinity LFA-1 and retain the recruited DC-derived vesicles at their cell surface for at least 24 h.[4,48] This suggests that stable receptor–ligand interactions are formed between vesicle-associated membrane proteins and T-cell plasma membrane proteins. In addition to vesicle binding, these receptor–ligand interactions may be involved in signaling toward T-cells and modulation of their function. In contrast, DC-derived vesicles captured by DCs rapidly fuse with the plasma membrane or are endocytosed by these cells.[8] Upon fusion with either the plasma membrane or the endosomal membrane, luminal contents of the vesicles, for example, miRNAs, can be released in the cytosol of DCs and can modulate DC functions.[8] Consequently, the functional outcome of crosstalk between DC-derived vesicles and their target cells depends not only on the molecular composition of the vesicles but also on the target cell type.

17.6.1 Immune-Activating Effects of DC-Derived Vesicles

17.6.1.1 Stimulation of $CD4^+$ T-cells by DC-derived vesicles

The presence of MHC molecules and costimulatory molecules on DC-derived vesicles sparked the idea that these vesicles could have a function in T-cell activation. Indeed several *in vitro* and *in vivo* studies demonstrated the T-cell-activating potential of these vesicles. In general mature DC-derived vesicles are more potent *in vitro* in inducing T-cell responses than vesicles derived from immature DCs.[50] Immature DC-derived vesicles could only activate $CD4^+$ T-cell clones weakly[14,51] and were not able to stimulate naive $CD4^+$ T-cells *in vitro*.[47] However, activation of naive $CD4^+$ T-cells could be induced when immature DC-derived vesicles were recruited by bystander DC.[47,52] The presence of ICAM-1 on DC vesicles was critical for the ability of these vesicles to induce $CD4^+$ T-cell proliferation and might be required for the efficient recruitment of these vesicles to bystander DCs.[45] Activation of naive T-cells was stronger when immature DC-derived vesicles were recruited on mature bystander DCs *versus* immature bystander DCs.[47,52] DC vesicles captured by MHC II–negative DCs could also induce naive $CD4^+$ T-cell proliferation, indicating the functional transfer of MHC II–peptide complexes to the surface of bystander DCs[47] (Fig. 17.2). On the other hand, immature DC vesicles captured by costimulation-deficient (CD80-/CD86-deficient) bystander DCs were unable to induce proliferation of naive $CD4^+$ T-cells.[47] This indicates that, besides MHC II complexes on the DC vesicles, additional costimulatory molecules on the bystander DCs were required to activate naive $CD4^+$ T-cells.[45,47]

In vivo, mature DC-derived vesicles injected into the footpath of mice were shown to bind to CD8α⁺ DCs present in the draining lymph nodes through LFA-1/ICAM-1 interactions. This allowed these CD8α⁺ DCs to prime naive CD4⁺ T-cells *in vivo*,[45,53] leading to rapid rejection of male skin grafts in female mice in an antigen-specific fashion[45] (Fig. 17.2).

Vesicles released by DCs can also induce humoral responses *in vivo*. Immunization of mice with vesicles from diphtheria toxoid (DT)-pulsed DCs induced primary immunoglobulin M (IgM) and IgG anti-DT responses in naive mice.[54] Vesicles derived from mature DT-pulsed DCs showed an enhanced ability to induce primary IgG responses as compared to vesicles from immature DCs.[54] Likewise, immunization with vesicles derived from whole protein (OVA)-pulsed mature DCs were shown to elicit potent Th1 and B-cell responses *in vivo*.[51] Whether the DC-derived vesicles eliciting strong humoral responses *in vivo* via DC targeting and/or B-cell targeting remains to be elucidated.

In conclusion these findings demonstrate that both immature and mature DC-derived vesicles can activate CD4⁺ T-cells but that the requirements with respect to the need for accessory DCs differ. Furthermore, DC-derived vesicle–induced activation of antigen-experienced CD4⁺ T-cells is more easily achieved than priming of naive CD4⁺ T-cells.

17.6.1.2 Stimulation of CD8⁺ T-cells by DC-derived vesicles

Similar to CD4⁺ T-cell activation, vesicles derived from mature antigen-pulsed DCs are more potent inducers of CD8⁺ T-cell activation than immature DC-derived vesicles. Indeed, vesicles from both immature and mature antigen-pulsed DCs were able induce proliferation of antigen-experienced CD8⁺ T-cell clones *in vitro*.[55] However, mature DC-derived vesicles induced a twofold higher proliferation of these CD8⁺ T-cell clones.[55] Mature DC-derived vesicles loaded with viral peptides were also shown to induce interferon-γ (IFN-γ) production during recall responses of freshly isolated human CD8⁺ T-cells, whereas immature DC-derived vesicles were poor inducers of IFN-γ production.[56] These studies indicate that antigen-experienced CD8⁺ T-cells can be directly activated by vesicles derived from antigen-pulsed DCs without the requirement of bystander DCs (Fig. 17.2). In contrast, priming of naive CD8⁺ T-cells by DC-derived vesicles requires recruitment of these vesicles onto bystander DCs.[52] Both peptide-loaded immature and mature DC-derived vesicles could activate naive $CD8^+$ T-cells upon transfer to bystander DC *in vitro*. *In vivo*, however, only mature DC-derived vesicles were able to promote the differentiation of naive $CD8^+$ T-cells into $CD8^+$ effector cells.[52]

In early studies, Zitvogel et al. showed that tumor-antigen-loaded DC-derived vesicles could promote antitumor $CD8^+$ T-cell responses, which led to the rejection of established tumors in mice.[57] On the basis of these results, clinical trials were initiated using vesicles derived from immature DCs pulsed with a tumor antigen.[58] However, antitumor T-cell responses were not efficiently induced with these immature DC-derived vesicle-based vaccines. With the novel insights regarding the potency of activated DC-derived vesicles, a phase II clinical trial is currently ongoing based on vesicles derived from IFN-γ-activated DCs, which have enhanced immunostimulatory properties.[59]

Since DC-derived vesicles recruited by $CD4^+$ T-cells remain at the cell surface for at least 24 h,[4,48] the acquired MHC–peptide loaded complexes could be involved in antigen presentation to other T-cells.[4,48] Indeed, it has been demonstrated that MHC–peptide complexes derived from vesicles released by OVA-protein-pulsed DCs can be presented *in vitro* by ConA-stimulated splenic $CD4^+$ T-cells, thereby inducing OVA-specific $CD8^+$ T-cell proliferation as potently as the peptide-pulsed DCs themselves.[49] Vaccination of these DC vesicle–decorated $CD4^+$ T-cells led to an increase in the number of OVA-specific $CD8^+$ T-cells in the spleen of mice. The magnitude of this response was again comparable to vaccination with OVA peptide–pulsed DC, while injection of the OVA-pulsed DC-derived vesicles alone induced only a slight increase in the number of OVA-specific $CD8^+$ T-cells.[49] It is, however, not clear whether the injected DC-derived vesicles were recruited to activated $CD4^+$ T-cells *in vivo* and whether the decoration of $CD4^+$ T-cells by DC-derived vesicles plays a physiological role in the stimulate of effector $CD8^+$ T-cell responses *in vivo*.

17.6.1.3 Activation of NK cells through immature DC-derived vesicles

NK cells have also been identified as targets for DC-derived vesicles (Fig. 17.2). Vesicles from immature DCs can carry ligands for

NKp30 and NKG2D, two activating receptors of the NK cell.[60,61] Upon incubation with immature DC-derived vesicles, NK cells were activated, as was shown by the release of TNFα and IFN-γ.[60] This release was further increased in response to vesicles derived from heat shock–treated immature DCs.[60] *In vivo* administration of immature DC-derived vesicles induced proliferation of NK cells in mice, which was dependent on vesicle-associated IL-15Rα and triggered NK cell–mediated cytotoxicity against tumor metastases in the lungs of mice.[61] It has been postulated that the presence of IL-15Rα on the surface of human immature DC-derived vesicles can synergize the effect of soluble IL-15 on NK cell activation, possibly via "transpresentation" of IL-15 to NK cells.

17.6.2 Immune-Suppressive Effects of DC-Derived Vesicles

DC-derived vesicles, especially those released by immature DCs, can also mediate immune-suppressive and/or tolerizing effects (Fig. 17.2). Although immature DC-derived vesicles can activate NK cells and induce minor levels of $CD4^+$ T-cell proliferation, they are mainly thought to play a role in tolerance induction. The immune-suppressive properties of immature DC-derived vesicles have primarily been investigated *in vivo*. It has been demonstrated that the administration of vesicles derived from immature DCs either before or after allograft transplantation can induce tolerance and thereby prolong allograft survival.[62,63] The combined administration of an immune-suppressive drug (LF 15-0195) and donor DC-derived vesicles after transplantation significantly improved survival of cardiac allografts.[62] Since LF can inhibit DC maturation *in vivo*, it has been proposed that donor vesicle-derived MHC complexes are presented by immature host DCs, thereby inducing donor-specific tolerance.[62] In another study, pretreatment of rats with vesicles derived from immature donor DCs prolonged survival of intestinal grafts.[63] *Ex vivo* analysis of splenic T-cells, derived from pretreated allograft recipients, showed higher levels of IL-10, lower levels of IFN-γ, and less proliferation in response to donor splenocytes as compared to allograft recipients that did not receive pretreatment. Additionally, total Foxp3 levels were higher in spleen homogenates from allograft recipients pretreated with donor DC-derived vesicles.[63] Altogether these findings indicate that immature donor

DC-derived vesicles could be exploited to induce donor-specific tolerance in allograft transplantation.

In an experimental sepsis model, administration of immature DC-derived vesicles improved the clearance of apoptotic cells and reduced the release of sepsis-related cytokines, thereby significantly reducing mortality.[64] The enhanced phagocytosis was abrogated by the inhibition of MFG-E8 before vesicle administration.[65] In this model, immature DC-derived vesicles might provide a source of MFG-E8, which is a crucial component for the uptake of apoptotic cells. This could also explain why mature DC-derived vesicles, which contain lower levels of MFG-E8[21,50,64], did not show these beneficial effects.[64]

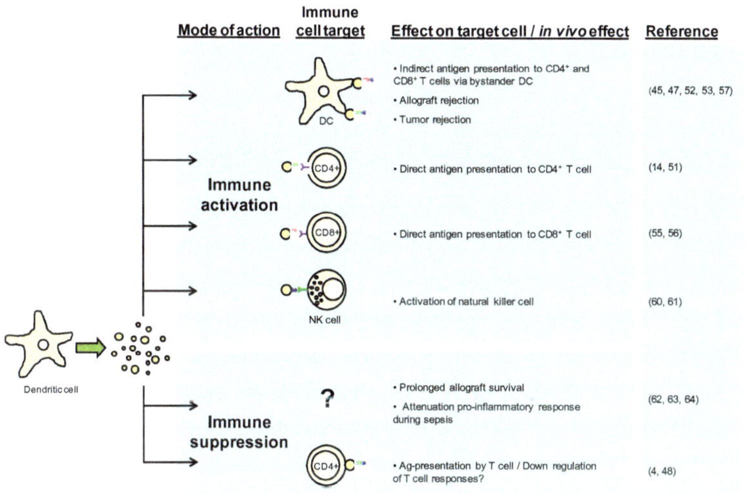

Figure 17.2 Targets and pleiotropic effects of DC-derived vesicles. This figure summarizes the main targets and effects of vesicles release by DCs on immune cells, both *in vitro* and *in vivo*.

In contrast to the strong immune-suppressive potential of immature DC-derived vesicles in the allograft transplantation and sepsis models, these vesicles could poorly interfere in already established immune responses, for example, collagen-induced arthritis.[66] This suggests that immature DC-derived vesicles contain tolerogenic properties but that these can be overruled during strong immune activation. Thus, although immature DC-derived vesicles are interesting candidates to be used as therapeutic agents for the treatment of diseases that require dampening of immune responses, the molecular composition of these vesicles should be optimized to retain the suppressive capacity in a primed environment. Several investigators tried to modify the molecular composition of these vesicles to increase their immune-suppressive properties. Robbins et al. generated tolerizing DCs by culturing them in the presence of recombinant IL-10 or by transducing them with vectors encoding IL-10, IL-4, or FasL.[67,68] Vesicles derived from these DCs could reduce *in vitro* proliferation in mixed lymphocyte reactions,[67] as well as delayed-type hypersensitivity responses and collagen-induced arthritis *in vivo*.[67] The potential of DC-derived vesicles to exert immune-suppressive functions, like their parental cell, and the fact that they are much more stable in circulation make them interesting therapeutic agents for the treatment of, for example, autoimmune diseases.

17.6.3 DC-Derived Vesicles as an Antigen Source

DC-derived vesicles can function as vehicles for the spreading of antigens. In many studies it is, however, unclear how an antigen is transferred to target cells via these vesicles and whether the antigen is further processed by the target cell. A first possibility is the functional transfer of MHC–peptide complexes or intact antigens to the surface of the target cell. This mechanism was illustrated by the ability of MHC II–negative DCs to stimulate $CD4^+$ T-cell responses after recruitment of MHC II–bearing vesicles, so-called "cross dressing."[47] These vesicle-derived MHC II–peptide complexes can be presented at the target cell surface directly after vesicle binding or after fusion of the vesicle with the plasma membrane. Vesicle-associated intact antigens exposed at the plasma membrane of the target cell can play a role in antigen presentation to B-cells.[51,54] Indeed, intact OVA protein on the surface of vesicles derived from mature DCs pulsed with OVA protein elicited production of OVA-specific IgG upon injection in mice, whereas vesicles directly loaded with OVA peptides did not.[51] However, vesicles derived from DCs pulsed with complete DT were also able to induce IgG responses, while no intact DT could be found in or on these vesicles.[54] Hence the exact mechanism via which DC-derived vesicles transfer intact antigens and induce humoral responses remains to be elucidated.

Alternatively, the transferred antigens can be processed by the target cell for antigen (cross-) presentation.[69] It was, for example, shown that DCs can process allogenic MHC-derived peptides transferred via vesicles and present these allopeptides via self-MHC.[69] Altogether, these findings demonstrate that DC-derived vesicles have a physiological role in antigen spreading.

17.7 DC- and CD4$^+$ T-Cell-Derived Vesicles for the Balancing of Immune Responses

Intense crosstalk between DCs and CD4$^+$ T-cells takes place during cognate interaction. This ensures a proper immune response to tackle infections and to prevent senescence and tumor development. These responses have to be strictly regulated to prevent immunopathology, autoimmune diseases, and chronic inflammation. Immune-stimulatory and immune-suppressive events therefore need to be carefully balanced in ongoing immune responses. We propose that vesicles released by DCs and T-cells during cognate interaction participate in the balancing of immune responses (Fig. 17.3). During cognate interactions, DCs release higher numbers of MHC II–containing vesicles.[4,21] These vesicles could amplify the antigen-presenting capacity of the DCs, both by direct antigen presentation to T-cells and by indirect antigen presentation after recruitment on bystander DCs.[47,51,52,55] Furthermore, the antigen-presenting capacity could be prolonged, since DC-derived vesicles can remain at the surface of CD4$^+$ T-cells after T-cells and DCs have dissociated. However, whether or not this antigen spreading will have an immune-activating effect depends on both the maturation status of the parental DC and the cell type and activation or maturation status of the target cell. Besides the targeting to bystander DCs, activated T-cells can also efficiently bind DC-derived vesicles, thereby recruiting MHC II onto their surface.[4,48] This may allow for T-cell-to-T-cell antigen presentation, which might lead to down-regulation of immune responses.[70,71] However, the recruitment of DC vesicles might also enable CD4$^+$ T-cells to activate CD8$^+$ T-cell responses.[49] In addition to these active immune modulatory properties, the respective blocking of TCR and MHC II complexes by

DCs and T-cell-derived vesicles could lead to suppression of immune responses.[31,33,36] The passive blocking limits the number of available specific TCR and/or MHC–peptide complexes on the surface of the CD4⁺ T-cell and DCs, respectively. Finally, T-cells can release vesicles carrying FasL, which can induce apoptosis of both DCs and bystander T-cells and possibly also of other immune cells.[24,25]

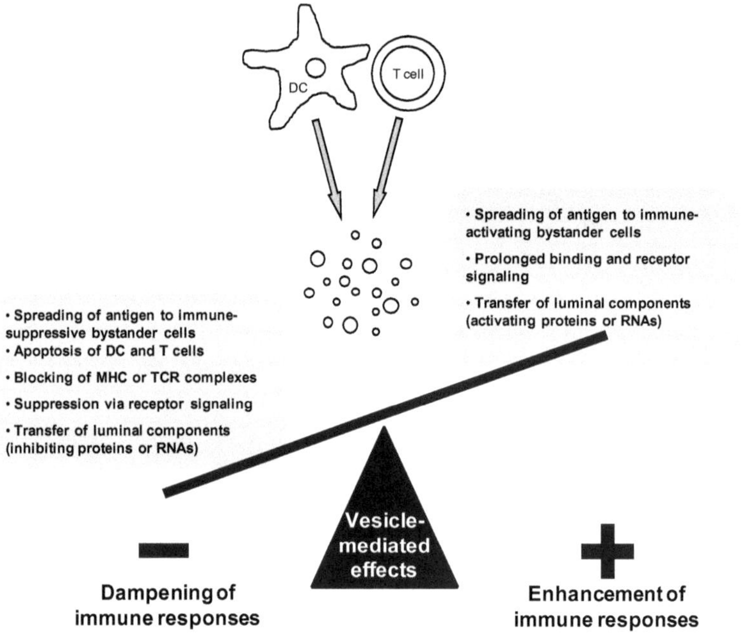

Figure 17.3 DC- and CD4⁺ T-cell-derived vesicles for balancing of immune responses. DCs and CD4⁺ T-cells release vesicles that can have immune-activating and/or immune-suppressive effects. These vesicles contribute to the spreading of both activating and suppressive signals that are involved in the regulation of immune responses.

The studies discussed here clearly indicate that DC- and T-cell-derived vesicles can play a major role in regulation of immune responses. The rapid expansion of this research field and the increasing interest in vesicle-mediated intercellular communication will catalyze progress in determining the physiological role and potential therapeutic applications of DC- and T-cell-derived vesicles.

Table 17.1 Characteristics of T-cell-derived vesicles, as described by different studies

	Vesicle-producing T-cell	T-cell activation protocol	Vesicle isolation protocol	Vesicle size	Vesicle composition	Target cell	Effect on target cell	Reference
$CD4^+$	OVA-specific T-cells generated by coculture of naive $CD4^+$ T-cells from OT-II mice with OVA-pulsed BMDCs	Cognate interaction with OVA-pulsed DCs 24 h	300 g (5 min) 1,200 g (20 min) 10,000 g (30 min) 100,000 g (1 h)	50–100 nm (based on EM)	CD4, TCR, LFA-1, CD25, and FasL (analyzed by conventional FC) LAMP1, TCR, and LFA-1, no calnexin (analyzed by WB)	OVA-pulsed BMDCs	Inhibition of $CD4^+$ T-cell proliferation *in vitro* Inhibition of $CD4^+$-dependent $CD8^+$ CTL responses (tumor killing and cytotoxicity assay *in vivo*)	33
	OT-II, DO11.10 $CD4^+$ T-cells Human peripheral $CD4^+$ T-lymphocytes	Cognate interaction with OVA-peptide/SEB-pulsed DCs 24 h	1,500 g (10 min) 4,000 g (20 min) 100,000 g (2.5 h)	<0.4 μm for CD3-positive vesicles (based on pore-size Transwell system)	CD2, CD3, CD27, and OX40 (based on transfer to DCs, analyzed by FC); CD28 not transferred	Spleen- (for mouse) or monocyte- (for human) derived DCs	Masking of MHC II complexes (by CD3-positive vesicles)	31

(*Continued*)

Table 17.1 (*Continued*)

Vesicle-producing T-cell	T-cell activation protocol	Vesicle isolation protocol	Vesicle size	Vesicle composition	Target cell	Effect on target cell	Reference
Murine $CD4^+$ T-cell clone KO4C1	Different levels of plate-bound anti-CD3 and anti-CD28 20 h	200 g (2 × 10 min) 500 g (2 × 10 min) 10,000 g (30 min) 100,000 g (65 min) Sucrose density gradient (192,000 g for <15 h)	50–250 nm	3 distinct subpopulations identified on the basis of light scattering and fluorescence of a general membrane dye (analyzed by high-resolution FC)	Not tested	Not tested	12
Rat $CD4^+$ T-cell clones A2b and Z1a	T-cell–T-cell Ag presentation 24 h	300 g (10 min) 1,200 g (20 min) 0.22 μm filter 100,000 g (1 h)	50–200 nm (based on EM)	MHC II (analyzed by EM) TCR, IL-2Rα, MHC II, ICAM-1, CD80, and CD86 (analyzed by conventional FC) Higher levels of IL-2Rα, MHC II, and ICAM-1 on vesicles from anergic T-cells	Spleen-derived B-cells and DCs	Responder T-cells stimulated by APCs exposed to vesicles derived from anergic T-cells showing a diminished response to subsequent presentation of their specific antigen	35

Vesicle-producing T-cell	T-cell activation protocol	Vesicle isolation protocol	Vesicle size	Vesicle composition	Target cell	Effect on target cell	Reference
Jurkat T-cells Human T-cell blasts	anti-CD3 30 min–24 h	300 g (6 min) 0.22 μm filter 100,000 g (1 h)	50–100 nm (based on EM)	CD3ε, CD2, and MHC I Upon activation (higher) levels of TCR, CD3ε, CXCR4, CD2, CD18, MHC I, MHC II, and CD63. No FasL, CD28, or CD45 (analyzed by WB, bead-based FC)	Not tested	Not tested	6
Jurkat T-cells Human T-cell blasts	PHA pulse (50 μg/mL) anti-CD59 1 h	800 g (10 min) 10,000 g (20 min) 100,000 g (18 h) or 100,000 g (8 h) alone	100–200 nm (based on EM)	FasL and APO2L (analyzed by WB, conventional FC, and EM)	Nonactivated Jurkat T-cells	Apoptosis of nonactivated Jurkat T-cells	24, 25

(*Continued*)

Table 17.1 (*Continued*)

Vesicle-producing T-cell	T-cell activation protocol	Vesicle isolation protocol	Vesicle size	Vesicle composition	Target cell	Effect on target cell	Reference
Jurkat E6.1 J-HM1-2.2 (Jurkat T-cells expressing HM1R) Human T-cell blasts	CCh (trigger HM1R) anti-CD3 PHA pulse (50 μg/mL) 10 h	800 g (10 min) 10,000 g (20 min) 100,000 g (12 h)	Not tested	CD63, FasL, no CD45 or CD28 (analyzed by WB)	Other Jurkat T-cells	Apoptosis of Jurkat T-cells	26
Jurkat T-cells Human peripheral blood T-lymphocytes	PMA (5 ng/mL) anti-CD3 / anti-CD28 coated Dynal beads 20 h	800 g (5 min) 4500 g (5 min) 100kDa Centricon cut-off (4,000 g, 10 min) or 20,000 g or 100,000 g (1 h)	100–800 nm (based on EM)	Positive for annexin V staining (analyzed by conventional FC), LFA-1 (analyzed by WB)	Mast cells (LAD2 or human cord blood mast cells)	B-hexosaminidase release (degranulation) Release of IL-8 and OSM ERK phosphorylation	29
Jurkat E6-1 T-cell Resting $CD4^+$ T-lymphocytes from peripheral blood	PHA (20 μg/mL) 3–4 d	200 g (10min) 0.22 μm filter 100,000 g (1.5 h)	50 nm (based on EM)	Positive for annexin V staining (analyzed by IF) CD81, no CD45 (analyzed by WB)	Monocytes isolated from peripheral blood	Accumulation of cholesterol Release of TNFα	30

Vesicle-producing T-cell	T-cell activation protocol	Vesicle isolation protocol	Vesicle size	Vesicle composition	Target cell	Effect on target cell	Reference
T-cell line HUT-78 Human peripheral blood T-lymphocytes	PHA (1 μg/nL) PMA (5 ng/mL) 6 h	800 g (5 min) 7,000 g (5 min) 20,000 g (45 min)	100–800 nm (based on EM)	Positive for annexin V staining (analyzed by conventional FC)	Monocytes isolated from peripheral blood	Release of proinflammatory cytokines TNFα, IL-1β, and sIL1-Ra (also IL-6, IL-8, CCL3, CCL4, and CCL2)	27, 28
Jurkat T-cells	No activation	2,000 g (5 min) 0.22 μm filter 10,000 g (30 min) 70,000 g (1 h) Sucrose density gradient (270,000 g for 16 h)	100–300 nm (based on EM)	CD81, CD63, no CD45 (in sucrose gradient, analyzed by WB)	Not tested	Not tested	10

(*Continued*)

Table 17.1 (*Continued*)

Vesicle-producing T-cell		T-cell activation protocol	Vesicle isolation protocol	Vesicle size	Vesicle composition	Target cell	Effect on target cell	Reference
$CD8^+$	CTL clone JS-132	Triggering vesicle release by interaction with target cell (within 10 min)	NA	40–70 nm (based on EM)	TCR, CD3, CD8, and MHC I (analyzed by EM)	Target cell of CTL	Not tested	34, 37
	OVA-specific T-cells generated by coculture of naive $CD8^+$ T-cells from OT-I or OT-I/Fas-ko mice with OVA-pulsed BMDCs	Cognate interaction with OVA-pulsed DCs 24 h	300 g (5 min) 1,200 g (20 min) 10,000 g (30 min) 100,000 g (1 h)	50–90 nm (based on EM)	CD8, CD28, ICAM-1, TCR, LFA-1, and FasL (analyzed by FC) TCR, LFA-1, ICAM-1, no calnexin (analyzed by WB)	OVA-pulsed BMDCs	Blocking of MHC Ip and induction of apoptosis by DCs Inhibition of both antitumor responses and autoimmunity *in vivo*	36

Vesicle-producing T-cell	T-cell activation protocol	Vesicle isolation protocol	Vesicle size	Vesicle composition	Target cell	Effect on target cell	Reference
$CD8^+$ T-cell line TG and primary $CD8^+$ T-cells from HIV-infected patients	anti-CD3	300 g (10 min) 800 g (30 min) 6,000 g (30 min) 15,000 g (30 min) or 60,000 g (1 h) 15,000 g pellet used for sucrose gradient at 90,000 g	30–100 nm (based on EM)	MHC II, CD9, CD63, CD81, CD3, no CD8 (captured on anti-MHC II beads, analyzed by FC)	TZM-bl cells (HeLa cells expressing CD4, CXCR4, and CCR5 and an additional reporter system) Expanded primary $CD4^+$ T-cells	Inhibition of HIV replication in both acute and chronically infected $CD4^+$ T-cells	32

Abbreviations: BMDC, bone marrow–derived DCs; FC, flow cytometry; HM1R, human muscarinic type 1 receptor; IF, immune fluorescence; WB, Western blot.

References

1. Thery C, et al. (2009) Membrane vesicles as conveyors of immune responses, *Nat. Rev.*, **9**, 581–593.
2. Cocucci E, et al. (2009) Shedding microvesicles: artefacts no more, *Trends Cell Biol.*, **19**, 43–51.
3. Simons M and Raposo G (2009) Exosomes—vesicular carriers for intercellular communication, *Curr. Opin. Cell Biol.*, **21**, 575–581.
4. Buschow SI, et al. (2009) MHC II in dendritic cells is targeted to lysosomes or T cell-induced exosomes via distinct multivesicular body pathways, *Traffic*, **10**, 1528–1542.
5. Thery C, et al. (1999) Molecular characterization of dendritic cell-derived exosomes. Selective accumulation of the heat shock protein hsc73, *J. Cell Biol.*, **147**, 599–610.
6. Blanchard N, et al. (2002) TCR activation of human T cells induces the production of exosomes bearing the TCR/CD3/zeta complex, *J. Immunol.*, **168**, 3235–3241.
7. Mittelbrunn M, et al. (2011) Unidirectional transfer of microRNA-loaded exosomes from T cells to antigen-presenting cells, *Nat. Commun.*, **2**, 282.
8. Montecalvo A, et al. (2012) Mechanism of transfer of functional microRNAs between mouse dendritic cells via exosomes, *Blood*, **119**, 756–766.
9. Sadallah S, et al. (2011) Ectosomes as immunomodulators, *Semin. Immunopathol.*, **33**, 487–495.
10. Booth AM, et al. (2006) Exosomes and HIV Gag bud from endosome-like domains of the T cell plasma membrane, *J. Cell Biol.*, **172**, 923–935.
11. de Gassart A, et al. (2003) Lipid raft-associated protein sorting in exosomes, *Blood*, **102**, 4336–4344.
12. van der Vlist EJ, et al. (2012) $CD4^+$ T cell activation promotes the differential release of distinct populations of nanosized vesicles, *JEV*, **1**, 18364.
13. Bobrie A, et al. (2012) Diverse subpopulations of vesicles secreted by different intracellular mechanisms are present in exosome preparations obtained by differential ultracentrifugation, *JEV*, **1**, 18397.
14. Raposo G, et al. (1996) B lymphocytes secrete antigen-presenting vesicles, *J. Exp. Med.*, **183**, 1161–1172.

15. Subra C, et al. (2010) Exosomes account for vesicle-mediated transcellular transport of activatable phospholipases and prostaglandins, *J. Lipid Res.*, **51**, 2105–2120.

16. Valadi H, et al. (2007) Exosome-mediated transfer of mRNAs and microRNAs is a novel mechanism of genetic exchange between cells, *Nat. Cell Biol.*, **9**, 654–659.

17. Gyorgy B, et al. (2011) Detection and isolation of cell-derived microparticles are compromised by protein complexes resulting from shared biophysical parameters, *Blood*, **117**, e39–e48.

18. Yuana Y, et al. (2010) Atomic force microscopy: a novel approach to the detection of nanosized blood microparticles, *J. Thromb. Haemost.*, **8**, 315–323.

19. Dragovic RA, et al. (2011) Sizing and phenotyping of cellular vesicles using nanoparticle tracking analysis, *Nanomedicine*, **7**, 780–788.

20. Lacroix R, et al. (2010) Overcoming limitations of microparticle measurement by flow cytometry, *Semin. Thromb. Haemost.*, **36**, 807–818.

21. Nolte-'t Hoen EN, et al. (2012) Quantitative and qualitative flow cytometric analysis of nano-sized cell-derived membrane vesicles, *Nanomedicine*, **8**, 712–720.

22. van der Vlist EJ, et al. (2012) Fluorescent labeling of nano-sized vesicles released by cells and subsequent quantitative and qualitative analysis by high resolution flow cytometry, *Nat. Prot.*, **7**, 1311–1326

23. Sallusto F and Lanzavecchia A (2009) Heterogeneity of $CD4^+$ memory T cells: functional modules for tailored immunity, *Eur. J. Immunol.*, **39**, 2076–2082.

24. Martinez-Lorenzo MJ, et al. (1999) Activated human T cells release bioactive Fas ligand and APO2 ligand in microvesicles, *J. Immunol.*, **163**, 1274–1281.

25. Monleon I, et al. (2001) Differential secretion of Fas ligand- or APO2 ligand/TNF-related apoptosis inducing ligand-carrying microvesicles during activation-induced death of human T cells, *J. Immunol.*, **167**, 6736–6744.

26. Alonso R, et al. (2005) Diacylglycerol kinase alpha regulates the secretion of lethal exosomes bearing Fas ligand during activation-induced cell death of T lymphocytes, *J. Biol. Chem.*, **280**, 28439–28450.

27. Carpintero R, et al. (2010) HDL interfere with the binding of T cell microparticles to human monocytes to inhibit pro-inflammatory cytokine production, *PLOS ONE*, **5**, e11869.

28. Scanu A, et al. (2008) Stimulated T cells generate microparticles, which mimic cellular contact activation of human monocytes: differential regulation of pro- and anti-inflammatory cytokine production by high-density lipoproteins, *J. Leukocyte Biol.*, **83**, 921–927.

29. Shefler I, et al. (2010) T cell-induced mast cell activation: a role for microparticles released from activated T cells, *J. Immunol.*, **185**, 4206–4212.

30. Zakharova L, et al. (2007) T cell exosomes induce cholesterol accumulation in human monocytes via phosphatidylserine receptor, *J. Cell Physiol.*, **212**, 174–181.

31. Busch A, et al. (2008) Transfer of T cell surface molecules to dendritic cells upon CD4$^+$ T cell priming involves two distinct mechanisms, *J. Immunol.*, **181**, 3965–3973.

32. Tumne A, et al. (2009) Noncytotoxic suppression of human immunodeficiency virus type 1 transcription by exosomes secreted from CD8$^+$ T cells, *J. Virol.*, **83**, 4354–4364.

33. Zhang H, et al. (2011) CD4$^+$ T cell-released exosomes inhibit CD8$^+$ cytotoxic T-lymphocyte responses and antitumor immunity, *Cell Mol. Immunol.*, **8**, 23–30.

34. Peters PJ, et al. (1989) Molecules relevant for T cell-target cell interaction are present in cytolytic granules of human T lymphocytes, *Eur J. Immunol.*, **19**, 1469–1475.

35. Nolte-'t Hoen EN, et al. (2004) Uptake of membrane molecules from T cells endows antigen-presenting cells with novel functional properties, *Eur. J. Immunol.*, **34**, 3115–3125.

36. Xie Y, et al. (2010) Dendritic cells recruit T cell exosomes via exosomal LFA-1 leading to inhibition of CD8$^+$ CTL responses through downregulation of peptide/MHC class I and Fas ligand-mediated cytotoxicity. *J. Immunol.*, **185**, 5268–5278.

37. Peters PJ, et al. (1991) Cytotoxic T lymphocyte granules are secretory lysosomes, containing both perforin and granzymes, *J. Exp. Med.*, **173**, 1099–1109.

38. Peters PJ, et al. (1990) A new model for lethal hit delivery by cytotoxic T lymphocytes, *Immunol. Today*, **11**, 28–32.

39. Zuccato E, et al. (2007) Sorting of Fas ligand to secretory lysosomes is regulated by mono-ubiquitylation and phosphorylation, *J. Cell Sci.*, **120**, 191–199.

40. Suda T, et al. (1995) Expression of the Fas ligand in cells of T cell lineage, *J. Immunol.*, **154**, 3806–3813.

41. Mariani SM, et al. (1995) Regulation of cell surface APO-1/Fas (CD95) ligand expression by metalloproteases, *Eur. J. Immunol.*, **25**, 2303–2307.

42. Reis e Sousa C (2006) Dendritic cells in a mature age, *Nat. Rev. Immunol.*, **6**, 476–483.

43. Guermonprez P, et al. (2002) Antigen presentation and T cell stimulation by dendritic cells, *Annu. Rev. Immunol.*, **20**, 621–667.

44. Viaud S, et al. (2010) Dendritic cell-derived exosomes for cancer immunotherapy: what's next? *Cancer Res.*, **70**, 1281–1285.

45. Segura E, et al. (2005) ICAM-1 on exosomes from mature dendritic cells is critical for efficient naive T-cell priming, *Blood*, **106**, 216–223.

46. Soo CY, et al. (2012) Nanoparticle tracking analysis monitors microvesicle and exosome secretion from immune cells, *Immunology*, **136**, 192–197.

47. Thery C, et al. (2002) Indirect activation of naive $CD4^+$ T cells by dendritic cell-derived exosomes, *Nat. Immunol.*, **3**, 1156–1162.

48. Nolte-'t Hoen EN, et al. (2009) Activated T cells recruit exosomes secreted by dendritic cells via LFA-1, *Blood*, **113**, 1977–1981.

49. Hao S, et al. (2007) Nonspecific $CD4^+$ T cells with uptake of antigen-specific dendritic cell-released exosomes stimulate antigen-specific $CD8^+$ CTL responses and long-term T cell memory, *J. Leukocyte Biol.*, **82**, 829–838.

50. Segura E, et al. (2005) Mature dendritic cells secrete exosomes with strong ability to induce antigen-specific effector immune responses, *Blood Cells Mol. Dis.*, **35**, 89–93.

51. Qazi KR, et al. (2009) Antigen-loaded exosomes alone induce Th1-type memory through a B-cell-dependent mechanism, *Blood*, **113**, 2673–2683.

52. Andre F, et al. (2004) Exosomes as potent cell-free peptide-based vaccine. I. Dendritic cell-derived exosomes transfer functional MHC class I/peptide complexes to dendritic cells, *J. Immunol.*, **172**, 2126–2136.

53. Segura E, et al. (2007) $CD8^+$ dendritic cells use LFA-1 to capture MHC-peptide complexes from exosomes *in vivo*, *J. Immunol.*, **179**, 1489–1496.

54. Colino J and Snapper CM (2006) Exosomes from bone marrow dendritic cells pulsed with diphtheria toxoid preferentially induce type

1 antigen-specific IgG responses in naive recipients in the absence of free antigen, *J. Immunol.*, **177**, 3757–3762.

55. Utsugi-Kobukai S, et al. (2003) MHC class I-mediated exogenous antigen presentation by exosomes secreted from immature and mature bone marrow derived dendritic cells, *Immunol. Lett.*, **89**, 125–131.
56. Admyre C, et al. (2006) Direct exosome stimulation of peripheral human T cells detected by ELISPOT, *Eur. J. Immunol.*, **36**, 1772–1781.
57. Zitvogel L, et al. (1998) Eradication of established murine tumors using a novel cell-free vaccine: dendritic cell-derived exosomes, *Nat. Med.*, **4**, 594–600.
58. Escudier B, et al. (2005) Vaccination of metastatic melanoma patients with autologous dendritic cell (DC) derived-exosomes: results of thefirst phase I clinical trial, *J. Transl. Med.*, **3**, 10.
59. Chaput N and Thery C (2011) Exosomes: immune properties and potential clinical implementations, *Semin. Immunopathol.*, **33**, 419–440.
60. Simhadri VR, et al. (2008) Dendritic cells release HLA-B-associated transcript-3 positive exosomes to regulate natural killer function, *PLOS ONE*, **3**, e3377.
61. Viaud S, et al. (2009) Dendritic cell-derived exosomes promote natural killer cell activation and proliferation: a role for NKG2D ligands and IL-15Ralpha, *PLOS ONE*, **4**, e4942.
62. Peche H, et al. (2006) Induction of tolerance by exosomes and short-term immunosuppression in a fully MHC-mismatched rat cardiac allograft model, *Am. J. Transplant.*, **6**, 1541–1550.
63. Yang X, et al. (2011) Exosomes derived from immature bone marrow dendritic cells induce tolerogenicity of intestinal transplantation in rats, *J. Surg. Res.*, **171**, 826–832.
64. Miksa M, et al. (2009) Immature dendritic cell-derived exosomes rescue septic animals via milk fat globule epidermal growth factor-factor VIII [corrected, *J. Immunol.*, **183**, 5983–5990.
65. Miksa M, et al. (2006) Dendritic cell-derived exosomes containing milk fat globule epidermal growth factor-factor VIII attenuate proinflammatory responses in sepsis, *Shock*, **25**, 586–593.
66. Kim SH, et al. (2006) Exosomes derived from genetically modified DC expressing FasL are anti-inflammatory and immunosuppressive, *Mol. Ther.*, **13**, 289–300.

67. Kim SH, et al. (2005) Exosomes derived from IL-10-treated dendritic cells can suppress inflammation and collagen-induced arthritis, *J. Immunol.*, **174**, 6440–6448.

68. Ruffner MA, et al. (2009) B7–1/2, but not PD-L1/2 molecules, are required on IL-10-treated tolerogenic DC and DC-derived exosomes for *in vivo* function, *Eur. J. Immunol.*, **39**, 3084–3090.

69. Morelli AE, et al. (2004) Endocytosis, intracellular sorting, processing of exosomes by dendritic cells, *Blood*, **104**, 3257–3266.

70. Helft J, et al. (2008) Antigen-specific T-T interactions regulate CD4 T-cell expansion, *Blood*, **112**, 1249–1258.

71. Taams LS, et al. (1998) Anergic T cells actively suppress T cell responses via the antigen-presenting cell, *Eur. J. Immunol.*, **28**, 2902–2912.

Index